OPTIMAL FILTERING

PRENTICE-HALL INFORMATION
AND SYSTEM SCIENCES SERIES
Thomas Kailath, *Editor*

OPTIMAL
FILTERING

Brian D. O. Anderson

John B. Moore

Professors of Electrical Engineering

University of Newcastle
New South Wales, Australia

PRENTICE-HALL, INC.
Englewood Cliffs, New Jersey 07632

Library of Congress Cataloging in Publication Data

ANDERSON, BRIAN D O
　Optimal filtering.

　(Information and system sciences series)
　Includes bibliographies and index.
　1. Signal processing.　2. Electric filters.
I. Moore, John Barratt, date　joint author.
II. Title.
TK5102.5.A53　　621.3815'32　　78–8938
ISBN　0–13–638122–7

© 1979 by Prentice-Hall, Inc., Englewood Cliffs, N.J.　07632

Printed in the United States of America

10　9　8　7　6　5　4

PRENTICE-HALL INTERNATIONAL, INC., *London*
PRENTICE-HALL OF AUSTRALIA PTY. LIMITED, *Sydney*
PRENTICE-HALL OF CANADA, LTD., *Toronto*
PRENTICE-HALL OF INDIA PRIVATE LIMITED, *New Delhi*
PRENTICE-HALL OF JAPAN, INC., *Tokyo*
PRENTICE-HALL OF SOUTHEAST ASIA PTE. LTD., *Singapore*
WHITEHALL BOOKS LIMITED, *Wellington, New Zealand*

CONTENTS

3 THE DISCRETE-TIME KALMAN FILTER

36

4 TIME-INVARIANT FILTERS

62

5 KALMAN FILTER PROPERTIES

90

6 COMPUTATIONAL ASPECTS

129

PREFACE

This book is a graduate level text which goes beyond and augments the undergraduate exposure engineering students might have to signal processing; particularly, communication systems and digital filtering theory. The material covered in this book is vital for students in the fields of control and communications and relevant to students in such diverse areas as statistics, economics, bioengineering and operations research. The subject matter requires the student to work with linear system theory results and elementary concepts in stochastic processes which are generally assumed at graduate level. However, this book is appropriate at the senior year undergraduate level for students with background in these areas.

Certainly the book contains more material than is usually taught in one semester, so that for a one semester or quarter length course, the first three chapters (dealing with the rudiments of Kalman filtering) can be covered first, followed by a selection from later chapters. The chapters following Chapter 3 build in the main on the ideas in Chapters 1, 2 and 3, rather than on all preceding chapters. They cover a miscellany of topics; for example, time-invariant filters, smoothing, and nonlinear filters. Although there is a significant benefit in proceeding through the chapters in sequence, this is not essential, as has been shown by the authors' experience in teaching this course.

The pedagogical feature of the book most likely to startle the reader is the concentration on discrete-time filtering. Recent technological developments as well as the easier path offered students and instructors are the two reasons for this course of action. Much of the material of the book has been with us in one form or another for ten to fifteen years, although again, much is relatively recent. This recent work has given new perspectives on the earlier material; for example, the notion of the innovations process provides helpful insights in deriving the Kalman filter.

We acknowledge the research support funding of the Australian Research Grants Committee and the Australian Radio Research Board. We are indebted also for specific suggestions from colleagues, Dr. G. Goodwin and Dr. A. Cantoni; joint research activities with former Ph.D. students Peter Tam and Surapong Chirarattananon; and to the typing expertise of Dianne Piefke. We have appreciated discussions in the area of optimal filtering with many scholars including Professors K. Ästrom, T. Kailath, D. Mayne, J. Meditch and J. Melsa.

B. D. O. ANDERSON

New South Wales, Australia

J. B. MOORE

CHAPTER *1*

INTRODUCTION

1.1 FILTERING

Filtering in one form or another has been with us for a very long time. For many centuries, man has attempted to remove the more visible of the impurities in his water by filtering, and one dictionary gives a first meaning for the noun filter as "a contrivance for freeing liquids from suspended impurities, especially by passing them through strata of sand, charcoal, etc."

Modern usage of the word filter often involves more abstract entities than fluids with suspended impurities. There is usually however the notion of something passing a barrier: one speaks of news filtering out of the war zone, or sunlight filtering through the trees. Sometimes the barrier is interposed by man for the purpose of sorting out something that is desired from something else with which it is contaminated. One example is of course provided by water purification; the use of an ultraviolet filter on a camera provides another example. When the entities involved are signals, such as electrical voltages, the barrier—in the form perhaps of an electric network—becomes a filter in the sense of signal processing.

It is easy to think of engineering situations in which filtering of signals might be desired. Communication systems always have unwanted signals, or noise, entering into them. This is a fundamental fact of thermodynamics. The user of the system naturally tries to minimize the inaccuracies caused by the presence of this noise—by filtering. Again, in many control systems the control is derived by feedback, which involves processing measurements derived from the system. Frequently, these measurements will contain random inaccuracies or be contaminated by unwanted signals, and filtering is necessary in order to make the control close to that desired.

1.2 HISTORY OF SIGNAL FILTERING

Filters were originally seen as circuits or systems with frequency selective behaviour. The series or parallel tuned circuit is one of the most fundamental such circuits in electrical engineering, and as a "wave trap" was a crucial ingredient in early crystal sets. More sophisticated versions of this same idea are seen in the IF strip of most radio receivers; here, tuned circuits, coupled by transformers and amplifiers, are used to shape a passband of frequencies which are amplified, and a stopband where attenuation occurs.

Something more sophisticated than collections of tuned circuits is necessary for many applications, and as a result, there has grown up an extensive body of filter design theory. Some of the landmarks are constant k and m-derived filters [1], and, later, Butterworth filters, Chebyshev filters, and elliptical filters [2]. In more recent years, there has been extensive development of numerical algorithms for filter design. Specifications on amplitude and phase response characteristics are given, and, often with the aid of sophisticated computer-aided design packages which allow interactive operation, a filter is designed to meet these specifications. Normally, there are also constraints imposed on the filter structure which have to be met; these constraints may involve impedance levels, types of components, number of components, etc.

Nonlinear filters have also been used for many years. The simplest is the AM envelope detector [3], which is a combination of a diode and a low-pass filter. In a similar vein, an automatic gain control (AGC) circuit uses a low-pass filter and a nonlinear element [3]. The phase-locked-loop used for FM reception is another example of a nonlinear filter [4], and recently the use of Dolby® systems in tape recorders for signal-to-noise ratio enhancement has provided another living-room application of nonlinear filtering ideas.

The notion of a filter as a device processing continuous-time signals and possessing frequency selective behaviour has been stretched by two major developments.

The first such development is digital filtering [5–7], made possible by recent innovations in integrated circuit technology. Totally different circuit modules from those used in classical filters appear in digital filters, e.g., analog-to-digital and digital-to-analog converters, shift registers, read-only memories, even microprocessors. Therefore, though the ultimate goals of digital and classical filtering are the same, the practical aspects of digital filter construction bear little or no resemblance to the practical aspects of, say, *m*-derived filter construction. In digital filtering one no longer seeks to minimize the active element count, the size of inductors, the dissipation of the reactive elements, or the termination impedance mismatch. Instead, one may seek to minimize the word length, the round-off error, the number of wiring operations in construction, and the processing delay.

Aside from the possible cost benefits, there are other advantages of this new approach to filtering. Perhaps the most important is that the filter parameters can be set and maintained to a high order of precision, thereby achieving filter characteristics that could not normally be obtained reliably with classical filtering. Another advantage is that parameters can be easily reset or made adaptive with little extra cost. Again, some digital filters incorporating microprocessors can be time-shared to perform many simultaneous tasks effectively.

The second major development came with the application of statistical ideas to filtering problems [8–14] and was largely spurred by developments in theory. The classical approaches to filtering postulate, at least implicitly, that the useful signals lie in one frequency band and unwanted signals, normally termed noise, lie in another, though on occasions there can be overlap. The statistical approaches to filtering, on the other hand, postulate that certain statistical properties are possessed by the useful signal and unwanted noise. Measurements are available of the sum of the signal and noise, and the task is still to eliminate by some means as much of the noise as possible through processing of the measurements by a filter. The earliest statistical ideas of Wiener and Kolmogorov [8, 9] relate to processes with statistical properties which do not change with time, i.e., to stationary processes. For these processes it proved possible to relate the statistical properties of the useful signal and unwanted noise with their frequency domain properties. There is, thus, a conceptual link with classical filtering.

A significant aspect of the statistical approach is the definition of a measure of suitability or performance of a filter. Roughly the best filter is that which, on the average, has its output closest to the correct or useful signal. By constraining the filter to be linear and formulating the performance measure in terms of the filter impulse response and the given statistical properties of the signal and noise, it generally transpires that a unique impulse response corresponds to the best value of the measure of performance or suitability.

As noted above, the assumption that the underlying signal and noise processes are stationary is crucial to the Wiener and Kolmogorov theory. It was not until the late 1950s and early 1960s that a theory was developed that did not require this stationarity assumption [11–14]. The theory arose because of the inadequacy of the Wiener-Kolmogorov theory for coping with certain applications in which nonstationarity of the signal and/or noise was intrinsic to the problem. The new theory soon acquired the name *Kalman filter theory.*

Because the stationary theory was normally developed and thought of in frequency domain terms, while the nonstationary theory was naturally developed and thought of in time domain terms, the contact between the two theories initially seemed slight. Nevertheless, there is substantial contact, if for no other reason than that a stationary process is a particular type of non-stationary process; rapprochement of Wiener and Kalman filtering theory is now easily achieved.

As noted above, Kalman filtering theory was developed at a time when applications called for it, and the same comment is really true of the Wiener filtering theory. It is also pertinent to note that the problems of implementing Kalman filters and the problems of implementing Wiener filters were both consistent with the technology of their time. Wiener filters were implementable with amplifiers and time-invariant network elements such as resistors and capacitors, while Kalman filters could be implemented with digital integrated circuit modules.

The point of contact between the two recent streams of development, digital filtering and statistical filtering, comes when one is faced with the problem of implementing a discrete-time Kalman filter using digital hardware. Looking to the future, it would be clearly desirable to incorporate the practical constraints associated with digital filter realization into the mathematical statement of the statistical filtering problem. At the present time, however, this has not been done, and as a consequence, there is little contact between the two streams.

1.3 SUBJECT MATTER OF THIS BOOK

This book seeks to make a contribution to the evolutionary trend in statistical filtering described above, by presenting a hindsight view of the trend, and focusing on recent results which show promise for the future. The basic subject of the book is the Kalman filter. More specifically, the book starts with a presentation of discrete-time Kalman filtering theory and then explores a number of extensions of the basic ideas.

There are four important characteristics of the basic filter:

1. Operation in discrete time
2. Optimality

3. Linearity
4. Finite dimensionality

Let us discuss each of these characteristics in turn, keeping in mind that derivatives of the Kalman filter inherit most but not all of these characteristics.

Discrete-time operation. More and more signal processing is becoming digital. For this reason, it is just as important, if not more so, to understand discrete-time signal processing as it is to understand continuous-time signal processing. Another practical reason for preferring to concentrate on discrete-time processing is that discrete-time statistical filtering theory is much easier to learn first than continuous-time statistical filtering theory; this is because the theory of random sequences is much simpler than the theory of continuous-time random processes.

Optimality. An optimal filter is one that is best in a certain sense, and one would be a fool to take second best if the best is available. Therefore, provided one is happy with the criterion defining what is best, the argument for optimality is almost self-evident. There are, however, many secondary aspects to optimality, some of which we now list. Certain classes of optimal filters tend to be robust in their maintenance of performance standards when the quantities assumed for design purposes are not the same as the quantities encountered in operation. Optimal filters normally are free from stability problems. There are simple operational checks on an optimal filter when it is being used that indicate whether it is operating correctly. Optimal filters are probably easier to make adaptive to parameter changes than suboptimal filters.

There is, however, at least one potential disadvantage of an optimal filter, and that is complexity; frequently, it is possible to use a much less complex filter with but little sacrifice of performance. The question arises as to how such a filter might be found. One approach, which has proved itself in many situations, involves approximating the signal model by one that is simpler or less complex, obtaining the optimal filter for this less complex model, and using it for the original signal model, for which of course it is suboptimal. This approach may fail on several grounds: the resulting filter may still be too complex, or the amount of suboptimality may be unacceptably great. In this case, it can be very difficult to obtain a satisfactory filter of much less complexity than the optimal filter, even if one is known to exist, because theories for suboptimal design are in some ways much less developed than theories for optimal design.

Linearity. The arguments for concentrating on linear filtering are those of applicability and sound pedagogy. A great many applications involve linear systems with associated gaussian random processes; it transpires that the optimal filter in a minimum mean-square-error sense is then linear. Of course, many applications involve nonlinear systems and/or nongaussian random processes, and for these situations, the optimal filter is nonlinear. However, the plain fact of the matter is that optimal nonlinear filter design and implementation are very hard, if not impossible, in many instances. For this reason, a suboptimal linear filter may often be used as a substitute for an optimal nonlinear filter, or some form of nonlinear filter may be derived which is in some way a modification of a linear filter or, sometimes, a collection of linear filters. These approaches are developed in this book and follow our discussion of linear filtering, since one can hardly begin to study nonlinear filtering with any effectiveness without a knowledge of linear filtering.

Finite dimensionality. It turns out that finite-dimensional filters should be used when the processes being filtered are associated with finite-dimensional systems. Now most physical systems are not finite dimensional; however, almost all infinite-dimensional systems can be approximated by finite-dimensional systems, and this is generally what happens in the modeling process. The finite-dimensional modeling of the physical system then leads to an associated finite-dimensional filter. This filter will be suboptimal to the extent that the model of the physical system is in some measure an inaccurate reflection of physical reality. Why should one use a suboptimal filter? Though one can without too much difficulty discuss infinite-dimensional filtering problems in discrete time, and this we do in places in this book, finite-dimensional filters are very much to be preferred on two grounds: they are easier to design, and far easier to implement, than infinite-dimensional filters.

1.4 OUTLINE OF THE BOOK

The book falls naturally into three parts.

The first part of the book is devoted to the formulation and solution of the basic Kalman filtering problem. By the end of the first section of Chapter 3, the reader should know the fundamental Kalman filtering result, and by the end of Chapter 3, have seen it in use.

The second part of the book is concerned with a deeper examination of the operational and computational properties of the filter. For example, there is discussion of time-invariant filters, including special techniques for computing these filters, and filter stability; the Kalman filter is shown to have a signal-to-noise ratio enhancement property.

In the third part of the book, there are a number of developments taking off from the basic theory. For example, the topics of smoothers, nonlinear and adaptive filters, and spectral factorization are all covered.

There is also a collection of appendices to which the reader will probably refer on a number of occasions. These deal with probability theory and random processes, matrix theory, linear systems, and Lyapunov stability theory. By and large, we expect a reader to know some, but not all, of the material in these appendices. They are too concentrated in presentation to allow learning of the ideas from scratch. However, if they are consulted when a new idea is encountered, they will permit the reader to learn much, simply by using the ideas.

Last, we make the point that there are many ideas developed in the problems. Many are not routine.

REFERENCES

[1] SKILLING, H. H., *Electrical Engineering Circuits*, John Wiley & Sons, Inc., New York, 1957.

[2] STORER, J. E., *Passive Network Synthesis*, McGraw-Hill Book Company, New York, 1957.

[3] TERMAN, F. E., *Electronic and Radio Engineering*, McGraw-Hill Book Company, New York, 1955.

[4] VITERBI, A. J., *Principles of Coherent Communication*, McGraw-Hill Book Company, New York, 1966.

[5] GOLD, B., and C. M. RADER, *Digital Processing of Signals*, McGraw-Hill Book Company, New York, 1969.

[6] RABINER, L. R., and B. GOLD, *Theory and Application of Digital Signal Processing*, Prentice-Hall, Inc., Englewood Cliffs, N.J., 1975.

[7] OPPENHEIM, A. V., and R. W. SCHAFER, *Digital Signal Processing*, Prentice-Hall, Inc., Englewood Cliffs, N.J., 1975.

[8] WIENER, N., *Extrapolation, Interpolation & Smoothing of Stationary Time Series*, The M.I.T. Press, Cambridge, Mass., 1949.

[9] KOLMOGOROV, A. N., "Interpolation and Extrapolation," *Bull. de l'académie des sciences de U.S.S.R.*, Ser. Math. 5, 1941, pp. 3–14.

[10] WAINSTEIN, L. A., and V. D. ZUBAKOV, *Extraction of Signals from Noise*, Prentice-Hall, Inc., Englewood Cliffs, N.J., 1962.

[11] KALMAN, R. E., and R. S. BUCY, "New Results in Linear Filtering and Prediction Theory," *J. of Basic Eng., Trans. ASME*, Series D, Vol. 83, No. 3, 1961, pp. 95–108.

[12] KALMAN, R. E., "A New Approach to Linear Filtering and Prediction Problems," *J. Basic Eng., Trans. ASME*, Series D, Vol. 82, No. 1, 1960, pp. 35–45.

[13] KALMAN, R. E., "New Methods in Wiener Filtering Theory," *Proc. Symp. Eng. Appl. Random Functions Theory and Probability* (eds. J. L. Bogdanoff and F. Kozin), John Wiley & Sons, Inc., New York, 1963.

[14] KAILATH, T., "A View of Three Decades of Linear Filtering Theory," *IEEE Trans. Inform. Theory*, Vol. IT-20, No. 2, March 1974, pp. 146–181.

CHAPTER 2

FILTERING, LINEAR SYSTEMS,
AND ESTIMATION

2.1 SYSTEMS, NOISE, FILTERING, SMOOTHING, AND PREDICTION

Our aim in this section is to give the reader some feel for the concepts of filtering, smoothing, and prediction. Later in this chapter we shall consider a specific filtering problem, and in the next chapter present its solution. This will provide the basis for the definition and solution of most of the other problems discussed in this book.

In order to have any sort of filtering problem in the first place, there must be a *system*, generally dynamic, of which measurements are available. Rather than develop the notion of a system with a large amount of mathematical formalism, we prefer here to appeal to intuition and common sense in pointing out what we mean. The system is some physical object, and its behaviour can normally be described by equations. It operates in real time, so that the independent variable in the equations is time. It is assumed to be causal, so that an output at some time $t = t_0$ is in no way dependent on inputs applied subsequent to $t = t_0$. Further, the system may operate in discrete or continuous time, with the underlying equations either difference equations or differential equations, and the output may change at discrete instants of time or on a continuous basis.

Later, we shall pose specific mathematical models for some systems, and even formally identify the system with the model.

In discussing filtering and related problems, it is implicit that the systems under consideration are *noisy*. The noise may arise in a number of ways. For example, inputs to the system may be unknown and unpredictable except for their statistical properties, or outputs from the system may be derived with the aid of a noisy sensor, i.e., one that contributes on a generally random basis some inaccuracy to the measurement of the system output. Again, outputs may only be observed via a sensor after transmission over a noisy channel.

In virtually all the problems we shall discuss here, it will be assumed that the output measurement process is noisy. On most occasions, the inputs also will be assumed to be noisy.

Now let us consider exactly what we mean by *filtering*. Suppose there is some quantity (possibly a vector quantity) associated with the system operation whose value we would like to know at each instant of time. For the sake of argument, assume the system in question is a continuous time system, and the quantity in question is denoted by $s(\cdot)$.* It may be that this quantity is not directly measurable, or that it can only be measured with error. In any case, we shall suppose that noisy measurements $z(\cdot)$ are available, with $z(\cdot)$ not the same as $s(\cdot)$.

The term filtering is used in two senses. First, it is used as a generic term: filtering is the recovery from $z(\cdot)$ of $s(\cdot)$, or an approximation to $s(\cdot)$, or even some information about $s(\cdot)$. In other words, noisy measurements of a system are used to obtain information about some quantity that is essentially internal to the system. Second, it is used to distinguish a certain kind of information processing from two related kinds, smoothing and prediction. In this sense, filtering means the recovery at time t of some information about $s(t)$ using measurements up till time t. The important thing to note is the *triple* occurrence of the time argument t. First, we are concerned with obtaining information about $s(\cdot)$ at time t, i.e., $s(t)$. Second, the information is available at time t, not at some later time. Third, measurements right up to, but not after, time t are used. [If information about $s(t)$ is to be available at time t, then causality rules out the use of measurements taken later than time t in producing this information.]

An example of the application of filtering in everyday life is in radio reception. Here the signal of interest is the voice signal. This signal is used to modulate a high frequency carrier that is transmitted to a radio receiver. The received signal is inevitably corrupted by noise, and so, when demodulated, it is filtered to recover as well as possible the original signal.

---

*Almost without exception throughout the book, $x(t)$ will denote the value taken by a function at time t, and $x(\cdot)$ will denote that function. Therefore, $x(t)$ is a number, and $x(\cdot)$ an infinite set of pairs, $\{t, x(t)\}$, for t ranging over all possible values.

Smoothing differs from filtering in that the information about $s(t)$ need not become available at time t, and measurements derived later than time t can be used in obtaining information about $s(t)$. This means there must be a delay in producing the information about $s(t)$, as compared with the filtering case, but the penalty of having a delay can be weighed against the ability to use more measurement data than in the filtering case in producing the information about $s(t)$. Not only does one use measurements up to time t, but one can also use measurements after time t. For this reason, one should expect the smoothing process to be more accurate in some sense than the filtering process.

An example of smoothing is provided by the way the human brain tackles the problem of reading hastily written handwriting. Each word is tackled sequentially, and when word is reached that is particularly difficult to interpret, several words after the difficult word, as well as those before it, may be used to attempt to deduce the word. In this case, the $s(\cdot)$ process corresponds to the sequence of correct words and the $z(\cdot)$ process to the sequence of handwritten versions of these words.

Prediction is the forecasting side of information processing. The aim is to obtain at time t information about $s(t + \lambda)$ for some $\lambda > 0$, i.e., to obtain information about what $s(\cdot)$ will be like subsequent to the time at which the information is produced. In obtaining the information, measurements up till time t can be used.

Again, examples of the application of prediction abound in many areas of information processing by the human brain. When attempting to catch a ball, we have to predict the future trajectory of the ball in order to position a catching hand correctly. This task becomes more difficult the more the ball is subject to random disturbances such as wind gusts. Generally, any prediction task becomes more difficult as the environment becomes noisier.

Outline of the Chapter

In Sec. 2.2, we introduce the basic system for which we shall aim to design filters, smoothers, and predictors. The system is described by linear, discrete-time, finite-dimensional state-space equations, and has noisy input and output.

In Sec. 2.3, we discuss some particular ways one might try to use noisy measurement data to infer estimates of the way internal variables in a system may be behaving. The discussion is actually divorced from that of Sec. 2.2, in that we pose the estimation problem simply as one of estimating the value of a random variable X given the value taken by a second random variable Y, with which X is jointly distributed.

Linkage of the ideas of Secs. 2.2 and 2.3 occurs in the next chapter. The ideas of Sec. 2.3 are extended to consider the problem of estimating the successive values of a random sequence, given the successive values of a second random sequence; the random sequences in question are those arising from the model discussed in Sec. 2.2, and the estimating device is the Kalman filter.

The material covered in this chapter can be found in many other places; see, e.g., [1] and [2].

Problem 1.1. Consider the reading of a garbled telegram. Defining signal-to-noise ratio as simply the inverse of the probability of error for the reception of each word letter, sketch on the one graph what you think might be reasonable plots of reading performance (probability of misreading a word) versus signal-to-noise ratio for the following cases.

1. Filtering—where only the past data can be used to read each word
2. One-word-ahead prediction
3. Two-words-ahead prediction
4. Smoothing—where the reader can look ahead one word
5. Smoothing—where the reader can look ahead for the remainder of the sentence

Problem 1.2. Interpret the following statement using the ideas of this section: "It is easy to be wise after the event."

2.2 THE GAUSS-MARKOV DISCRETE-TIME MODEL

System Description

We shall restrict attention in this book primarily to discrete-time systems, or, equivalently, systems where the underlying system equations are difference equations rather than differential equations.

The impetus for the study of discrete-time systems arises because frequently in a practical situation system observations are made and control strategies are implemented at discrete time instants. An example of such a situation in the field of economics arises where certain statistics or economic indices may be compiled quarterly and budget controls may be applied yearly. Again, in many industrial control situations wherever a digital computer is used to monitor and perhaps also to control a system, the discrete-time framework is a very natural one in which to give a system model description—even in the case where the system is very accurately described by differential equations. This is because a digital computer is intrinsically a discrete-time system rather than a continuous-time system.

The class of discrete-time systems we shall study in this section has as a prototype the linear, finite-dimensional system depicted in Fig. 2.2-1. The system depicted may be described by state-space equations*

$$x_{k+1} = F_k x_k + G_k w_k \qquad (2.1)$$

$$z_k = y_k + v_k = H'_k x_k + v_k \qquad (2.2)$$

The subscript is a time argument; for the moment we assume that the initial time at which the system commences operating is finite. Then by shift of the time origin, we can assume that (2.1) and (2.2) hold for $k \geq 0$. Further, we shall denote successive time instants without loss of generality by integer k.

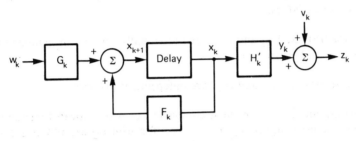

Fig. 2.2-1   Finite-dimensional linear system serving as signal model.

Equations of the type just given can arise in considering a continuous-time linear system with sampled measurements, as described in Appendix C in some detail.

To denote the set $\{(x_k, k) \,|\, k \geq 0\}$, we shall use the symbol $\{x_k\}$. As usual, x_k will be a value taken by $\{x_k\}$ at time k. In (2.1) and (2.2), x_k is, of course, the *system state* at time k. Under normal circumstances, $y_k = H'_k x_k$ would be the corresponding system output, but in this case there is added to $\{y_k\}$ a *noise process* $\{v_k\}$, which results in the *measurement process* $\{z_k\}$. The *input process* to the system is $\{w_k\}$, and like $\{v_k\}$, it is a noise process. Further details of $\{v_k\}$ and $\{w_k\}$ will be given shortly, as will some motivation for introducing the whole model of Fig. 2.2-1.

Of course, the processes $\{v_k\}$, $\{w_k\}$, $\{x_k\}$, $\{y_k\}$, and $\{z_k\}$ in general will be vector processes. Normally we shall not distinguish between scalar and vector quantities.

Our prime concern in this and the next chapter will be to pose in precise terms, and solve, a filtering problem for the system depicted. In loose terms, the filtering problem is one of producing an estimate at time k of the system state x_k using measurements up till time k; i.e., the aim is to use the measured quantities $z_0, z_1, \ldots, z_k$ to intelligently guess at the value of x_k. Further, at each time instant k, we want the guess to be available.

*These and other key equations are summarized at the end of the section.

As those with even the most elementary exposure to probability theory will realize, almost nothing can be done unless some sort of probabilistic structure is placed on the input noise process $\{w_k\}$ and output noise process $\{v_k\}$. Here, we shall make the following assumptions:

ASSUMPTION 1. $\{v_k\}$ and $\{w_k\}$ are individually *white* processes. Here we speak of a white process* as one where, for any k and l with $k \neq l$, v_k and v_l are independent random variables and w_k and w_l are independent random variables.

ASSUMPTION 2. $\{v_k\}$ and $\{w_k\}$ are individually zero mean, gaussian processes with known covariances.

ASSUMPTION 3. $\{v_k\}$ and $\{w_k\}$ are independent processes.

Assumption 2 under normal circumstances would mean that the joint probability density of, say, $v_{k_1}, v_{k_2}, \ldots, v_{k_m}$ for arbitrary m and k_i is gaussian. In view of the whiteness of $\{v_k\}$ guaranteed by Assumption 1, the joint probability density is simply the product of the individual densities, and is therefore gaussian if the probability density of v_k for each single k is gaussian.

Here, we remind the reader that the probability density of a gaussian random variable† v is entirely determined by the mean m and covariance R of v, which are defined by

$$m = E[v] \qquad R = E[(v - m)(v - m)'] \qquad (2.3)$$

When v has dimension n and R is nonsingular, the probability density is

$$p_V(v) = \frac{1}{(2\pi)^{n/2} |R|^{1/2}} \exp\left[-\frac{1}{2}(v - m)' R^{-1}(v - m)\right] \qquad (2.4)$$

If R is singular, $p_V(v)$ is now no longer well defined, and probabilistic properties of v are more easily defined by its *characteristic function*, viz.

$$\phi_V(s) = E[\exp js'v] = \exp\left(jm's - \tfrac{1}{2}s'Rs\right) \qquad (2.5)$$

Since $\{v_k\}$ is white, we can arrange for Assumption 2 to be fulfilled if we specify that v_k has zero mean for each k and the covariance $E[v_k v_k']$ is known for all k. (For note that the *process* covariance is the *set* of values of $E[v_k v_l']$

---

*Historically, white processes have often been defined as those uncorrelated from instant to instant, as being also stationary, and usually having zero mean. The definition above is less restrictive.

†See Appendix A for a review of a number of ideas of probability theory and stochastic processes.

for all k and l. However, we see that for $k \neq l$

$$E[v_k v_l'] = E[v_k]E[v_l'] \quad \text{by the whiteness assumption}$$

$$= 0 \qquad \qquad \text{by the zero mean assumption)}$$

Consequently, if we know that $E[v_k v_k'] = R_k$ say, then the covariance of the $\{v_k\}$ process is given by

$$E[v_k v_l'] = R_k \delta_{kl} \tag{2.6}$$

for all k and l, where δ_{kl} is the Kronecker delta, which is 1 for $k = l$ and 0 otherwise. Likewise, the covariance of the $\{w_k\}$ process, which is zero mean, is completely specified by a sequence of matrices $\{Q_k\}$ such that

$$E[w_k w_l'] = Q_k \delta_{kl} \tag{2.7}$$

Of course, Q_k and R_k are nonnegative definite symmetric for all k. Note also that Assumption 3 and the zero mean assumption imply

$$E[v_k w_l'] = 0 \tag{2.8}$$

for all k and l. [Later in the book, we shall frequently relax (2.8). Prob. 2.6 considers the case of dependent v_k and w_k.]

For convenience, we can sum up the assumptions on $\{w_k\}$ and $\{v_k\}$ as follows:

The processes $\{v_k\}$ and $\{w_k\}$ are zero mean, independent gaussian processes with covariances given by (2.6) and (2.7).

Initial State Description

So far, we have not specified an initial condition for the difference equation (2.1). Under normal circumstances, one might expect to be told that at the initial time $k = 0$, the state x_0 was some prescribed vector. Here, however, we prefer to leave our options more open. From the practical point of view, if it is impossible to measure x_k exactly for arbitrary k, it is unlikely that x_0 will be available. This leads us to the adoption of a random initial condition for the system. In particular, we shall assume that x_0 is a gaussian random variable of known mean $\bar{x}_0$ and known covariance P_0, i.e.,

$$E[x_0] = \bar{x}_0 \qquad E\{[x_0 - \bar{x}_0][x_0 - \bar{x}_0]'\} = P_0 \tag{2.9}$$

Further, we shall assume that x_0 is independent of v_k and w_k for any k.

At this stage, the reader is probably wondering to what extent all the assumptions made are vital, and to what extent they are suggested by physical reality. Why shouldn't one, for example, choose an arbitrary probability density for x_0? Why should x_0 have to be independent of v_k?

We hope to indicate in subsequent parts of the book relaxations of many assumptions, and we hope also to indicate by way of examples the reasonableness of many. At this point we simply comment that experiments establish that many naturally occurring processes are gaussian; that by modelling certain natural processes as resulting from the sum of a number of individual, possibly nongaussian, processes, the central limit theorem of probability theory [3] suggests an approximately gaussian character for the sum; and finally, that the filtering problem is generally easier to solve with the gaussian assumption. So it is a combination of experiment, analysis, and pragmatism that suggests the gaussian assumption.

> EXAMPLE 2.1. In order to illustrate where a state-space signal model might be used in an engineering situation, consider the problem of prediction of air pollution levels to control the pollutants in an industrial district. The control would perhaps be to limit or close down an industrial process when certain pollutant levels were expected to rise beyond a certain threshold without such control. To meet such a requirement in Tokyo, a research team in a study reported in [4] selected five kinds of pollutants—OX, NO, NO_2, SO_2, and CO—as state variables in a state model and Kalman predictor. The meteorological conditions such as temperature, wind speed, and humidity could also have been included in the state vector, but their effects as studied in a factor analysis were shown to be relatively less significant.
>
> The structure of one of the models considered in [4] is simply $x_{k+1} = F_k x_k + b_k + w_k$ (regarded in [4] as a linear multiple regression model). Here, x_k is the state vector consisting of the concentrations of the pollutants, F_k is a system matrix, b_k is a bias vector and w_k is a vector of model errors. The observations are $y_k = x_k + v_k$, where v_k is the measurement noise. The system parameters including bias are not in this case obtained from a study of the physics and chemistry of the pollution dispersion, but are nonphysical parameters derived from an identification procedure not discussed here. In order to improve the model so as to whiten w_k, an increased order model
>
> $$x_{k+1} = A_1 x_k + A_2 x_{k-1} + \cdots + A_m x_{k-m} + b_k + w_k$$
>
> is also studied in [4]. This is readily reorganized as a higher order state model where now the state vector is $[x_k', x_{k-1}', \ldots, x_{k-m}']'$ and the measurements are again $y_k = x_k + v_k$.
>
> We see that the model errors are treated as system noise, and the measurement errors are taken into account as observation noise. The dimension of the model is not necessarily the dimension of the measurement vector; and in estimating the states from the measurements it may well be that in a model one or more of the measurements may contribute very little to the estimation (or prediction). In [4], predictors are developed based on Kalman filtering ideas, and the model is validated on actual data. The model is shown to be more useful than simpler autoregressive models, which are but a special class of the models discussed in this section, as pointed out in the next example.
>
> EXAMPLE 2.2. Provided that correlation of input and output noise is allowed, we can represent scalar autoregressive (AR), moving average (MA), and

ARMA processes in the standard form. Let $\{w_k\}$ be a sequence of independent $N(0, 1)$ random variables (i.e., each w_k is gaussian with mean zero and variance 1). Firstly, the process $\{z_k\}$ defined by

$$z_k = w_k + c_1 w_{k-1} + \cdots + c_n w_{k-n}$$

for constants c_i is a moving average process of order n. We set, with x_k an n-vector,

$$x_{k+1} = \begin{bmatrix} 0 & 1 & 0 & \cdots & 0 \\ 0 & 0 & 1 & \cdots & 0 \\ \cdot & \cdot & & & \cdot \\ \cdot & \cdot & & & \cdot \\ \cdot & \cdot & & & \cdot \\ 0 & 0 & 0 & \cdots & 1 \\ 0 & 0 & 0 & \cdots & 0 \end{bmatrix} x_k + \begin{bmatrix} 0 \\ 0 \\ \cdot \\ \cdot \\ \cdot \\ 0 \\ 1 \end{bmatrix} w_k$$

(Then $x_k = [w_{k-n} \quad w_{k-n+1} \quad \cdots \quad w_{k-1}]'$.) Also

$$z_k = [c_n \quad c_{n-1} \quad \cdots \quad c_1] x_k + v_k$$

where $v_k = w_k$. Secondly, with $\{w_k\}$ as above, the process $\{z_k\}$ defined by

$$z_k + a_1 z_{k-1} + \cdots + a_n z_{k-n} = w_k$$

with the a_i constants is called an autoregressive process of order n. (Usually, the zeros of $z^n + a_1 z^{n-1} + \cdots + a_n$ lie in $|z| < 1$, since this actually ensures that $\{z_k\}$ is stationary if $k_0 = -\infty$; this point need not, however, concern us here.) The process can be obtained in the following way:

$$x_{k+1} = \begin{bmatrix} -a_1 & -a_2 & \cdots & -a_{n-1} & -a_n \\ 1 & 0 & \cdots & 0 & 0 \\ 0 & 1 & \cdots & 0 & 0 \\ \cdot & \cdot & & \cdot & \cdot \\ \cdot & \cdot & & \cdot & \cdot \\ \cdot & \cdot & & \cdot & \cdot \\ 0 & 0 & \cdots & 1 & 0 \end{bmatrix} x_k + \begin{bmatrix} 1 \\ 0 \\ 0 \\ \cdot \\ \cdot \\ \cdot \\ 0 \end{bmatrix} w_k$$

$$z_k = -[a_1 \quad a_2 \quad \cdots \quad a_n] x_k + v_k$$

with $v_k = w_k$. (To see this, observe that from the state equation, with $\alpha_k = x_k^{(1)}$, we have

$$x_k = [\alpha_k \quad \alpha_{k-1} \quad \cdots \quad \alpha_{k-n}]'$$

and

$$\alpha_{k+1} = -a_1 \alpha_k - a_2 \alpha_{k-1} \cdots - a_n \alpha_{k-n} + w_k$$

From the output equation, we have

$$z_k = -a_1 \alpha_k - a_2 \alpha_{k-1} \cdots - a_n \alpha_{k-n} + w_k$$

Thus $z_k = \alpha_{k+1}$, and the autoregressive equation is immediate.) The ARMA equation is examined in the problems.

Gaussian and Markov Properties of the System State

We wish to note three important properties of the random process $\{x_k\}$ of the fundamental system of Fig. 2.2-1.

First, x_k *is a gaussian random variable.* (In other words, the gaussian character of the system input $\{w_k\}$ and the initial state x_0 propagates so as to make x_k gaussian for arbitrary k.) Why is this so? Observe from (2.1) that

$$x_k = \Phi_{k,0}x_0 + \sum_{l=0}^{k-1} \Phi_{k,l+1}G_l w_l \qquad (2.10)$$

where

$$\Phi_{k,l} = F_{k-1}F_{k-2}\cdots F_l \qquad (k > l) \qquad \Phi_{k,k} = I \qquad (2.11)$$

(Recall that $\Phi_{k,l}$ for all k and l constitutes the transition matrix for the homogeneous equation $x_{k+1} = F_k x_k$. As such, it has the important properties $\Phi_{k,k} = I$ and $\Phi_{k,l}\Phi_{l,m} = \Phi_{k,m}$ for all k, l, and m with $k \geq l \geq m$.)

Equation (2.10) expresses x_k as a linear combination of the jointly gaussian random vectors x_0, w_0, $w_1, \ldots, w_{k-1}$. (Note that the variables are jointly gaussian as a result of their being individually gaussian and independent.) Now since linear transformations of gaussian random variables preserve their gaussian character,* it follows that x_k is a gaussian random variable.

The second property is: $\{x_k\}$ *is a gaussian random process.* Of course, this property is simply an extension of the first. In effect, one has to show that for arbitrary m and k_i, $i = 1, \ldots, m$, the set of random variables x_{k_i} is jointly gaussian. The details will be omitted.

Finally, we claim that $\{x_k\}$ *is a Markov process.* In other words, if $k_1 < k_2 < \cdots < k_m < k$, the probability density of x_k conditioned on x_{k_1}, $x_{k_2}, \ldots x_{k_m}$ is simply the probability density of x_k conditioned on x_{k_m}:

$$p(x_k \,|\, x_{k_1}, x_{k_2}, \ldots, x_{k_m}) = p(x_k \,|\, x_{k_m})$$

This property is essentially a consequence of two factors: the whiteness of w_k and the causality of the system (2.1). In outline, one can argue the Markov property as follows. From (2.1), one has

$$x_k = \Phi_{k,k_m}x_{k_m} + \sum_{l=k_m}^{k-1} \Phi_{k,l+1}G_l w_l \qquad (2.12)$$

Now the summation in this equation involves w_l for $l \geq k_m$, with $k_m > k_i$ for all $i < m$. Hence the particular w_l are all independent of $x_{k_1}, x_{k_2}, \ldots, x_{k_m}$. Knowing $x_{k_1}, x_{k_2}, \ldots, x_{k_m}$, therefore, conveys no information whatsoever about these w_l. Of the sequence $x_{k_1}, x_{k_2}, \ldots, x_{k_m}$, only x_{k_m} can be relevant as a conditioning variable for x_k. This is the essence of the Markov property.

What of the measurement process $\{z_k\}$? Certainly, it too is a gaussian process, for essentially the same reasons as $\{x_k\}$. *In fact, $\{x_k\}$ and $\{z_k\}$ are jointly gaussian.* But $\{z_k\}$ is, perhaps surprisingly at first glance, no longer a Markov process (except in isolated instances). Roughly speaking, the reason is that the process $\{y_k\}$ is not white, nor is the correlation between y_k and y_l

---

*Once more, we remind the reader of the existence of Appendix A, summarizing many results of probability theory and stochastic processes.

normally zero if $|k - l| > 1$. This means that y_{k-2} and y_{k-1} may convey more information jointly about y_k than y_{k-1} alone. Consequently, $\{y_k\}$ is not usually Markov and so neither is $\{z_k\}$.

Propagation of Means and Covariances

As just noted, $\{x_k\}$ and $\{z_k\}$ are jointly gaussian processes. Therefore, their probabilistic properties are entirely determined by their means and covariances. We wish here to indicate what these means and covariances are. The means are easily dealt with. From (2.10), the linearity of the expectation operator, and the fact that $E[w_k] = 0$ for all k, we obtain

$$E[x_k] = \Phi_{k,0}\bar{x}_0 \tag{2.13}$$

Equivalently,

$$E[x_{k+1}] = F_k E[x_k] \tag{2.14}$$

which equally follows from (2.1) as from (2.13). From (2.2), we have

$$E[z_k] = H'_k E[x_k] \tag{2.15}$$

Now let us consider the covariance functions. For ease of notation, we shall write $\bar{x}_k$ for $E[x_k]$ and shall compute the quantity

$$P_{k,l} = E\{[x_k - \bar{x}_k][x_l - \bar{x}_l]'\} \tag{2.16}$$

for $k \geq l$. (The case $k < l$ can be recovered by matrix transposition.) The calculation is straightforward. From (2.10) and (2.13), we have

$$P_{k,l} = E\left\{\left[\Phi_{k,0}(x_0 - \bar{x}_0)\right.\right.$$
$$\left.\left. + \sum_{m=0}^{k-1} \Phi_{k,m+1}G_m w_m\right]\left[\Phi_{l,0}(x_0 - \bar{x}_0) + \sum_{n=0}^{l-1} \Phi_{l,n+1}G_n w_n\right]'\right\}$$

Next, we use the fact that the random variables $x_0 - \bar{x}_0, w_0, \ldots, w_{k-1}$ are all independent. This means that when expectations are taken in the above expression, many terms disappear. One is left with

$$P_{k,l} = \Phi_{k,0}E\{[x_0 - \bar{x}_0][x_0 - \bar{x}_0]'\}\Phi'_{l,0}$$
$$+ \sum_{m=0}^{l-1} \Phi_{k,m+1}G_m Q_m G'_m \Phi'_{l,m+1}$$
$$= \Phi_{k,l}\left\{\Phi_{l,0}P_0\Phi'_{l,0} + \sum_{m=0}^{l-1} \Phi_{l,m+1}G_m Q_m G'_m \Phi'_{l,m+1}\right\} \tag{2.17}$$

In obtaining the first equality in (2.17), we have used the independence property, and $E[w_m w'_n] = Q_m \delta_{mn}$, see (2.7). The second equality follows from the assumptions on x_0 and from Eq. (2.11) for $\Phi_{k,l}$. Equations (2.13) and (2.17) together provide all the probabilistic information there is to know about the gaussian process $\{x_k\}$. However, an alternative formulation is sometimes helpful: just as the mean $\bar{x}_k$ satisfies a difference equation [see (2.14)], so we can see how to obtain $P_{k,l}$ from a difference equation. First, specializing (2.17) to the case $k = l$, we obtain

$$P_{k,k} = \Phi_{k,0} P_0 \Phi'_{k,0} + \sum_{m=0}^{k-1} \Phi_{k,m+1} G_m Q_m G'_m \Phi'_{k,m+1} \qquad (2.18)$$

Let us adopt the notation P_k in lieu of $P_{k,k}$ in view of the frequent future reoccurrence of the quantity. Observe that this notation is consistent with the use of P_0 to denote $E\{[x_0 - \bar{x}_0][x_0 - \bar{x}_0]'\}$ since

$$P_k = P_{k,k} = E\{[x_k - \bar{x}_k][x_k - \bar{x}_k]'\} \qquad (2.19)$$

Now using either (2.18) or the state equation (2.1), it is straightforward to show that

$$P_{k+1} = F_k P_k F'_k + G_k Q_k G'_k \qquad (2.20)$$

Equation (2.20) constitutes a difference equation for P_k, allowing computation of this quantity recursively, starting with the known quantity P_0. Once P_k is obtained for all k, the matrix $P_{k,l}$ for all k and l follows. Reference to (2.17) (which is valid for $k \geq l$) and (2.18) shows that

$$P_{k,l} = \Phi_{k,l} P_l \qquad k \geq l \qquad (2.21)$$

From (2.16), it is evident that $P_{k,l} = P'_{l,k}$; so for $k \leq l$ we must obtain

$$P_{k,l} = P_{k,k} \Phi'_{l,k} \qquad k \leq l \qquad (2.22)$$

Equations (2.20) through (2.22) together give another way of obtaining the state covariance.

The mean of the $\{z_k\}$ process has already been studied. The covariance essentially follows from that for $\{x_k\}$. Recall (2.2):

$$z_k = H'_k x_k + v_k \qquad (2.2)$$

Let us write $\bar{z}_k$ for $E[z_k]$. It follows that

$$\begin{aligned}
\text{cov}\,[z_k, z_l] &= E\{[z_k - \bar{z}_k][z_l - \bar{z}_l]'\} \\
&= E\{H'_k[x_k - \bar{x}_k][x_l - \bar{x}_l]'H_l\} \\
&\quad + E\{H'_k[x_k - \bar{x}_k]v'_l\} \\
&\quad + E\{v_k[x_l - \bar{x}_l]'H_l\} \\
&\quad + E\{v_k v'_l\}
\end{aligned}$$

Evidently the first summand in this expression can be written as

$$\begin{aligned}
E\{H'_k[x_k - \bar{x}_k][x_l - \bar{x}_l]'H_l\} &= H'_k E\{[x_k - \bar{x}_k][x_l - \bar{x}_l]'\}H_l \\
&= H'_k P_{k,l} H_l
\end{aligned}$$

Noting that $\{v_k\}$ must be independent of $\{x_k - \bar{x}_k\}$ (the latter process being determined by x_0 and $\{w_k\}$, which are independent of $\{v_k\}$), we see that the second and third summands are zero. The fourth summand has been noted to be $R_k \delta_{kl}$. Therefore

$$\begin{aligned}
\mathrm{cov}\,[z_k, z_l] &= H_k' \Phi_{k,l} P_l H_l + R_k \delta_{kl} && k \geq l \\
&= H_k' P_k \Phi_{l,k}' H_l + R_k \delta_{kl} && k \leq l
\end{aligned} \tag{2.23}$$

Dropping the Gaussian Assumption

Hitherto, x_0, $\{v_k\}$, and $\{w_k\}$ have been assumed gaussian. If this is not the case, but they remain described by their first order and second order statistics, then *all the calculations still carry through in the sense that formulas for the mean and covariance of the* $\{x_k\}$ *and* $\{z_k\}$ *sequences are precisely as before.* Of course, in the gaussian case, knowledge of the mean and covariance is sufficient to deduce density functions of any order. In the nongaussian case, knowledge of the mean and covariance does not provide other than incomplete information about higher order moments, let alone probability density functions.

Main Points of the Section

The most important points to understand in this section are the form of the model—its linearity and finite dimensionality and the assumptions regarding the input noise, measurement noise, and initial state. The fact that the mean and covariance of the $\{x_k\}$ and $\{z_k\}$ process can be calculated is important, as is their gaussian nature. While the particular formulas for these means and covariances are not of prime importance, the student could well remember that these formulas can be obtained by solving difference equations.

The particular formulas are as follows:

Signal model. With x_0, $\{v_k\}$, $\{w_k\}$ independent and gaussian,

$$x_{k+1} = F_k x_k + G_k w_k \qquad k \geq 0$$
$$z_k = y_k + v_k = H_k' x_k + v_k$$

with

$$E[w_k] = 0,\ E[w_k w_l'] = Q_k \delta_{kl},\ E[v_k] = 0,\ E[v_k v_l'] = R_k \delta_{kl}$$
$$E[x_0] = \bar{x}_0,\ E\{[x_0 - \bar{x}_0][x_0 - \bar{x}_0]'\} = P_0$$

State statistics. $\{x_k\}$ is gaussian and Markov with

$$\bar{x}_k = E[x_k] = \Phi_{k,0}\bar{x}_0 \qquad \bar{x}_{k+1} = F_k \bar{x}_k$$
$$E\{[x_k - \bar{x}_k][x_k - \bar{x}_k]'\} = P_k = \Phi_{k,0} P_0 \Phi_{k,0}' + \sum_{m=0}^{k-1} \Phi_{k,m+1} G_m Q_m G_m' \Phi_{k,m+1}'$$
$$P_{k+1} = F_k P_k F_k' + G_k Q_k G_k'$$
$$E\{[x_k - \bar{x}_k][x_l - \bar{x}_l]'\} = P_{k,l} = \Phi_{k,l} P_l \qquad k \geq l$$

Output statistics. $\{z_k\}$ is gaussian (in fact jointly gaussian with $\{x_k\}$), but not normally Markov, with

$$\bar{z}_k = E[z_k] = H_k'\bar{x}_k$$

$$E\{[z_k - \bar{z}_k][z_l - \bar{z}_l]'\} = H_k'\Phi_{k,l}P_lH_l + R_k\delta_{kl} \qquad k \geq l$$

Expressions for the mean and covariance of the $\{x_k\}$ and $\{z_k\}$ processes are still valid if x_0, $\{v_k\}$, and $\{w_k\}$ fail to be gaussian.

Problem 2.1. (This problem shows that for jointly gaussian random variables, the properties of zero correlation and independence are equivalent.) Let a and b be two jointly gaussian vector random variables, with $E[a]E[b'] = E[ab']$. Show that the variables are independent. (Since the probability densities of a and b may not exist, prove independence by showing that the joint characteristic function of a and b is the product of the separate characteristic functions of a and b.)

Problem 2.2. In setting up a description of the system of Fig. 2.2-1, (See p. 13), we assumed x_0 to be a gaussian random variable of known mean and covariance. Can one retain this notion and at the same time cover the case when x_0 takes a prescribed value?

Problem 2.3. Establish the recursive formula (2.20) for P_k both from (2.18) and directly from (2.1) with $k = l$. How is the formula initialized?

Problem 2.4. Adopt the same model as described in this section, save that the noise processes $\{v_k\}$ and $\{w_k\}$ possess nonzero known means. Find a new equation for the evolution of the mean of x_k, and show that the covariance of the $\{x_k\}$ process is unaltered.

Problem 2.5. Show for the model described in this section that cov $[z_k, z_l]$ can be written in the form

$$A_k'B_l1(k - l) + B_k'A_l1(l - k) + C_k\delta_{kl}$$

where $1(k - l)$ is 0 for $k - l \leq 0$, and 1 otherwise, and A_k and B_k are matrices for each k.

Problem 2.6. With $\{w_k\}$ a sequence of independent $N(0, 1)$ random variables, define a process z_k by

$$z_k + a_1z_{k-1} + \cdots + a_nz_{k-n} = w_k + c_1w_{k-1} + \cdots + c_nw_{k-n}$$

Establish that a signal model like that of this section can be obtained from the AR signal model of Example 2.2 by varying only the output equation.

Problem 2.7. What variations should be made to the state and output statistics calculations in case $E[v_kw_l'] = C_k\delta_{kl}$ for some $C_k \neq 0$?

2.3 ESTIMATION CRITERIA

In this section, we wish to indicate how knowledge of the value taken by one random variable can give information about the value taken by a second random variable. Particularly, we shall note how an *estimate* can be made of the value taken by this second random variable. The detail of this section is independent of that of the previous section, but there is a conceptual link to be explored in the next chapter. In the last section, we introduced, amongst other things, two random sequences $\{x_k\}$ and $\{z_k\}$ and posed the filtering problem as one of finding at time k some information about x_k from $z_0, z_1, \ldots, z_k$. Now think of $z_0, z_1, \ldots, z_k$ as one (vector) random variable Z_k. Then the task of filtering at any one fixed time instant which is to find information about the random variable x_k given the value of the random variable Z_k, is a particular instance of the general problem considered in this section.

Much of this section could logically have been placed in Appendix A; it is, however, so crucial to what follows that we have elected to give it separate treatment.

Notation. As in Appendix A, we shall use in the remainder of this section an upper-case letter to denote a random variable, and a lowercase letter to denote a value taken by that variable; i.e., if the random variable is X and the underlying probability space Ω has elements ω, the symbol x will in effect be used in place of $X(\omega)$ and the symbol X in place of $X(\cdot)$, or the set of pairs $\{\omega, X(\omega)\}$ as ω ranges over Ω.

If X and Y are two vector random variables, what does the knowledge that $Y = y$ tell us about X? The answer to this question is summed up in the concept of *conditional probability density*. Suppose that before we know that $Y = y$, the random variable X has a probability density $p_X(x)$. Being told that $Y = y$ has the effect of modifying the probability density. The modified density, termed the conditional probability density, is

$$p_{X|Y}(x|y) = \frac{p_{X,Y}(x, y)}{p_Y(y)}$$

assuming that $p_Y(y) \neq 0$.

We shall present a number of examples to point up this and later ideas, and we caution the reader that these examples form an integral part of the text, and reference in the future will be made to them.

EXAMPLE 3.1. Consider the simple relationship

$$y = x + n$$

for the values of scalar random variables Y, X, and N. Suppose a value $Y = y$ is measured. For the case when X and N are independent and gaussian with

23

zero means and variances Σ_x and Σ_n, the conditional probability density $p_{X|Y}(x|y)$ can be evaluated as follows:

$$p_{X|Y}(x|y) = \frac{p_{Y|X}(y|x)p_X(x)}{p_Y(y)} = \frac{p_{Y|X}(y|x)p_X(x)}{\int_{-\infty}^{+\infty} p_{Y|X}(y|x)p_X(x)\,dx}$$

$$= \frac{p_{X+N|X}(x+n|x)p_X(x)}{\int_{-\infty}^{\infty} p_{X+N|X}(x+n|x)p_X(x)\,dx}$$

$$= \frac{p_N(y-x)p_X(x)}{\int_{-\infty}^{\infty} p_N(y-x)p_X(x)\,dx}$$

$$= \frac{(2\pi\Sigma_n)^{-1/2}\exp\left[-\tfrac{1}{2}(y-x)^2\Sigma_n^{-1}\right](2\pi\Sigma_x)^{-1/2}\exp\left(-\tfrac{1}{2}x^2\Sigma_x^{-1}\right)}{(2\pi)^{-1}(\Sigma_x\Sigma_n)^{-1/2}\int_{-\infty}^{\infty}\exp\left[-\tfrac{1}{2}(y-x)^2\Sigma_n^{-1}-\tfrac{1}{2}x^2\Sigma_x^{-1}\right]dx}$$

$$= \frac{(2\pi)^{-1}(\Sigma_x\Sigma_n)^{-1/2}\exp\left\{-\tfrac{1}{2}[(y-x)^2\Sigma_n^{-1}+x^2\Sigma_x^{-1}]\right\}}{(2\pi)^{-1/2}(\Sigma_x+\Sigma_n)^{-1/2}\exp\left\{-\tfrac{1}{2}y^2(\Sigma_x+\Sigma_n)^{-1}\right\}}$$

$$= \left(2\pi\frac{\Sigma_n\Sigma_x}{\Sigma_x+\Sigma_n}\right)^{-1/2}\exp\left\{-\frac{1}{2}\left(x-\frac{y\Sigma_x}{\Sigma_x+\Sigma_n}\right)^2\left(\frac{\Sigma_n\Sigma_x}{\Sigma_x+\Sigma_n}\right)^{-1}\right\}$$

The first of the above steps makes use of Bayes' rule. The remaining steps are straightforward from the probability point of view, except perhaps for the evaluation of the integral

$$\int_{-\infty}^{+\infty}\exp\left\{-\tfrac{1}{2}[(y-x)^2\Sigma_n^{-1}+x^2\Sigma_x^{-1}]\right\}dx$$

By writing this integral in the form

$$k_1\int_{-\infty}^{+\infty}\exp\left[-k_2(x-k_3)^2\right]dx$$

and using

$$\int_{-\infty}^{+\infty}\exp\left(-\tfrac{1}{2}x^2\right)dx = \sqrt{2\pi}$$

one can evaluate the integral.

Notice that in the above example $p_{X|Y}(x|y)$ *is itself a gaussian density,* with mean $y\Sigma_x(\Sigma_x+\Sigma_n)^{-1}$ and variance $\Sigma_n\Sigma_x(\Sigma_x+\Sigma_n)^{-1}$. *It is actually a general property that for two jointly gaussian random variables X and Y, the conditional density $p_{X|Y}(x|y)$ is also gaussian.* (Note that in Example 3.1, Y is gaussian, being the sum of two gaussian variables.) This is shown in the following example.

EXAMPLE 3.2. Let the pair of vectors X and Y be jointly gaussian, i.e., with $Z = [X'\quad Y']'$; Z is gaussian with mean and covariance

$$m = \begin{bmatrix} \bar{x} \\ \bar{y} \end{bmatrix} \quad \text{and} \quad \Sigma = \begin{bmatrix} \Sigma_{xx} & \Sigma_{xy} \\ \Sigma_{yx} & \Sigma_{yy} \end{bmatrix}$$

respectively. We shall show that X is conditionally gaussian.

Assuming nonsingularity of Σ and Σ_{yy}, we have

$$p_{X|Y}(x|y) = \frac{p_{XY}(x, y)}{p_Y(y)}$$

$$= \frac{1}{(2\pi)^{N/2}} \frac{|\Sigma_{yy}|^{1/2}}{|\Sigma|^{1/2}} \frac{\exp\{-\frac{1}{2}[x' - \bar{x}' \vdots y' - \bar{y}']\Sigma^{-1}[x' - \bar{x}' \vdots y' - \bar{y}']'\}}{\exp[-\frac{1}{2}(y - \bar{y})'\Sigma_{yy}^{-1}(y - \bar{y})]}$$

(Here, N is the dimension of X.) Now we shall rewrite this density, using the easily checked formula

$$\begin{bmatrix} I & -\Sigma_{xy}\Sigma_{yy}^{-1} \\ 0 & I \end{bmatrix} \Sigma \begin{bmatrix} I & 0 \\ -\Sigma_{yy}^{-1}\Sigma_{xy}' & I \end{bmatrix} = \begin{bmatrix} \Sigma_{xx} - \Sigma_{xy}\Sigma_{yy}^{-1}\Sigma_{yx} & 0 \\ 0 & \Sigma_{yy} \end{bmatrix}$$

First, taking determinants in this formula, we have

$$|\Sigma| = |\Sigma_{xx} - \Sigma_{xy}\Sigma_{yy}^{-1}\Sigma_{yx}||\Sigma_{yy}|$$

Second, it yields

$$[x' - \bar{x}' \vdots y' - \bar{y}']\Sigma^{-1}[x' - \bar{x}' \vdots y' - \bar{y}']'$$

$$= [x' - \bar{x} \vdots y' - \bar{y}'] \begin{bmatrix} I & 0 \\ -\Sigma_{yy}^{-1}\Sigma_{xy}' & I \end{bmatrix} \begin{bmatrix} (\Sigma_{xx} - \Sigma_{xy}\Sigma_{yy}^{-1}\Sigma_{yx})^{-1} & 0 \\ 0 & \Sigma_{yy}^{-1} \end{bmatrix}$$

$$\times \begin{bmatrix} I & -\Sigma_{xy}\Sigma_{yy}^{-1} \\ 0 & I \end{bmatrix} [x' - \bar{x}' \vdots y' - \bar{y}']'$$

$$= (x' - \bar{\bar{x}}')(\Sigma_{xx} - \Sigma_{xy}\Sigma_{yy}^{-1}\Sigma_{yx})^{-1}(x - \bar{\bar{x}}) + (y' - \bar{y}')\Sigma_{yy}^{-1}(y - \bar{y})$$

where

$$\bar{\bar{x}} = \bar{x} + \Sigma_{xy}\Sigma_{yy}^{-1}(y - \bar{y})$$

Therefore, we have

$$p_{X|Y}(x|y) = \frac{1}{(2\pi)^{N/2}|\Sigma_{xx} - \Sigma_{xy}\Sigma_{yy}^{-1}\Sigma_{yx}|^{1/2}}$$

$$\times \exp[-\frac{1}{2}(x' - \bar{\bar{x}}')(\Sigma_{xx} - \Sigma_{xy}\Sigma_{yy}^{-1}\Sigma_{yx})^{-1}(x - \bar{\bar{x}})]$$

As claimed then, X is indeed conditionally gaussian. In fact, this is true even when Σ or Σ_{yy} are singular; in this case one must eschew the use of probability densities and work with characteristic functions. Further, one must use the next best thing to an inverse of Σ_{yy}, and this is the pseudo-inverse, described in Appendix B; when an inverse exists, it equals the pseudo-inverse. The result is that X *conditioned on* $Y = y$ *has conditional mean* $\bar{x} + \Sigma_{xy}\Sigma_{yy}^{\#}(y - \bar{y})$ *and conditional covariance* $\Sigma_{xx} - \Sigma_{xy}\Sigma_{yy}^{\#}\Sigma_{xy}'$. Thus the inverse in the usual formula is simply replaced by a pseudo-inverse. Notice that the conditional covariance is independent of y; this is a special property associated with the particular form of density assumed—one cannot expect it in general.

Estimates of X Given Y = y

The conditional probability density $p_{X|Y}(x|y)$ with a particular value substituted for y and with x regarded as a variable sums up all the information which knowledge that $Y = y$ conveys about X. Since it is a function rather than a single real number or vector of real numbers, it makes sense to ask if one can throw away some of the information to obtain a simpler entity. One might, for example, seek a single estimate of the value taken by X given the knowledge that $Y = y$. How might one intelligently generate such an estimate?

Obviously, one such estimate would be the value of x maximizing $p_{X|Y}(x|y)$, that is, the maximum a posteriori estimate. However, we shall find it helpful to introduce a different kind of estimate, namely the minimum variance estimate (more properly, a conditional minimum variance estimate).

Minimum Variance Estimate

Let us denote an estimate of the value taken by X as $\hat{x}$ when we know that $Y = y$. Then, in general, $\hat{x}$ will not equal x, the actual value taken by X. On occasions, $x - \hat{x}$ may be small, on other occasions large. An average measure of the error is provided by

$$E\{\| X - \hat{x} \|^2 | Y = y\}$$

where $\| a \|^2 = a'a$. Remember that $\hat{x}$, the estimate, has to be determined somehow from y. It is therefore a fixed number (or vector of numbers) in the above expectation. On the other hand, knowledge of y does not pin down X, and so X remains a random variable in the expectation.

We define a *minimum variance estimate* $\hat{x}$ as one for which

$$E\{\| X - \hat{x} \|^2 | Y = y\} \leq E\{\| X - z \|^2 | Y = y\} \tag{3.1}$$

for all vectors z, determined in some way from y. As we shall see in a moment, $\hat{x}$ is unique. Other names for $\hat{x}$ are: least squares estimate, minimum mean-square estimate, and recognizable variants on these terms.*

A major property of the minimum variance estimate is contained in the following theorem; as a study of the theorem statement shows, the theorem also serves to establish uniqueness and to point out another reason that makes $\hat{x}$ an intelligent choice of estimate.

THEOREM 3.1. Let X and Y be two jointly distributed random vectors, and let Y be measured as taking the value y. Let $\hat{x}$ be a minimum variance estimate of X as defined above. Then $\hat{x}$ is also uniquely specified as *the conditional mean of X given that $Y = y$*, i.e.,

$$\hat{x} = E[X | Y = y] = \int_{-\infty}^{+\infty} x p_{X|Y}(x|y)\, dx \tag{3.2}$$

---

*Some of these names involve some abuse of nomenclature, since they may fail to suggest the conditional nature of the estimate and the fact that a priori probability information is used.

Proof. Observe that, with z possibly dependent on y but not depending on x,

$$E\{\|X - z\|^2 \,|\, Y = y\}$$

$$= \int_{-\infty}^{+\infty} (x - z)'(x - z)p_{X|Y}(x|y)\,dx$$

$$= \int_{-\infty}^{\infty} x'x p_{X|Y}(x|y)\,dx - 2z' \int_{-\infty}^{\infty} x p_{X|Y}(x|y)\,dx + z'z$$

$$= \left[z' - \int_{-\infty}^{\infty} x' p_{X|Y}(x|y)\,dx \right]\left[z - \int_{-\infty}^{\infty} x p_{X|Y}(x|y)\,dx \right]$$

$$+ \int_{-\infty}^{+\infty} x'x p_{X|Y}(x|y)\,dx - \left\| \int_{-\infty}^{+\infty} x p_{X|Y}(x|y)\,dx \right\|^2$$

The expression on the right side, regarded as a function of z, has a unique minimum when $z = E[X \,|\, Y = y]$.

As a byproduct, the proof of the above theorem contains the value of the average mean square error associated with the estimate $\hat{x}$. On setting $z = \hat{x}$ in the last equality, we obtain

$$E\{\|X - \hat{x}\|^2 \,|\, Y = y\} = E\{\|X\|^2 \,|\, Y = y\} - \|\hat{x}\|^2$$

Note that the theorem provides yet another descriptive term for this type of estimate, viz., *conditional mean estimate*.

EXAMPLE 3.3. As in Example 3.1, let X and N be two independent, zero mean, gaussian random variables of variances Σ_x and Σ_n. Let $Y = X + N$, and suppose Y is measured in an experiment as having the value y. Then the conditional mean estimate of the value taken by X is

$$\hat{x} = \int_{-\infty}^{+\infty} x p_{X|Y}(x|y)\,dx$$

$$= \left(2\pi \frac{\Sigma_x \Sigma_n}{\Sigma_x + \Sigma_n} \right)^{-1/2} \int_{-\infty}^{+\infty} x \exp\left[-\frac{1}{2}\left(x - \frac{y\Sigma_x}{\Sigma_x + \Sigma_n} \right)^2 \left(\frac{\Sigma_n \Sigma_x}{\Sigma_x + \Sigma_n} \right)^{-1} \right] dx$$

We are using the expression for $p_{X|Y}(x|y)$ computed in Example 3.1. The integral can be evaluated to yield

$$\hat{x} = \frac{\Sigma_x}{\Sigma_x + \Sigma_n} y$$

Alternatively, this follows immediately from the fact that $p_{X|Y}(x|y)$ is

$$N\left(\frac{\Sigma_x}{\Sigma_x + \Sigma_n} y, \frac{\Sigma_n \Sigma_x}{\Sigma_x + \Sigma_n} \right)$$

as noted earlier. The conditional error variance $E\{\|X - \hat{x}\|^2 \,|\, Y = y\}$ is simply the variance associated with the density $p_{X|Y}(x|y)$ since $\hat{x}$ is the mean of this density, and is accordingly $\Sigma_n \Sigma_x / (\Sigma_x + \Sigma_n)$. Of course if some other form of estimate were used, one would not expect the error variance to be the same as the variance associated with the conditional density.

EXAMPLE 3.4. As in Example 3.2, let X and Y be jointly distributed, gaussian random vectors with mean and covariance

$$\begin{bmatrix} \bar{x} \\ \bar{y} \end{bmatrix}, \quad \begin{bmatrix} \Sigma_{xx} & \Sigma_{xy} \\ \Sigma'_{xy} & \Sigma_{yy} \end{bmatrix}$$

Suppose Y is measured as having the value y. Then the conditional mean estimate of the value taken by X is known from the parameters of the gaussian density $p_{X|Y}(x|y)$ (or the gaussian characteristic function in case the density does not exist). Thus

$$\hat{x} = E[X| Y = y] = \bar{x} + \Sigma_{xy}\Sigma_{yy}^{-1}(y - \bar{y})$$

Thus $\hat{x}$ is derived from y by an affine transformation, i.e., *one of the form* $y \longrightarrow Ay + b = \hat{x}$. As for the scalar case, the conditional error covariance is the same as the covariance of the conditioned random variable when the estimate is the conditional mean, namely $\Sigma_{xx} - \Sigma_{xy}\Sigma_{yy}^{-1}\Sigma'_{xy}$. The average mean square error is the trace of this quantity. (Show this in two lines!)

Estimates and Estimators*

Hitherto, we have described a procedure which involves the use of a known vector of numbers, namely y, to produce with the aid of a conditional density another known vector of numbers, namely $\hat{x}$, termed the estimate of x. But clearly, what we have done is to have given *a general rule for passing from any vector of numbers y to the associated $\hat{x}$.* In other words, we have defined a function. The domain of this function is the set of values y, or the random variable Y. *As a function of a random variable, it is itself a random variable, which we shall call $\hat{X}$.* A particular value of $\hat{X}$ is $\hat{x}$, given by $E\{X| Y = y\}$, i.e.,

$$\hat{X}(y) = \hat{x} = E\{X| Y = y\}$$

So evidently,

$$\hat{X} = E\{X| Y\} \tag{3.3}$$

Since we wish to use the term estimate for a particular value $\hat{x}$ of $\hat{X}$ taken as a result of a particular value y taken by Y, we shall term $\hat{X}$ an *estimator* of X in terms of Y. *Thus the estimator is a rule—or a function, in the sense that a function is a rule—for associating particular values of two variables.* In contrast, an estimate is a value taken by the estimator, regarded as a function. The distinction is illustrated in Fig. 2.3-1.

*This material is not essential to the first and simplest derivation of the Kalman filter given in the text, but is essential for the more sophisticated proof in Chaps. 3 and 5.

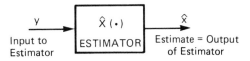

Fig. 2.3-1 The estimator is a function, or a device for assigning a number, given a measurement.

EXAMPLE 3.5. In estimating the value of a random variable X given a measurement of a random variable $Y = X + N$, with X and N having densities as described in Example 3.1, we found

$$\hat{x} = \frac{\Sigma_x}{\Sigma_x + \Sigma_n} y$$

The associated estimator is

$$\hat{X} = \frac{\Sigma_x}{\Sigma_x + \Sigma_n} Y$$

Note that $\hat{X}$, as a function of a random variable, is itself a random variable; and thus *it has its own mean*, viz.,

$$E[\hat{X}] = E\left[\frac{\Sigma_x Y}{\Sigma_x + \Sigma_n}\right] = \frac{\Sigma_x}{\Sigma_x + \Sigma_n} E[Y] = 0$$

and its own variance

$$E[\hat{X}^2] = \frac{\Sigma_x^2}{(\Sigma_x + \Sigma_n)^2} E[Y^2] = \frac{\Sigma_x^2}{\Sigma_x + \Sigma_n}$$

Minimum Variance Estimator Property

As we have seen, $\hat{x}$ is, in the sense of minimum error variance, the best estimate of the value taken by X, given that $Y = y$. We would then imagine that $\hat{X}$ would be, in the sense of minimum error variance, the best estimator. In other words, if $Z(\cdot)$ is an arbitrary function mapping values taken by Y into a space of the same dimension as X, we might have

$$E_{X,Y}\{\|X - \hat{X}(Y)\|^2\} \le E_{X,Y}\{\|X - Z(Y)\|^2\} \tag{3.4}$$

(the subscripts on the expectation operator indicate the variables with respect to which expectation is being taken.) Here the expectation is *not* a conditional one, but is over all possible values of X and Y.

This conjecture is nontrivially different from (3.1), but may be verified as follows. For the verification, we recall the following properties of the conditional expectation operator:

$$E_{X|Y}\{h(X, Y) \mid Y = y\} = E_{X|Y}\{h(X, y) \mid Y = y\} \tag{3.5}$$

and

$$E_Y\{E_{X|Y}\{h(X, Y) \mid Y = y\}\} = E_{X,Y}\{h(X, Y)\} \tag{3.6}$$

Now to verify the conjecture, we have from (3.1) that

$$E_{x|Y}\{\|X - \hat{X}(y)\|^2 \mid Y = y\} \leq E_{x|Y}\{\|X - Z(y)\|^2 \mid Y = y\}$$

and by (3.5),

$$E_{x|Y}\{\|X - \hat{X}(Y)\|^2 \mid Y = y\} \leq E_{x|Y}\{\|X - Z(Y)\|^2 \mid Y = y\}$$

Now take expectations with respect to Y. The inequality is preserved, and by (3.6) we have, as required,

$$E_{x,Y}\{\|X - \hat{X}(Y)\|^2\} \leq E_{x,Y}\{\|X - Z(Y)\|^2\}$$

In effect, we have proved the following theorem:

THEOREM 3.2. Let X and Y be two jointly distributed random vectors. Then the minimum variance estimator $\hat{X}$ of X in terms of Y is

$$\hat{X} = E\{X \mid Y\} \tag{3.3}$$

Equation (3.4) may sometimes be written loosely as

$$E\{\|X - \hat{x}\|^2\} \leq E\{\|X - z\|^2\} \tag{3.7}$$

where the expectation is over X and Y. This equation is inconsistent with our notational convention, in that $\hat{x}$ represents a particular value taken by a random variable, so that taking its expectation is a meaningless operation in the above context. The meaning which (3.7) is intended to convey should however be clear enough. It is precisely the meaning conveyed by the statement "$\hat{X}$ is a minimum variance estimator of X given Y", where the word "conditional" does *not* appear.

EXAMPLE 3.6. Let X, Y be jointly gaussian with mean and covariance as in Example 3.2. Then

$$\hat{X}(Y) = E[X \mid Y] = \bar{x} + \Sigma_{xy}\Sigma_{yy}^{-1}(Y - \bar{y})$$

and

$$E\{[X - \hat{X}(y)][X - \hat{X}(y)]' \mid Y = y\} = \Sigma_{xx} - \Sigma_{xy}\Sigma_{yy}^{-1}\Sigma_{yx}$$

Also

$$E_{x,Y}\{[X - \hat{X}(Y)][X - \hat{X}(Y)]'\} = E_Y[E_{x|Y}\{[X - \hat{X}(Y)][X - \hat{X}(Y)]' \mid Y = y\}]$$
$$= E_Y[\Sigma_{xx} - \Sigma_{xy}\Sigma_{yy}^{-1}\Sigma_{xy}']$$
$$= \Sigma_{xx} - \Sigma_{xy}\Sigma_{yy}^{-1}\Sigma_{xy}'$$

and

$$E_{x,Y}\{\|X - \hat{X}(Y)\|^2\} = \text{trace}\,[\Sigma_{xx} - \Sigma_{xy}\Sigma_{yy}^{-1}\Sigma_{xy}']$$

In formal terms, we have shown that *the (unconditioned) error variance associated with the conditional mean estimate is the same as the conditional error covariance stemming from a particular $Y = y$; but note that this would not normally be the case in the absence of the gaussian assumption.*

Unbiased Estimates and Estimator Properties

As a further illustration of the above remarks distinguishing the conditional nature of a property of an estimate from the unconditioned nature of the associated property of an estimator, we consider the question of bias. We can talk of $\hat{x}$ as being an unbiased estimate in that the conditional expected error in using $\hat{x}$ as an estimate of X, given y, is zero:

$$E_{X|Y}\{X - \hat{x} \mid Y = y\} = E_{X|Y}\{X \mid Y = y\} - \hat{x}$$
$$= 0 \qquad (3.8)$$

We can talk of $\hat{Y}$ as being an unbiased estimator in that both

$$E_{X|Y}\{X - \hat{X}(Y) \mid Y = y\} = 0 \qquad (3.9)$$

and

$$E_{X,Y}\{X - \hat{X}(Y)\} = 0 \qquad (3.10)$$

In later sections and chapters, we shall often use the same symbol to denote a random variable and the value taken by it; in particular, the same symbol may be used to denote an estimator and an estimate. There should be no confusion if the above distinctions are kept in mind.

EXAMPLE 3.7. Let X and Y be jointly distributed random variables as in Example 3.2. Then

$$E[X \mid Y = y] = \bar{x} + \Sigma_{xy}\Sigma_{yy}^{-1}(y - \bar{y})$$

and

$$\hat{X}(Y) = \bar{x} + \Sigma_{xy}\Sigma_{yy}^{-1}(Y - \bar{y})$$

It follows that

$$E[\hat{X}(Y)] = \bar{x} + \Sigma_{xy}\Sigma_{yy}^{-1}[E(Y) - \bar{y}]$$
$$= \bar{x}$$
$$= E[X]$$

as expected.

Other Estimation Criteria

As we have shown, the conditional mean estimate is that which minimizes the average value of $\|x - \hat{x}\|^2$. It is somewhat arbitrary that one chooses to measure the error associated with an estimate as $\|x - \hat{x}\|^2$, rather than, say, $\|x - \hat{x}\|$, $\|x - \hat{x}\|^4$, or cosh $\|x - \hat{x}\|^3$. Two alternative error measures and the associated estimates are described in the problems of this section. The estimates both have some intuitive significance, being the conditional median and conditional mode (also termed the maximum a posteriori estimate).

For a gaussian density with x scalar, the median is identical with the mean in view of the symmetry of the density. *For any gaussian density the mode is identical with the mean; thus maximum a posteriori estimates are precisely conditional mean estimates.* Other types of estimates also agree with the conditional mean estimate for gaussian densities and indeed certain other densities (see [1, 5, 6]).

It is also possible to demand that estimators have a certain structure, e.g., that they define an affine function; one then seeks the best estimator within the class defined by this structure. Some development along these lines is given in a later chapter; it turns out that the error measure provided by the average value of $||x - \hat{x}||^2$ is particularly suited to the development of estimators which are constrained to be linear.

Main Points of the Section

For arbitrary densities,

1. $p_{X|Y}(x|y)$ for fixed y and variable x, sums up the information that the equality $Y = y$ provides about X.

2. The conditional mean estimate $\hat{x} = \int_{-\infty}^{+\infty} x p_{X|Y}(x|y)$ is also the conditional minimum variance estimate:

$$E\{||X - \hat{x}||^2 \mid Y = y\} \leq E\{||X - z(y)||^2 \mid Y = y\}$$

for all functions z of y, and

$$E\{||X - \hat{x}||^2 \mid Y = y\} = E\{||X||^2 \mid Y = y\} - ||\hat{x}||^2$$

3. The estimator $\hat{X} = E[X \mid Y]$ is a function of Y, with $\hat{X}(y) = E[X \mid Y = y] = \hat{x}$, and is a minimum variance estimator. That is,

$$E_{X,Y}\{||X - \hat{X}(Y)||^2\} \leq E_{X,Y}\{||X - Z(Y)||^2\}$$

for all functions $Z(\cdot)$, and is unbiased, i.e., $E_{X,Y}[\hat{X}(Y)] = E[X]$

For x, y jointly gaussian with mean and covariance

$$\begin{bmatrix} \bar{x} \\ \bar{y} \end{bmatrix} \quad \text{and} \quad \begin{bmatrix} \Sigma_{xx} & \Sigma_{xy} \\ \Sigma_{yx} & \Sigma_{yy} \end{bmatrix}$$

x conditioned on y is gaussian, with conditional mean and conditional covariance

$$\bar{x} + \Sigma_{xy}\Sigma_{yy}^{-1}(y - \bar{y}) \quad \text{and} \quad \Sigma_{xx} - \Sigma_{xy}\Sigma_{yy}^{-1}\Sigma_{yx}$$

The conditional mean is also a maximum a posteriori estimate, and the conditional covariance is the conditional error covariance associated with use of a conditional mean estimate and, being independent of the measurement, is also an unconditional error covariance associated with the conditional mean estimator.

Problem 3.1. Suppose X and Y are jointly distributed random variables. When Y is unknown, an intelligent estimate of the value of X is $\bar{x} = E[X]$. This estimate has the property that $E\{\|X - \bar{x}\|^2\} \leq E\{\|X - z\|^2\}$ for all z, and has average error $E\{\|X - \bar{x}\|^2\}$. Now suppose that one is told that $Y = y$. Let $\hat{x} = E[X \mid Y = y]$. Show that

$$E\{\|X - \hat{x}\|^2 \mid Y = y\} = E\{\|X - \bar{x}\|^2 \mid Y = y\} - \|\hat{x} - \bar{x}\|^2$$

Conclude the intuitively reasonable result that the mean estimation error $E\{\|X - \hat{x}\|^2\}$ averaged over all values of X and Y will be bounded above by $E\{\|X - \bar{x}\|^2\}$. When will the bound not be attained, i.e., when will there be a strict improvement in the knowledge of X? Extend the argument to cover the case when X, Y, and Z are three jointly distributed random variables and one knows that $Y = y$ and $Z = z$.

Problem 3.2. The conditional mean estimate may not always be a reasonable estimate. Construct an example where X can only take discrete values and $E\{X \mid Y = y\}$ may not equal any of these values. In this case, a maximum a posteriori estimate may be appropriate.

Problem 3.3. Let X and Y be two jointly distributed random variables with X scalar, and let Y take the value y. Let $\hat{x}$ be an estimate chosen so that

$$E\{|X - \hat{x}| \mid Y = y\} \leq E\{|X - z| \mid Y = y\}$$

In other words, $\hat{x}$ is chosen to minimize the average value of the *absolute error* between $\hat{x}$ and the actual value taken by X. Show that $\hat{x}$ is the median of the conditional density $p_{X\mid Y}(x \mid y)$. [The median of a continuous density $p_A(a)$ is that value of a, call it α, for which $P(A \leq \alpha) = P(A \geq \alpha)$.]

Problem 3.4. Let X and Y be two jointly distributed random variables, and let Y take the value y. Weight the error between the value taken by X and an estimate $\hat{x}$ uniformly for the region I_ϵ defined by $\|x - \hat{x}\| \geq \epsilon$, and give it zero weight for $\|x - \hat{x}\| < \epsilon$. In other words, define a performance of the estimator by

$$P = \int_{I_\epsilon} p_{X\mid Y}(x \mid y)\, dx$$

Show that in the limit as $\epsilon \longrightarrow 0$, the best estimate (in the sense of maximizing P) is the maximum a posteriori estimate, or the conditional mode. [Assume $p_{X\mid Y}(x \mid y)$ is continuous in x.]

Problem 3.5. * Suppose $p_{X\mid Y}(x \mid y)$ is gaussian. Show that the conditional mean estimate is the same as the conditional modal and maximum a posteriori estimates, using the results of Probs. 3.3 and 3.4.

Problem 3.6. * Let $L(\cdot)$ be a scalar function with $L(0) = 0$, $L(y) \geq L(z)$ for $\|y\| \geq \|z\|$, $L(y) = L(-y)$, and with $L(\cdot)$ convex. Let $p_{X\mid Y}(x \mid y)$ be symmetric

*These problems refer to material which may have been omitted at a first reading.

about $\hat{x} = E\{X | Y = y\}$. Prove that for all z,

$$E\{L(X - \hat{x}) | Y = y\} \leq E\{L(X - z) | Y = y\}$$

[*Hint:* Set $\tilde{x} = x - \hat{x}$, $\tilde{z} = z - \hat{x}$, and show that

$$E\{L(X - z) | Y = y\} = \int L(\tilde{z} - \tilde{x}) p_{\tilde{X}|Y}(\tilde{x} | y) \, d\tilde{x} = \int L(\tilde{z} + \tilde{x}) p_{\tilde{X}|Y}(\tilde{x} | y) \, d\tilde{x}$$

$$= \int \frac{1}{2} [L(\tilde{z} - \tilde{x}) + L(\tilde{z} + \tilde{x})] p_{\tilde{X}|Y}(\tilde{x} | y) \, d\tilde{x}$$

Then use the evenness and convexity of $L(\cdot)$.]

Problem 3.7. Assume X and N are independent gaussian random variables of means $\bar{x}$, $\bar{n}$ and covariance matrices Σ_x and Σ_n. Then $Y = X + N$ is gaussian and $p_{X|Y}(x | y)$ is gaussian. Show that the associated conditional mean and covariance are

$$\Sigma_n (\Sigma_x + \Sigma_n)^{-1} \bar{x} + \Sigma_x (\Sigma_x + \Sigma_n)^{-1} (y - \bar{n})$$

and

$$\Sigma_x - \Sigma_x (\Sigma_x + \Sigma_n)^{-1} \Sigma_x = \Sigma_x (\Sigma_x + \Sigma_n)^{-1} \Sigma_n = (\Sigma_x^{-1} + \Sigma_n^{-1})^{-1}$$

[Assume the various inverses exist, and first find the joint density $p_{X,Y}(x, y)$.]

Problem 3.8. Let X and Y be jointly gaussian random vectors and let $\hat{X}(Y) = E[X | Y]$. Show that $\hat{X}(Y)$, thought of as a random variable, is itself gaussian. [*Note:* This is quite a different statement from the fact that $p_{X|Y}(x | y)$ is gaussian.] Find its mean and covariance in terms of the associated mean and covariance of $[X' \quad Y'']$.

REFERENCES

[1] MEDITCH, J. S., *Stochastic Optimal Linear Estimation and Control*, McGraw-Hill Book Company, New York, 1969.

[2] KALMAN, R. E., "A New Approach to Linear Filtering and Prediction Problems," *J. Basic Eng., Trans. ASME*, Series D. Vol. 82, No. 1, March 1960, pp. 35–45.

[3] TUCKER, M. G., *A Graduate Course in Probability*, Academic Press, Inc., New York, 1967.

[4] SAWARAGI, Y., *et al.*, "The Prediction of Air Pollution Levels by Nonphysical Models Based on Kalman Filtering Method," *J. Dynamic Systems, Measurement and Control*, Vol. 98, No. 4, December 1976, pp. 375–386.

[5] WELLS, C. H., and LARSON, R. E., "Application of Combined Optimum Control and Estimation Theory to Direct Digital Control," *Proc. IEEE*, Vol. 58, No. 1, January 1970, pp. 16–22.

[6] DEUTSCH, R., *Estimation Theory*, Prentice-Hall, Inc., Englewood Cliffs, N. J., 1965.

[7] VAN TREES, H. L., *Detection, Estimation and Modulation Theory*, John Wiley & Sons, Inc., New York, 1968.

CHAPTER 3

THE DISCRETE-TIME

KALMAN FILTER

3.1 THE KALMAN FILTER

Outline of Chapter

In this section, we shall tie together the ideas of the previous chapter to state the Kalman filter problem, and we shall indicate what the solution to the problem is. We offer a derivation of the filter that is simple and direct but to some degree uninspiring, and we offer a number of preliminary comments about the filter. In the next section, we illustrate a major property of the Kalman filter. The final two sections of the chapter present some motivating applications.

In a later chapter, we give an alternative derivation of the Kalman filter, and in the process derive many important properties of the filter. The reason for the early presentation of the filter, with a concomitant delay in the presentation of an interesting proof, is solely to let the student see first the simplicity and strength of the Kalman filter as an engineering tool.

The Filtering Problem

Recall that in the last chapter we introduced the system depicted in Fig. 3.1-1 and described for $k \geq 0$ by the following equations:

$$x_{k+1} = F_k x_k + G_k w_k \tag{1.1}$$

$$z_k = H_k' x_k + v_k \tag{1.2}$$

Below, we shall recall the probabilistic descriptions of $\{v_k\}$, $\{w_k\}$, and x_0. More general models will be considered later. One generalization involving external inputs is studied in Prob. 1.1.

The filtering problem has been stated in broad terms to have been one requiring the deduction of information about x_k, using measurements up till time k. In order to obtain the simplest filtering equations, we shall initially modify the filtering problem slightly by seeking to deduce information about x_k using measurements up till time $k - 1$. In effect then, we are considering a *one-step prediction problem*. Convention has it, however, that this is still termed a filtering probelm.

Bearing in mind the material of the last chapter dealing with estimation, we can refine this one-step prediction problem to *one of requiring computation of the sequence* $E\{x_k | z_0, z_1, \ldots, z_{k-1}\}$ *for* $k = 0, 1, 2, \ldots$. We shall denote this quantity by $\hat{x}_{k/k-1}$ and shall use the symbol Z_{k-1} to denote the set $\{z_0, z_1, \ldots, z_{k-1}\}$. This use of a capital letter and lower-case letters is a variation on the notation used earlier.

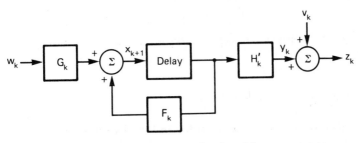

Fig. 3.1-1 Basic signal model.

At the same time as knowing $\hat{x}_{k/k-1}$, it is obviously of interest to know how good the estimate $\hat{x}_{k/k-1}$ is. We shall measure this estimate by the error covariance matrix $\Sigma_{k/k-1}$, where

$$\Sigma_{k/k-1} = E\{[x_k - \hat{x}_{k/k-1}][x_k - \hat{x}_{k/k-1}]' | Z_{k-1}\} \tag{1.3}$$

We shall aim to calculate this quantity also. Notice that, in view of the formula trace $(AB) = $ trace (BA) for two matrices A and B,

$$\text{trace } \Sigma_{k/k-1} = E\{\text{tr } [(x_k - \hat{x}_{k/k-1})(x_k - \hat{x}_{k/k-1})'] | Z_{k-1}]\}$$
$$= E\{\|x_k - \hat{x}_{k/k-1}\|^2 | Z_{k-1}\}$$

is the conditional error variance associated with the estimate $\hat{x}_{k/k-1}$ and the conditional mean estimate minimizes this error variance.

Plainly, it will sometimes be relevant to aim to compute the true filtered estimate $E[x_k | Z_k]$, which we shall denote by $\hat{x}_{k/k}$, instead of $\hat{x}_{k/k-1}$. At the same time, we would seek to know the associated error covariance matrix $\Sigma_{k/k}$. It turns out that the estimate $\hat{x}_{k/k}$ can be obtained in essentially the same way as $\hat{x}_{k/k-1}$, as can its error covariance, save that the formulas are more complicated.

Evidently the notation $E\{x_k \mid Z_{k-1}\}$ suggests that $\hat{x}_{k/k-1}$ can only be computed when $k - 1 \geq 0$, or $k \geq 1$. By convention, we shall define $\hat{x}_{k/k-1}$ for $k = 0$ (i.e., $\hat{x}_{0/-1}$) to be $\bar{x}_0 = E\{x_0\}$, i.e., the expected value of x_0 given no measurements. For the same reason, we take $\Sigma_{0/-1}$ to be P_0.

Now let us combine all the above ideas with those of the last chapter. We can state the basic filtering problem as follows.

> ***Discrete-time Kalman filtering problem.*** For the linear, finite-dimensional, discrete-time system of (1.1) and (1.2) defined for $k \geq 0$, suppose that $\{v_k\}$ and $\{w_k\}$ are independent, zero mean, gaussian white processes with
>
> $$E[v_k v'_k] = R_k \delta_{kl} \qquad E[w_k w'_l] = Q_k \delta_{kl} \qquad (1.4)$$
>
> Suppose further that the initial state x_0 is a gaussian random variable with mean $\bar{x}_0$ and covariance P_0, independent of $\{v_k\}$ and $\{w_k\}$. Determine the estimates
>
> $$\hat{x}_{k/k-1} = E[x_k \mid Z_{k-1}] \qquad \hat{x}_{k/k} = E[x_k \mid Z_k] \qquad (1.5)$$
>
> and the associated error covariance matrices $\Sigma_{k/k-1}$ and $\Sigma_{k/k}$.

Solution of the Kalman Filter Problem

Let us now note what the solution to the filtering problem is. Then we shall give a proof.

The Kalman filter comprises the system depicted in Fig. 3.1-2 and is described for $k \geq 0$ by the equations

$$\hat{x}_{k+1/k} = [F_k - K_k H'_k]\hat{x}_{k/k-1} + K_k z_k \qquad (1.6)$$

with

$$\hat{x}_{0/-1} = \bar{x}_0 \qquad (1.7)$$

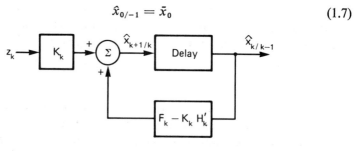

Fig. 3.1-2   Structure of filter.

The gain matrix K_k is determined from the error covariance matrix by*

$$K_k = F_k \Sigma_{k/k-1} H_k [H_k' \Sigma_{k/k-1} H_k + R_k]^{-1} \qquad (1.8)$$

assuming the inverse exists, and the conditional error covariance matrix is given recursively by a so-called discrete-time Riccati equation

$$\Sigma_{k+1/k} = F_k [\Sigma_{k/k-1} - \Sigma_{k/k-1} H_k (H_k' \Sigma_{k/k-1} H_k + R_k)^{-1} H_k' \Sigma_{k/k-1}] F_k' + G_k Q_k G_k'$$

$$(1.9)$$

This equation is initialized by

$$\Sigma_{0/-1} = P_0 \qquad (1.10)$$

One obtains $\hat{x}_{k/k}$ and $\Sigma_{k/k}$ as follows:

$$\hat{x}_{k/k} = \hat{x}_{k/k-1} + \Sigma_{k/k-1} H_k (H_k' \Sigma_{k/k-1} H_k + R_k)^{-1} (z_k - H_k' \hat{x}_{k/k-1}) \quad (1.11)$$

$$\Sigma_{k/k} = \Sigma_{k/k-1} - \Sigma_{k/k-1} H_k (H_k' \Sigma_{k/k-1} H_k + R_k)^{-1} H_k' \Sigma_{k/k-1} \qquad (1.12)$$

"First-Principles" Derivation of the Kalman Filter Equations

Recall from the last chapter (see end of Sec. 2.3) that if X and Y are jointly gaussian, with $Z = [X' \quad Y']'$ possessing mean and covariance

$$\begin{bmatrix} \bar{x} \\ \bar{y} \end{bmatrix} \quad \text{and} \quad \begin{bmatrix} \Sigma_{xx} & \Sigma_{xy} \\ \Sigma_{yx} & \Sigma_{yy} \end{bmatrix}$$

then the random variable X, when conditioned on the information that $Y = y$, is gaussian, with mean and covariance,

$$\bar{x} + \Sigma_{xy} \Sigma_{yy}^{-1} (y - \bar{y}) \quad \text{and} \quad \Sigma_{xx} - \Sigma_{xy} \Sigma_{yy}^{-1} \Sigma_{yx}$$

respectively. (A pseudo-inverse can replace the inverse if the inverse fails to exist.) There is an obvious extension of this result which we shall use, and which we state here without proof. Suppose that there is introduced another random variable W so that X, Y, and W are jointly gaussian. Also suppose that the mean and covariance of Z above are not those given a priori but are, in fact, the mean and covariance *conditioned on the fact that $W = w$*. Then the a posteriori mean and covariance stated apply to the random variable X conditioned on the information *that $Y = y$ and also that $W = w$*. In short, one can condition all variables without affecting the result.

Returning to the signal model of interest [see (1.1), (1.2) and (1.4)], we proceed as follows.

1. The random variable $[x_0' \quad z_0']'$ has mean $[\bar{x}_0' \quad \bar{x}_0' H_0]'$ and covariance

$$\begin{bmatrix} P_0 & P_0 H_0 \\ H_0' P_0 & H_0' P_0 H_0 + R_0 \end{bmatrix}$$

---

*We shall assume nonsingularity of $H_k' \Sigma_{k/k-1} H_k + R_k$. This normally holds, and is guaranteed if R_k is positive definite.

Hence x_0 conditioned on z_0 has mean

$$\hat{x}_{0/0} = \bar{x}_0 + P_0 H_0 (H_0' P_0 H_0 + R_0)^{-1}(z_0 - H_0' \bar{x}_0)$$

and covariance

$$\Sigma_{0/0} = P_0 - P_0 H_0 (H_0' P_0 H_0 + R_0)^{-1} H_0' P_0$$

2. From (1.1) and the various independence assumptions, it follows that x_1 conditioned on z_0 is gaussian with mean and covariance

$$\hat{x}_{1/0} = F_0 \hat{x}_{0/0} \quad \text{and} \quad \Sigma_{1/0} = F_0 \Sigma_{0/0} F_0' + G_0 Q_0 G_0'$$

3. From these equations and (1.2) it follows that z_1 conditioned on z_0 is gaussian with mean and covariance

$$\hat{z}_{1/0} = H_1' \hat{x}_{1/0} \quad \text{and} \quad H_1' \Sigma_{1/0} H_1 + R_1$$

It also follows that

$$E\{[x_1 - \hat{x}_{1/0}][z_1 - \hat{z}_{1/0}]' \,|\, z_0\} = \Sigma_{1/0} H_1$$

so that the random variable $[x_1' \quad z_1']$ conditioned on z_0 has mean and covariance

$$\begin{bmatrix} \hat{x}_{1/0} \\ H_1' \hat{x}_{1/0} \end{bmatrix} \quad \text{and} \quad \begin{bmatrix} \Sigma_{1/0} & \Sigma_{1/0} H_1 \\ H_1' \Sigma_{1/0} & H_1' \Sigma_{1/0} H_1 + R_1 \end{bmatrix}$$

4. Applying the basic result, we conclude that x_1 conditioned on z_0 and z_1 has mean

$$\hat{x}_{1/1} = \hat{x}_{1/0} + \Sigma_{1/0} H_1 (H_1' \Sigma_{1/0} H_1 + R_1)^{-1}(z_1 - H_1' \hat{x}_{1/0})$$

and covariance

$$\Sigma_{1/1} = \Sigma_{1/0} - \Sigma_{1/0} H_1 (H_1' \Sigma_{1/0} H_1 + R_1)^{-1} H_1' \Sigma_{1/0}$$

5. With updating of time indices, step 2 now applies to yield

$$\hat{x}_{2/1} = F_1 \hat{x}_{1/1}$$
$$\Sigma_{2/1} = F_1 \Sigma_{1/1} F_1' + G_1 Q_1 G_1'$$

6. More generally, repetition of steps 2 through 4 yields

$$\hat{x}_{k/k} = \hat{x}_{k/k-1} + \Sigma_{k/k-1} H_k (H_k' \Sigma_{k/k-1} H_k + R_k)^{-1}(z_k - H_k' \hat{x}_{k/k-1})$$
$$\hat{x}_{k+1/k} = F_k \hat{x}_{k/k}$$
$$\Sigma_{k/k} = \Sigma_{k/k-1} - \Sigma_{k/k-1} H_k (H_k' \Sigma_{k/k-1} H_k + R_k)^{-1} H_k' \Sigma_{k/k-1}$$
$$\Sigma_{k+1/k} = F_k \Sigma_{k/k} F_k' + G_k Q_k G_k'$$

(When the inverse fails to exist, a pseudo-inverse can be used.) The equations taken together yield (1.6) through (1.12). We remark that the equations yielding $\hat{x}_{k/k}$ and $\Sigma_{k/k}$ from $\hat{x}_{k/k-1}$ and $\Sigma_{k/k-1}$ are sometimes termed *measurement-update equations*, while the equations yielding $\hat{x}_{k+1/k}$ and $\Sigma_{k+1/k}$ from $\hat{x}_{k/k}$ and $\Sigma_{k/k}$ are known as *time-update equations*.

While the above proof is perhaps quick and easy, it has a number of drawbacks. Thus a gaussian assumption is necessary, and it proves unsuited to developing deeper properties of the Kalman filter, e.g., those relating to innovations and smoothing, discussed in later chapters. It is harder to cope with correlated $\{v_k\}$ and $\{w_k\}$ sequences, as we shall want to do.

Obvious Properties of the Filter

We now list a number of properties of the Kalman filter, the importance of which is in no way lessened by the ease with which some can be seen.

1. *The Kalman filter is a linear, discrete-time, finite-dimensional system.* From the point of view of someone wishing to build a filter, this is a marvelous stroke of good fortune, even if it is the only logical outcome of the problem specification. After all, one might have conjectured that the filter was nonlinear, or infinite dimensional.

2. The input of the filter is the process $\{z_k\}$, the output is $\{\hat{x}_{k/k-1}\}$. Obviously, the particular set of numbers appearing as $\hat{x}_{k/k-1}$ depends on the particular set of numbers appearing at the input as $z_0, z_1, \ldots, z_{k-1}$. On the other hand, the conditional error covariance matrix equation (1.9) shows that

$$\Sigma_{k/k-1} = E\{[x_k - \hat{x}_{k/k-1}][x_k - \hat{x}_{k/k-1}]' \,|\, Z_{k-1}\} \qquad (1.3)$$

is *actually independent* of Z_{k-1}. No one set of measurements helps any more than any other to eliminate some uncertainty about x_k. The gain K_k is also independent of Z_{k-1}. Because of this, *the error covariance $\Sigma_{k/k-1}$ and gain matrix K_k can be computed before the filter is actually run.* (Such phenomena are usually not observed in nonlinear filtering problems.)

3. The filter equation (1.6) can be thought of either as an equation yielding the estimator (the *rule* for passing from any sequence $\{z_k\}$ to the associated conditional mean estimate) or as an equation yielding the estimate (the *value* $\hat{x}_{k/k-1}$, expressed in terms of a set of values taken by the $\{z_k\}$ process). We do not distinguish in notation between the conditional mean estimator and estimate. The conditional covariance definition of (1.3) however identifies $\Sigma_{k/k-1}$ as the covariance associated with a particular *estimate*. Note too that because $\Sigma_{k/k-1}$ is independent of Z_{k-1}, we may take the expectation of both sides of (1.3) over all possible Z_{k-1} to conclude that

$$\Sigma_{k/k-1} = E\{[x_k - \hat{x}_{k/k-1}][x_k - \hat{x}_{k/k-1}]'\} \qquad (1.13)$$

This equation means that $\Sigma_{k/k-1}$ is an unconditional error covariance matrix associated with the Kalman filter; i.e., $\Sigma_{k/k-1}$ is also the covariance matrix associated with the *estimator*.

4. Consider the redrawing of Fig. 3.1-2 in the form of Fig. 3.1-3. Suppose also for the moment that the input and output additive noises in the system of Fig. 3.1-1 are not present, so that we have

$$x_{k+1} = F_k x_k \tag{1.14}$$

$$z_k = H_k' x_k \tag{1.15}$$

Then we can argue that the arrangement of Fig. 3.1-3 represents a logical form of state estimator for (1.14) and (1.15). We argue first on qualitative grounds. If at some time instant k it is true that $\hat{x}_{k/k-1} = x_k$, then it is evident from Fig. 3.1-3 that the input to the gain block K_k (the output of the differencing element) will be zero. Accordingly, we will then have

$$\hat{x}_{k+1/k} = F_k \hat{x}_{k/k-1} = F_k x_k = x_{k+1}$$

At time $k + 1$, the input to the gain block K_{k+1} will again be zero. This will lead to

$$\hat{x}_{k+2/k+1} = x_{k+2}$$

and so on. Precisely because part of the Fig. 3.1-3 scheme is a copy of the original system, tracking will occur. Now in the event that $x_k \neq \hat{x}_{k/k-1}$, there will be an input to the K_k block. Hopefully, this affects $\hat{x}_{k+1/k}$ to steer it toward x_{k+1}. Quantitatively, we have

$$\hat{x}_{k+1/k} = F_k \hat{x}_{k/k-1} + K_k(z_k - H_k' \hat{x}_{k/k-1}) \tag{1.16}$$

with $z_k - H_k' \hat{x}_{k/k-1}$ a measure of the estimation error. This equation may be rewritten as

$$\hat{x}_{k+1/k} = F_k \hat{x}_{k/k-1} + K_k H_k'(x_k - \hat{x}_{k/k-1})$$

and, together with (1.14), implies

$$(x_{k+1} - \hat{x}_{k+1/k}) = (F_k - K_k H_k')(x_k - \hat{x}_{k/k-1}) \tag{1.17}$$

Let $\Psi_{k,l}$ denote the transition matrix associated with (1.17). If

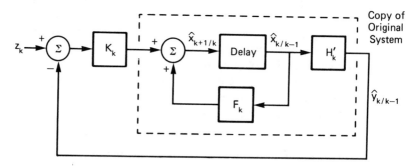

Fig. 3.1-3 Filter redrawn to emphasize its structure as a copy of original system driven by estimation error.

$\Psi_{k+m,k} \rightarrow 0$ as $m \rightarrow \infty$, we can expect $\hat{x}_{k+m/k+m-1}$ to ultimately track x_{k+m}, because

$$x_{k+m} - \hat{x}_{k+m/k+m-1} = \Psi_{k+m,k}(x_k - \hat{x}_{k/k-1}) \qquad (1.18)$$

Estimator design in the noiseless case amounts, therefore, to correct selection of $\{K_i\}$ to ensure that $\Psi_{k,l}$ has the requisite property.

When (1.14) and (1.15) do not hold and the original noisy arrangement of Fig. 3.1-1 applies, it is reasonable to conjecture that the arrangement of Fig. 3.1-3 functions as a state estimator. However, presence of the noise renders unlikely the possibility of $x_k - \hat{x}_{k/k-1}$ approaching zero as $k \rightarrow \infty$.

5. Because $\{x_k\}$ and $\{z_k\}$ are jointly gaussian processes as discussed in the last chapter, it follows that x_k conditioned on Z_{k-1} is gaussian. The conditional density of x_k is, therefore, in effect defined by the conditional mean, which is $\hat{x}_{k/k-1}$, and the conditional covariance, which is $\Sigma_{k/k-1}$ [see (1.3)]. It follows that the Kalman filter equations provide a procedure for *updating the entire conditional probability density of x_k*.

6. In Eqs (1.8) and (1.9), the inverse of the matrix $H_k'\Sigma_{k/k-1}H_k + R_k$ occurs. This matrix may not be nonsingular, although it is nonnegative definite, since $\Sigma_{k/k-1}$ and R_k, being covariance matrices, are individually nonnegative definite. One way to force positive definiteness of $H_k'\Sigma_{k/k-1}H_k + R_k$ is to demand a priori that R_k be positive definite. This has the significance that no measurement is exact (see Prob. 1.3), and is therefore often reasonable on physical grounds. In the event that the $H_k'\Sigma_{k/k-1}H_k + R_k$ is singular, however, its inverse may be replaced in (1.8) and (1.9) by its pseudo-inverse.

7. Suppose the underlying signal model is time invariant and the input and output noise processes are stationary. Thus F_k, G_k, H_k, Q_k, and R_k are constant. In general, $\Sigma_{k/k-1}$ and therefore K_k will not be constant, so *the Kalman filter will normally still be time varying despite time invariance and stationarity in the signal model*.

8. Throughout this section almost all our discussion has been in terms of the quantities $\hat{x}_{k/k-1}$ and $\Sigma_{k/k-1}$ rather than $\hat{x}_{k/k}$ and $\Sigma_{k/k}$. Some of the problems explore what can be said about $\hat{x}_{k/k}$ and $\Sigma_{k/k}$.

A Generalization

As we shall see in a later section, it is possible to have a situation in which one or more of the matrices F_k, G_k, H_k, Q_k, and R_k take values depending on Z_{k-1}. In this case some, but not all, of the previous statements hold true. Most importantly, $\hat{x}_{k/k-1}$ and $\Sigma_{k/k-1}$ are still given by Eqs. (1.6) through (1.10). But now, the gain matrix K_k and error covariance matrix $\Sigma_{k/k-1}$ are not precomputable, and they depend on Z_{k-1}. This means also that while $\Sigma_{k/k-1}$ is a conditional error covariance, it is not an unconditioned one. (In contrast, we shall later encounter situations in which the same equations yield $\Sigma_{k/k-1}$ as an unconditioned error covariance matrix which is *not* also a conditional error covariance matrix.) The "first-principles" derivation offered earlier works with little or no change if F_k, G_k, etc. depend on Z_{k-1}.

Main Points of the Section

The Kalman filter equations should be committed to memory, and the following points remembered. The Kalman filter is a linear, discrete-time, finite-dimensional system. Normally, the covariance matrix $\Sigma_{k/k-1}$ is both a conditional error covariance matrix associated with the state estimate, and an unconditional error covariance matrix associated with the filter qua estimator; it can be precomputed, as can the filter gain. The filter has the same structure as a class of deterministic estimators. The Kalman filter equations define the evolution of the gaussian conditional probability density of the state. The Kalman filter equations are also valid for the case when F_k, G_k, H_k, Q_k, and R_k are not necessarily independent of Z_{k-1}, in which case the covariance matrix $\Sigma_{k/k-1}$, though still a conditional error covariance matrix, is not an unconditional error covariance. The equations are summarized for convenience:

SIGNAL MODEL:

$$x_{k+1} = F_k x_k + G_k w_k$$
$$z_k = y_k + v_k = H'_k x_k + v_k$$

x_0, $\{v_k\}$, $\{w_k\}$ are jointly gaussian and mutually independent; x_0 is $N(\bar{x}_0, P_0)$; $\{v_k\}$ is zero mean, covariance $R_k \delta_{kl}$; $\{w_k\}$ is zero mean, covariance $Q_k \delta_{kl}$.

KALMAN FILTER:

$$\hat{x}_{k+1/k} = (F_k - K_k H'_k)\hat{x}_{k/k-1} + K_k z_k \qquad \hat{x}_{0/-1} = \bar{x}_0$$
$$K_k = F_k \Sigma_{k/k-1} H_k (H'_k \Sigma_{k/k-1} H_k + R_k)^{-1}$$
$$\Sigma_{k+1/k} = F_k [\Sigma_{k/k-1} - \Sigma_{k/k-1} H_k (H'_k \Sigma_{k/k-1} H_k + R_k)^{-1} H'_k \Sigma_{k/k-1}]F'_k + G_k Q_k G'_k$$
$$= E\{[x_{k+1} - \hat{x}_{k+1/k}][x_{k+1} - \hat{x}_{k+1/k}]' | Z_k\} \qquad \Sigma_{0/-1} = P_0$$
$$\hat{x}_{k/k} = \hat{x}_{k/k-1} + \Sigma_{k/k-1} H_k (H'_k \Sigma_{k/k-1} H_k + R_k)^{-1}(z_k - H'_k \hat{x}_{k/k-1})$$
$$\Sigma_{k/k} = \Sigma_{k/k-1} - \Sigma_{k/k-1} H_k (H'_k \Sigma_{k/k-1} H_k + R_k)^{-1} H'_k \Sigma_{k/k-1}$$

Problem 1.1. Suppose that the arrangement of Fig. 3.1-1 is augmented by the insertion of a known input sequence $\{u_k\}$, as shown in Fig. 3.1-4. The equations describing this arrangement are

$$x_{k+1} = F_k x_k + G_k w_k + \Gamma_k u_k$$
$$z_k = H'_k x_k + v_k$$

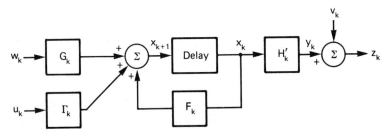

Fig. 3.1-4 Signal model with addition of external, known input.

The processes $\{v_k\}$ and $\{w_k\}$ and the initial state x_0 are as given earlier. By using the deterministic state estimator ideas discussed toward the end of the section, conjecture a filter structure.

Problem 1.2. The matrix $\Sigma_{k+1/k}$ is nonnegative definite symmetric because it is a covariance matrix. Show by an induction argument that (1.9) implies, on purely algebraic grounds, that $\Sigma_{k+1/k}$ is nonnegative definite symmetric for all k; use the fact that Q_k is nonnegative definite symmetric for all k and $\Sigma_{0/-1} = P_0$ is nonnegative definite symmetric. (*Hint for proving nonnegativity:* Show that

$$\Sigma_{k+1/k} = F_k[I \; \vdots \; -\Sigma_{k/k-1}H_k(H_k'\Sigma_{k/k-1}H_k + R_k)^{-1}]$$

$$\times \left\{ \begin{bmatrix} I \\ H_k' \end{bmatrix} \Sigma_{k/k-1}[I \quad H_k] + \begin{bmatrix} 0 & 0 \\ 0 & R_k \end{bmatrix} \right\}$$

$$\times \begin{bmatrix} I \\ -(H_k'\Sigma_{k/k-1}H_k + R_k)^{-1}H_k'\Sigma_{k/k-1} \end{bmatrix} F_k' + G_kQ_k'G_k)$$

Problem 1.3. The "measurement equation" we have used is $z_k = H_k'x_k + v_k$, with v_k gaussian, of zero mean, and covariance R_k. Show that if R_k is singular, some linear functional of x_k is determined by z_k with zero error; i.e., there exists a vector a of dimension equal to that of x_k such that z_k determines $a'x_k$.

Problem 1.4. In the previous chapter, it was shown that the quantity $P_k = E[x_kx_k']$ associated with the system of (1.1) with $E[x_0] = 0$ could be obtained from $P_{k+1} = F_kP_kF_k' + G_kQ_kG_k'$. Using this equation and the recursive equation for $\Sigma_{k+1/k}$, show that $P_{k+1} - \Sigma_{k+1/k} \geq 0$. Give an interpretation of this result.

Problem 1.5. Find recursive equations expressing $\hat{x}_{k+1/k+1}$ in terms of $\hat{x}_{k/k}$ and $\Sigma_{k+1/k+1}$ in terms of $\Sigma_{k/k}$. Observe the structural complexity in this latter equation in comparison with that associated with the recursive equation for $\Sigma_{k+1/k}$ in terms of $\Sigma_{k/k-1}$.

Problem 1.6. (a) Does the recursive equation for $\hat{x}_{k/k}$ computed in solving Prob. 1.5 show that $\hat{x}_{k/k}$ is obtainable as the output of a linear, discrete-time, finite-dimensional system?
(b) Is $\Sigma_{k/k}$ both a conditional and unconditional error covariance?
(c) Is there an interpretation of the $\hat{x}_{k/k}$ filter as a copy of the original system driven by an estimation error?

3.2 BEST LINEAR ESTIMATOR PROPERTY OF THE KALMAN FILTER

In this section, we shall take a completely different viewpoint of the Kalman filter. We *drop* the assumption that x_0, $\{v_k\}$, and $\{w_k\}$ are jointly gaussian, and we seek that particular filter *among a limited class of linear filters* which will produce an estimate minimizing a mean square error. We find that the Kalman filter defined in the last section is the optimal filter in this class of linear filters. The major tool will be the type of analysis of the last chapter, allowing the calculation of the mean and covariance of the state and output of a linear system. The proof is patterned after one in [1].

As before, we shall consider the signal model

$$x_{k+1} = F_k x_k + G_k w_k \tag{2.1}$$

$$z_k = H'_k x_k + v_k \tag{2.2}$$

and assume that $E[v_k] = 0$, $E[w_k] = 0$, $E[x_0] = \bar{x}_0$, $E[v_k v'_l] = R_k \delta_{kl}$ $E[w_k w'_l] = Q_k \delta_{kl}$, $E\{[x_0 - \bar{x}_0][x_0 - \bar{x}_0]'\} = P_0$, and that $\{v_k\}$, $\{w_k\}$, and x_0 are uncorrelated. We shall study the particular class of filters defined by

$$x^e_{k+1/k} = (F_k - K^e_k H'_k) x^e_{k/k-1} + K^e_k z_k \tag{2.3}$$

Here, the set of matrices $\{K^e_k\}$ is arbitrary. The general importance of this class of filters in tackling the problem of designing a state estimator for a noiseless plant was discussed in the last section. Also, the Kalman filter is a member of the class. Actually, a widening of the class is described in Prob. 2.2.

We shall show that by taking

$$x^e_{0/-1} = \bar{x}_0 \qquad K^e_k = K_k \text{ for all } k \tag{2.4}$$

with K_k as defined in the previous section, we minimize, in a sense to be made precise below,

$$\Sigma^e_{k/k-1} = E\{[x_k - x^e_{k/k-1}][x_k - x^e_{k/k-1}]'\} \tag{2.5}$$

for all k. At the same time, we have that

$$E[x_k - x^e_{k/k-1}] = 0 \tag{2.6}$$

for all k. The equalities (2.4), of course, serve to make the filter (2.3) coincide with the Kalman filter defined in the last section.

The quantity $x_k - x^e_{k/k-1}$ is the error associated with the estimate $x^e_{k/k-1}$ of x_k, and (2.6) implies that this estimation error, which is a random variable for any one k, has zero mean. Note that this is indeed a property possessed by the estimate $\hat{x}_{k/k-1}$ of the previous section, precisely because it is a conditional mean:

$$E[x_k - \hat{x}_{k/k-1}] = E[x_k] - E[E[x_k \,|\, Z_{k-1}]]$$
$$= E[x_k] - E[x_k]$$
$$= 0$$

Equation (2.5) is a matrix measure of the error associated with the *estimator* $x^e_{k/k-1}$ defined by (2.3). [Were the expectation in (2.5) conditioned on Z_{k-1}, the matrix would measure the error associated with the *estimate* $x^e_{k/k-1}$ resulting from a particular Z_{k-1}.]

We shall show that for arbitrary K^e_k, we have*

$$\Sigma^e_{k/k-1} \geq \Sigma_{k/k-1} \tag{2.7}$$

for all k, with equality being attained if and only if $K^e_k = K_k$. Computations defining $\Sigma_{k/k-1}$ were set out in the last section, where it was also claimed that $\Sigma_{k/k-1}$ was both the conditional error covariance matrix associated with the Kalman filter estimate, and the unconditioned error covariance matrix associated with the Kalman filter regarded as an estimator,† provided that $x_0, \{w_k\}$, and $\{v_k\}$ are jointly gaussian.

To obtain additional significance for (2.7), we recall the following properties of the trace operator:

1. $A = A' \geq 0$ implies tr $A \geq 0$, with equality if and only if $A = 0$.
2. tr $AB =$ tr BA.
3. $E[\text{tr } A] = \text{tr } E[A]$, where the entries of A are random variables.

Then (2.7) implies

$$\text{tr } E\{[x_k - x^e_{k/k-1}][x_k - x^e_{k/k-1}]'\} \geq \text{tr } E\{[x_k - \hat{x}_{k/k-1}][x_k - \hat{x}_{k/k-1}]'\}$$

or

$$E\{\| x_k - x^e_{k/k-1} \|^2\} \geq E\{\| x_k - \hat{x}_{k/k-1} \|^2\} \tag{2.8}$$

As noted, we shall show that (2.7), and therefore (2.8), holds with equality if and only if $K^e_k = K_k$. Therefore, the mean square error on the left side of (2.8) will be minimized precisely when $K^e_k = K_k$. *At least among the set of filters of the form (2.3), that which was defined in the previous section is the minimum variance filter, whether or not certain variables are gaussian.*

Let us now sum up the main result in a theorem statement.

THEOREM 2.1. Consider the system defined for $k \geq 0$ by (2.1) and (2.2), with $\{v_k\}$ and $\{w_k\}$ uncorrelated, zero mean processes with

$$E\{v_k v'_l\} = R_k \delta_{kl} \qquad E\{w_k w'_l\} = Q_k \delta_{kl}$$

Suppose also that x_0 has mean $\bar{x}_0$ and covariance matrix P_0, and is uncorrelated with $\{v_k\}$ and $\{w_k\}$.

---

*For symmetric matrices A and B, the notation $A \geq B$ means $A - B \geq 0$, or $A - B$ is nonnegative definite.

†Actually, if one or more of $F_k, G_k, \ldots$ depends on Z_{k-1}, the second interpretation is not valid.

Let quantities K_k and $\Sigma_{k/k-1}$ be defined recursively by

$$K_k = F_k \Sigma_{k/k-1} H_k [H_k' \Sigma_{k/k-1} H_k + R_k]^{-1} \tag{2.9}$$

$$\Sigma_{k+1/k} = F_k [\Sigma_{k/k-1} - \Sigma_{k/k-1} H_k (H_k' \Sigma_{k/k-1} H_k + R_k)^{-1} H_k' \Sigma_{k/k-1}] F_k' + G_k Q_k G_k' \tag{2.10}$$

with $\Sigma_{0/-1} = P_0$. Let a filter estimating $\{x_k\}$ be defined by (2.3). Then the estimation error matrix

$$\Sigma_{k/k-1}^e = E\{[x_k - x_{k/k-1}^e][x_k - x_{k/k-1}^e]'\} \tag{2.5}$$

satisfies

$$\Sigma_{k/k-1}^e \geq \Sigma_{k/k-1} \tag{2.7}$$

and the minimum possible value of $\Sigma_{k/k-1}^e$, namely $\Sigma_{k/k-1}$, is attained if and only if $x_{0/-1}^e = \bar{x}_0$ and $K_k^e = K_k$ for all k. Moreover, the estimator is unbiased, i.e.,

$$E[x_k - x_{k/k-1}^e] = 0 \tag{2.6}$$

Proof. We present an outline proof; see [1] for details. We note also that material in Chap. 5 will provide an alternative approach.

1. The error $x_k - x_{k/k-1}^e$ satisfies the following recursive equation.

$$x_{k+1} - x_{k+1/k}^e = (F_k - K_k^e H_k')(x_k - x_{k/k-1}^e) + [G_k \quad -K_k^e] \begin{bmatrix} w_k \\ v_k \end{bmatrix}$$

2. The error covariance matrix $\Sigma_{k+1/k}^e$ satisfies

$$\Sigma_{k+1/k}^e = (F_k - K_k^e H_k') \Sigma_{k/k-1}^e (F_k - K_k^e H_k')' + G_k Q_k G_k' + K_k^e R_k K_k^{e'} \tag{2.11}$$

3. $\Sigma_{0/-1}^e \geq P_0$ (basis step of an induction).
4. $\Sigma_{k/k-1}^e \geq \Sigma_{k/k-1}$ implies $\Sigma_{k+1/k}^e \geq \Sigma_{k+1/k}$ (recursion step of induction).
5. Equation (2.6) is almost immediate.

It is important to realize that the proof outlined above is of itself rigorous, but fails to establish the *overall* optimality of the Kalman filter, whether or not gaussian assumptions are made. Among the set of filters of a *restricted* class, the Kalman filter is without question optimal in the sense that it is a minimum error variance estimator, whether or not gaussian assumptions are made.

We comment also that in this section we have given a formal demonstration that $\Sigma_{k/k-1}$ is in actual fact the *unconditional* error covariance associated with the Kalman filter. It is not, in general, also a conditional error covariance, although, as we know from the last section, with the assumption that x_0, $\{v_k\}$ and $\{w_k\}$ are jointly gaussian it is true that $\Sigma_{k/k-1}$ is a conditional error covariance. (In fact, we also described briefly in the last section a special situation in which $\Sigma_{k/k-1}$ was a conditional but *not* an unconditional error covariance!) Note, though, that in proving here that the Kalman filter is a minimum variance estimator, we are implicitly proving it provides a minimum variance estimate for each set of measurements Z_{k-1}. In other words, we know that

$$E\{\|x_k - \hat{x}_{k/k-1}\|^2 \,|\, Z_{k-1}\}$$

is minimized, even though we have not shown in this section that its minimum value is tr $\Sigma_{k/k-1}$ for all Z_{k-1} or that

$$E\{[x_k - x_{k/k-1}^e][x_k - x_{k/k-1}^e]' \,|\, Z_{k-1}\} = \Sigma_{k/k-1} \tag{2.12}$$

for all Z_{k-1}.

The reader will have noticed that although the filter (2.3) has the form

$$x_{k+1/k}^e = A_k x_{k/k-1}^e + K_k^e z_k \tag{2.13}$$

with K_k^e arbitrary, the matrix A_k was not arbitrary, but forced to be $F_k - K_k^e H_k'$. One of the problems explores the possibility of choosing the best possible matrix A_k in (2.13) in order to minimize the error covariance matrix. This has been done for the corresponding continuous time problem in [2].

Main Points of the Section

Among a class of linear estimators, the Kalman filter produces the smallest unconditional error covariance matrix whether or not x_0, $\{v_k\}$ and $\{w_k\}$ are gaussian. The notion of smallest is a technical one of matrix theory, but also implying here smallest minimum mean square error. The unconditional error covariance matrix is $\Sigma_{k/k-1}$ as defined in the previous section.

Problem 2.1. Show that, among the set of linear filters considered in this section, the Kalman filter determines that estimate $x_{k/k-1}^e$ of x_k which minimizes $E\{[x_k - x_{k/k-1}^e]'A[x_k - x_{k/k-1}^e]\}$ for any nonnegative definite symmetric A. (*Hint:* Begin by writing the matrix A as $B'B$ for some B. This is always possible when A is nonnegative definite.)

Problem 2.2. In lieu of an assumed filter structure of the form (2.3), with K_k^e to be chosen, assume a structure of the form (2.13), with A_k and K_k^e to be chosen. Suppose that for all initial means $\bar{x}_0$ of x_0, the same filter is to be used. Show that the requirement that $x_{k/k-1}^e$ be an unbiased estimator of x_k, i.e., $E[x_{k/k-1}^e - x_k] = 0$ for all $\bar{x}_0$, and an assumption that F_k is nonsingular for all k imply that

$$A_k = F_k - K_k^e H_k'$$

Problem 2.3. Two equivalent equations for $\Sigma_{k+1/k}$ are

$$\Sigma_{k+1/k} = F_k[\Sigma_{k/k-1} - \Sigma_{k/k-1}H_k(H_k'\Sigma_{k/k-1}H_k + R_k)^{-1}H_k'\Sigma_{k/k-1}]F_k' + G_k Q_k G_k'$$

$$= (F_k - K_k H_k')\Sigma_{k/k-1}(F_k - K_k H_k')' + K_k R_k K_k' + G_k Q_k G_k'$$

where K_k is given by the usual formula. Compare from the computational point of view the implementation of these two equations.

3.3 IDENTIFICATION AS A KALMAN FILTERING PROBLEM

In this and the next section, we aim to introduce the reader to engineering-type applications of the Kalman filter. Hopefully, this will engender confidence as to its wide applicability.

Kalman filtering can be applied to provide a technique for the identification of the coefficients in a scalar ARMA equation of the form

$$y_k + a^{(1)}y_{k-1} + \cdots + a^{(n)}y_{k-n} = a^{(n+1)}u_{k-1} + \cdots + a^{(n+m)}u_{k-m} \quad (3.1)$$

Measurements of the system input $\{u_k\}$ and output $\{y_k\}$ become available in real time, and the aim is to estimate the values of the coefficients $a^{(1)}, \ldots, a^{(n+m)}$ using these measurements, as discussed in [1] and [3, 4].

Equations of the form (3.1) can arise in the study of systems controlled and measured on a sampled data basis. In reference [1], there is a description of a paper mill via an equation like (3.1); the problem is to identify the coefficients and then develop control strategies.

If (3.1) is taken as the equation describing the system and the measurement process, and if the $a^{(i)}$ are constant, the identification problem is almost trivial and, with sufficient measurements, the coefficients can be found by solving a set of linear equations. It is, however, more realistic to model the $a^{(i)}$ as being subject to random perturbations and to model the measurements as being noisy. So we suppose that for each i

$$a^{(i)}_{k+1} = a^{(i)}_k + w^{(i)}_k \quad (3.2)$$

where $\{w^{(i)}_k\}$ is a zero mean, white, gaussian random process, independent of $\{w^{(j)}_k\}$ for $i \neq j$. Also, we assume that (3.1) is replaced by

$$y_k + a^{(1)}_k y_{k-1} + \cdots + a^{(n)}_k y_{k-n} = a^{(n+1)}_k u_{k-1} + \cdots + a^{(n+m)}_k u_{k-m} + v_k$$

$$(3.3)$$

where $\{v_k\}$ is a zero mean, white gaussian random process, independent of the processes $\{w^i_k\}$.

We need to assume values for the variances of $w^{(i)}_k$ and v_k, and in assigning these values, the fullest possible knowledge must be used of the physical arrangement of which (3.2) and (3.3) constitute a representation. In other words, a variance for v_k should be assigned on the basis of our knowledge of the noise introduced by measurement sensors, and we should assign a variance to $w^{(i)}_k$ after an assessment, possibly subjective, of the way the $a^{(i)}_k$ are likely to vary.

Finally, we need to assume an a priori mean and variance for each $a^{(i)}$, reflecting our estimate before measurements are taken of the value of these coefficients and the likely error in the estimate respectively. To apply the Kalman filtering theory, we assume too that the $a_0^{(i)}$ are gaussian random variables. (We could alternatively drop the gaussian assumptions and still obtain the Kalman filter as a best linear estimator, as argued in the last section.)

Now we can pose the identification problem in Kalman filter terms. Define an $(n + m)$-dimensional state vector x_k by

$$x_k^{(1)} = a_k^{(1)}, \qquad x_k^{(2)} = a_k^{(2)}, \qquad \ldots, \qquad x_k^{(n+m)} = a_k^{(n+m)} \qquad (3.4)$$

Define also the $(n + m)$-dimensional, white, zero mean, gaussian process $\{w_k\}$ as the vector process formed from the $\{w_k^i\}$. Then Eqs. (3.2) and (3.4) lead to the state equation

$$x_{k+1} = x_k + w_k \qquad (3.5)$$

Next, define the matrix, actually a row vector,

$$H_k' = [-y_{k-1} \quad -y_{k-2} \quad \cdots \quad -y_{k-n} \quad u_{k-1} \quad u_{k-2} \quad \cdots \quad u_{k-m}] \qquad (3.6)$$

and the process $\{z_k\}$ by $z_k = y_k$. Then (3.3) and (3.6) yield

$$z_k = H_k' x_k + v_k \qquad (3.7)$$

Notice that at time 0, we cannot say what H_k is for $k > 0$. However, by the time z_k is received, the value of H_k is known. This is sufficient for the purposes of defining the Kalman filter. The filter in this case becomes

$$\hat{x}_{k+1/k} = [I - K_k H_k'] \hat{x}_{k/k-1} + K_k z_k \qquad (3.8)$$

with

$$K_k = \Sigma_{k/k-1} H_k [H_k' \Sigma_{k/k-1} H_k + R_k]^{-1} \qquad (3.9)$$

and

$$\Sigma_{k+1/k} = \Sigma_{k/k-1} - \Sigma_{k/k-1} H_k [H_k' \Sigma_{k/k-1} H_k + R_k]^{-1} H_k' \Sigma_{k/k-1} + Q_k \qquad (3.10)$$

Here, $R_k = E[v_k^2]$ and $Q_k = E[w_k w_k']$. Equation (3.8) is initialized with $\hat{x}_{0/-1}$ set equal to the vector of a priori estimates of the coefficients, and Eq. (3.10) is initialized with $\Sigma_{0/-1}$ set equal to the a priori covariance matrix of the coefficients.

Three important comments on the above material follow.

1. Because of the dependence of H_k on the actual system measurements, $\Sigma_{k/k-1}$ and K_k cannot be computed a priori. Since $\Sigma_{k/k-1}$ is no longer independent of the measurements, it loses its interpretation as an unconditional error covariance matrix of the estimator, though it is still a conditional error covariance matrix. For the case when $\{w_k\}$ and $\{v_k\}$ are not gaussian, even this conditional error covariance interpretation for $\Sigma_{k/k-1}$, as calculated in (3.10), is lost. We note too that $\hat{x}_{k+1/k}$ is not derived by simple *linear* processing of the measurements.

2. In case the $a_k^{(i)}$ are known or thought to be constant, one might attempt to replace (3.5) by

$$x_{k+1} = x_k$$

This has the effect of setting $Q_k = 0$ in (3.10). It turns out that this procedure is fraught with peril for the uninitiated, as is argued in a later chapter. In broad terms, there is the possibility that the smallest of modeling errors can lead, in time, to overwhelming errors—not predicted from the error covariance formula—in the estimates of $a_k^{(i)}$. (A simple example is contained in Prob. 3.2.) The solution is simply to take each $w_k^{(i)}$ as having a small, nonzero covariance.

3. To the extent that the error covariance matrix $\Sigma_{k/k-1}$ depends on the measurements via H_k, it is evident that poor identification may result with some sets of measurements. [In particular, if $u_k = 0$ for all k, no identification of $a^{(n+1)}, \ldots, a^{(n+m)}$ can take place.] Effectively, what one wants is $\Sigma_{k/k-1} \leq p_k I$ for (almost) all measurement sequences, with p_k a sequence of scalars approaching zero, or at least a small number p, as $k \longrightarrow \infty$. Then for almost all measurement sequences the mean square parameter estimation error will approach zero, or some small quantity. It is possible to lay down some criteria on the excitation u_k which guarantee effective identification. Basically, u_k cannot be too small for too many k, and should persistently excite all modes of the system (3.1) [5].

Besides offering several formal results, this section illustrates an important point: with judicious modeling assumptions, it is possible to bring to bear the exact, mathematical Kalman filter theory—admittedly in an ad hoc fashion—onto a situation to which, strictly speaking, it is not applicable. The engineer should constantly be alert to such a possibility. On the other hand, it would only be fair to point out that trouble can frequently arise in trying to stretch the theory; this point is illustrated at greater length in a later chapter dealing with computational aspects and modeling errors.

Problem 3.1. An alternative model for the variation of the coefficients in (3.1) is provided by

$$a_{k+1}^{(i)} = f_i a_k^{(i)} + w_k^{(i)}$$

with $w_k^{(i)}$ as before and f_i a scalar constant. Show that if $|f_i| < 1$, $E[(a_k^{(i)})^2]$ approaches a limit as $k \longrightarrow \infty$. Argue that this model for the variation of the $a_k^{(i)}$ may then be more relevant than that of (3.2). How might the f_i be selected in practice?

Problem 3.2. Consider the identification of $a^{(1)}$ in the equation $y_k = a^{(1)} u_{k-1} + v_k$. Model the variation of $a^{(1)}$ by $a_{k+1}^{(1)} = a_k^{(1)}$; i.e., assume $a^{(1)}$ is strictly constant. Assume v_k has constant variance. Argue, making assumptions on u_{k-1} as required, that $\Sigma_{k+1/k} \longrightarrow 0$ as $k \longrightarrow \infty$, and obtain the limiting form of the filter. Then consider how this filter performs if $E[v_k] = \epsilon$, where ϵ is an arbitrarily small but nonzero constant.

Problem 3.3. For the signal model (3.3), with $a_k^{(i)} = 0$ for $i = n + 1, n + 2$, $\ldots$, it is clear that the input sequence $\{u_k\}$ is not required for the identification.

Show that knowledge of R_k is also not required. [The signal model of this problem is termed an autoregressive (AR) model.]

3.4 APPLICATION OF KALMAN FILTERS

In this section, we mention a number of "real world" problems that have been successfully solved using Kalman filtering ideas. Invariably, assumptions are introduced to manipulate the problem to a form amenable to the application of the Kalman filtering results of the previous sections. The intention is to achieve in each case a near optimum yet workable solution to the original problem. The details for one of the applications are then further explored.

A pollution estimation (prediction) and control application [6] has been mentioned in Section 2.2. Other chemical process applications abound, for example [7, 8], which require extended Kalman filtering theory as developed in Chap. 8.

Kalman filtering (and smoothing) has been applied in filtering noise from two-dimensional images. Early attempts [9] employed a low (fifth) order state vector with an assumption that the scanned picture is a stationary process (obviously such an assumption is not especially well founded since it overlooks the periodic discontinuities associated with jumping from one line or field to the next.) More recent attempts have designated the entire scanned, digitized picture (or suboptimally, a portion of the picture) as the state vector, but the results are perhaps no better than using a state consisting of the four or so pixels (picture elements) in the immediate vicinity of the pixel being processed and applying simple adaptive schemes for the parameters of the digital filter. In picture enhancement we see clearly the costs and limitations of optimal estimation techniques. One alternative is to work with high order state vectors and the consequent high computational burden with possible sensitivity problems. The other alternative is to select low order suboptimal models which may be inappropriate for some situations.

In the previous section, model identification via Kalman filtering ideas was discussed. A further application and extension of these methods to a civil engineering application is discussed in [10]. In [10], stream flow model identification via Kalman smoothing from very few data points is employed to achieve stream flow prediction.

Later in the text we will explore an application of Kalman filtering ideas (in particular the extended Kalman filter of Chap. 8) to the demodulation of frequency modulated signals in communication systems. This is but one application of Kalman filtering ideas to demodulation and detection schemes in communication system design. Demodulation is simply a state or signal estimation problem. As it turns out, for detection, too, the crucial step is often the design of a filter, termed a whitening filter; and in turn (and as will be later shown), whitening filter design is equivalent to Kalman filter design. One of the earliest references in this field is [11]. Examples of developments of the ideas of [11] are given in [12, 13]. Detection problems including application to radar signal processing are discussed in [14, 15, 16]. More recently, adaptive equalization for communication channels has been approached using Kalman filtering [17, 18] (see also Chap. 10).

Another area of application of state-estimation techniques as expounded in this text is to the area of determining the state of a power system. (See, for example, [19].)

Much of the early impetus for the developments of Kalman filter theory and application came from problems in the aerospace industry, as for example in [20, 21]. The state variables in such applications are frequently the position (usually three state variables) and the velocity (a further three variables). We now move on to explore in more detail an application involving such state variables.

The basic task is to estimate as accurately as possible the position and the velocity of a moving object from noisy measurements of its range and bearing. The moving object could be a vehicle such as a ship, aircraft, or tractor; or it could be a whale, school of fish, etc. The measuring equipment could involve radar, sonar, or optical equipment.

In order to keep the problem within manageable proportions, we restrict attention to the case where the movement is constrained to two dimensions. Actually this constraint is not too severe, since movement is frequently in two dimensions, at least to a first order of approximation.

Associated with the tracking problem, there will frequently be a control problem, or a differential game problem. For example, when aircraft in the vicinity of an airport are tracked, there will also be some control; general pursuit-evasion problems exemplify situations in which tracking and control (on both sides) are involved. However, we shall not consider the control aspects further here.

As a first step in tackling the tracking problem, we derive a discrete-time signal model with a state vector consisting of both the position and the velocity of the moving vehicle. The general problem of deriving a discrete-time model by sampling a continuous-time system is discussed in Appendix C. Here we shall proceed more from a "first-principles" approach in deriving the model. Using the two-dimensional rectangular xy coordinate system, we select as state vector

$$\bar{x}_k = \begin{bmatrix} \dot{x}_k \\ x_k \\ \dot{y}_k \\ y_k \end{bmatrix} \tag{4.1}$$

Here, x_k and y_k are the position coordinates and $\dot{x}_k$ and $\dot{y}_k$ are the components of velocity in the x and y directions. The discrete time instants are $k = 0, 1, 2, \ldots$.

In order to express the measurement data as a linear combination of the components of the state vector, we choose as a data vector

$$z_k = \begin{bmatrix} r_k \sin \theta_k \\ r_k \cos \theta_k \end{bmatrix} \tag{4.2}$$

where r_k denotes the range measurement and θ_k denotes the bearing measurement at the discrete time instant k. With this selection of data vector, we have an equation for z_k as

$$z_k = H' \bar{x}_k + v_k \tag{4.3}$$

where

$$H' = \begin{bmatrix} 0 & 1 & 0 & 0 \\ 0 & 0 & 0 & 1 \end{bmatrix} \tag{4.4}$$

and v_k denotes the noise perturbations on the measurement of $H' \bar{x}_k$.

Before proceeding with a description of the noise perturbations v_k, we comment on what might appear at this stage to be a more suitable selection of a state vector, namely $[\dot{r}_k \quad r_k \quad \dot{\theta}_k \quad \theta_k]'$. This selection of state vector would allow the data vector to be simply $[r_k \quad \theta_k]'$, rather than one involving sine and cosine terms as in (4.2). Unfortunately, the full state equations for even the very simple case of a vehicle moving at a constant speed on a fixed course are more complicated if we use this vector, as opposed to the state vector of (4.1). It is this fact which has influenced the choice of the vector.

Returning to our descriptions of the noise perturbations v_k, we comment that the statistical characteristics of the noise v_k depend to a large extent on the measuring equipment. Sonar, radar, and optical equipment each have their own error characteristics. For the purpose of this analysis we assume that measurement noises on the range and bearing are independent, and each is of zero mean. The respective variances are known quantities σ_r^2 and σ_θ^2. We make no further assumptions about the probability density functions of the noise at this point. It follows that the mean $E[v_k]$ is zero and the covariance matrix of our measurement noise vector v_k is approximately

$$R_k = E[v_k v_k'] \simeq \begin{bmatrix} \sigma_r^2 \sin^2 \theta_k + r_k^2 \sigma_\theta^2 \cos^2 \theta_k & (\sigma_r^2 - r_k^2 \sigma_\theta^2) \sin \theta_k \cos \theta_k \\ (\sigma_r^2 - r_k^2 \sigma_\theta^2) \sin \theta_k \cos \theta_k & \sigma_r^2 \cos^2 \theta_k + r_k^2 \sigma_\theta^2 \sin^2 \theta_k \end{bmatrix}$$

$$\tag{4.5}$$

The approximation arises from an assumption that the fluctuations are small, i.e., $\sigma_\theta \ll 1$ and $\sigma_r \ll r_k$. Next, rather than determining the probability density function associated with the measurement noise v_k from the corresponding functions governing range and bearing measurement noise, and keeping in mind that the Kalman filter theory is optimum only for the case of gaussian noise, we use our engineering judgment and tentatively assume that the measurement noise v_k is nearly gaussian. More precisely, we assume the noise is sufficiently near gaussian that when a filter is designed on the basis that the noise is gaussian and then used with the actual noise, the resulting performance is not far from that predicted by the gaussian assumption. This may lead us into error. On the other hand, it may represent the only feasible solution to the filtering problem, given the present status of theory and application. Alternatively, we could adopt the viewpoint that we know nothing about the noise statististics, save R_k, and are simply seeking the best filter within a certain class of filters.

A second problem arising with the use of the covariance matrix R_k is that it depends on the state vector which we are trying to estimate or, at least, on the positional coordinates of the state vector. Since the state vector is unknown, it makes sense, at least from a heuristic point of view, to replace the formula for R_k given above by one involving the current estimates of all the relevant quantities. Thus we would have the 1-2 entry of R as

$$[\sigma_r^2 - (\hat{x}^2 + \hat{y}^2)\sigma_\theta^2]\frac{\hat{x}\hat{y}}{\hat{x}^2 + \hat{y}^2}$$

Again, we are making an engineering assumption which can only be validated by seeing what happens in practice.

In our discussion of the signal model we have so far said nothing about the evolution of the state vector $\bar{x}$. The use of sonar, radar, and the like generally means that the measurements are made in discrete time, and it then makes sense to look for a discrete-time model for the evolution of the state vector. It is here that we again mould our problem somewhat in order to be able to apply theoretical results. We consider a state-space model for the vehicle motion as

$$\bar{x}_{k+1} = F\bar{x}_k + w_k \tag{4.6}$$

where

$$F = \begin{bmatrix} \alpha & 0 & 0 & 0 \\ \Delta & 1 & 0 & 0 \\ 0 & 0 & \alpha & 0 \\ 0 & 0 & \Delta & 1 \end{bmatrix} \tag{4.7}$$

The quantity Δ is the time interval between measurements. The zero mean, gaussian random vector w_k allows us to consider random manoeuvering of the vehicle under observation. The quantity α will be discussed below.

To understand (4.6), and particularly the reason why F has the form of (4.7), consider first a target moving with constant speed on a fixed course. Then we should have $\dot{x}_{k+1} = \dot{x}_k$ and $x_{k+1} = x_k + \Delta\dot{x}_k$, with like equations for $\dot{y}$ and y. This would lead to w_k in (4.6) being zero and an F matrix as in (4.7) with $\alpha = 1$.

Now suppose the target is maneuvering. The entries of w_k will be used to account for the randomness in its motion. First, we can examine the capability of the target for speed change and course change during the interval Δ; from a knowledge of the performance capabilities and operating patterns of the target, we could derive quantities σ_s^2 and σ_c^2, representing the mean square change in forward speeds and in the course θ which occur in any interval Δ. It would be reasonable to expect that the mean changes in forward speed and course would be zero and to assume that changes in speed and course over an interval Δ occur independently.

Suppose (temporarily) that α in (4.7) is unity, so that (4.6) implies

$$\dot{x}_{k+1} = \dot{x}_k + w_k^{(1)}$$

Then all the change in $\dot{x}$ from time k to $k + 1$ is associated with $w_k^{(1)}$. Likewise, all the change in $\dot{y}$ is associated with $w_k^{(3)}$.

Using the relations $\dot{x} = s \cos \theta$ and $\dot{y} = s \sin \theta$, it is possible to show that zero mean changes in s and θ imply $E[w_k^{(1)}] = E[w_k^{(3)}] = 0$, as we would expect, and that the expected mean square changes in s and θ lead to

$$E[(w_k^{(1)})^2] = \sigma_s^2 \sin^2 \theta_k + \sigma_c^2 (\dot{x}_k^2 + \dot{y}_k^2) \cos^2 \theta_k$$

$$E[(w_k^{(3)})^2] = \sigma_s^2 \cos^2 \theta_k + \sigma_c^2 (\dot{x}_k^2 + \dot{y}_k^2) \sin^2 \theta_k \qquad (4.8)$$

$$E[w_k^{(1)} w_k^{(3)}] = [\sigma_s^2 - \sigma_c^2 s_k^2] \sin \theta_k \cos \theta_k$$

provided $\sigma_s^2 \ll \dot{x}_k^2 + \dot{y}_k^2$ and σ_c^2 is small. Note that σ_s^2 and σ_c^2 are the mean square speed and course changes over the interval Δ. Accordingly, the smaller Δ is, the smaller will these quantities be, and therefore the better the above approximation. Now consider the effect of maneuvering on the positional coordinates. We have (at least approximately)

$$x_{k+1} = x_k + \tfrac{1}{2}\Delta(\dot{x}_k + \dot{x}_{k+1}) = x_k + \Delta(\dot{x}_k + \tfrac{1}{2}w_k^{(1)}) \qquad (4.9)$$

[The approximation is perfect if the average velocity between the sampling instants is $\tfrac{1}{2}(\dot{x}_k + \dot{x}_{k+1})$.] Proceeding likewise for y_{k+1} leads to equations

$$\begin{bmatrix} \dot{x}_{k+1} \\ x_{k+1} \\ \dot{y}_{k+1} \\ y_{k+1} \end{bmatrix} = \begin{bmatrix} 1 & 0 & 0 & 0 \\ \Delta & 1 & 0 & 0 \\ 0 & 0 & 1 & 0 \\ 0 & 0 & \Delta & 1 \end{bmatrix} \begin{bmatrix} \dot{x}_k \\ x_k \\ \dot{y}_k \\ y_k \end{bmatrix} + \begin{bmatrix} w_k^{(1)} \\ \dfrac{\Delta}{2} w_k^{(1)} \\ w_k^{(3)} \\ \dfrac{\Delta}{2} w_k^{(3)} \end{bmatrix} \qquad (4.10)$$

with (4.8) holding, and $E[w_k^{(1)}] = E[w_k^{(3)}] = 0$. This model, however, can still be improved slightly. Our assumptions on the target maneuvering implied that the speed $\{s_k\}$ of the target obeyed an equation of the form

$$s_{k+1} = s_k + u_k \qquad (4.11)$$

with $E[u_k] = 0$, $E[u_k^2] = \sigma_s^2$. If $\{u_k\}$ is white and if one sets $s_0 = 0$, one can derive $E[s_k^2] = k\sigma_s^2$, which implies that the mean square speed is unbounded.

Clearly this is an unrealistic assumption for any physical object. It would be more reasonable to have $E[s_k^2] = S$, where S is a constant independent of k and depending on the speed capabilities of the target. It can be shown that this would result if (4.11) were replaced by

$$s_{k+1} = \sqrt{\frac{S^2 - \sigma_s^2}{S^2}} \, s_k + u_k \tag{4.12}$$

(See Prob. 4.3.) In turn, this implies

$$\dot{x}_{k+1} = \sqrt{\frac{S^2 - \sigma_s^2}{S^2}} \, \dot{x}_k + w_k^{(1)}$$

and

$$\dot{y}_{k+1} = \sqrt{\frac{S^2 - \sigma_s^2}{S^2}} \, \dot{y}_k + w_k^{(3)}$$

So, finally, we are led to

$$\bar{x}_{k+1} = \begin{bmatrix} \sqrt{\dfrac{S^2 - \sigma_s^2}{S^2}} & 0 & 0 & 0 \\ \Delta & 1 & 0 & 0 \\ 0 & 0 & \sqrt{\dfrac{S^2 - \sigma_s^2}{S^2}} & 0 \\ 0 & 0 & \Delta & 1 \end{bmatrix} \bar{x}_k + w_k \tag{4.13}$$

with

$$E[w_k] = 0 \qquad w_k' = \left[w_k^{(1)} \quad \frac{\Delta}{2} w_k^{(1)} \quad w_k^{(3)} \quad \frac{\Delta}{2} w_k^{(3)} \right]$$

and (4.8) defining the covariance of w_k. To design the filter, we take the same conceptual jump for w_k as we did for v_k, i.e., we assume for design purposes that w_k is gaussian and that the state estimate can be used in defining the covariance of w_k.

One further matter which must be considered is a selection of the mean $E[\bar{x}_0]$ and covariance P_0 of the initial state, which is assumed gaussian for design purposes. If there is doubt about what the value of P_0 should be, an arbitrarily large value can be selected, since after a few iterations the estimation process is usually reasonably independent of the value chosen for P_0.

Having formulated the model in a form amenable to application of the earlier theory, the remaining work in determining the filter is straightforward.

As a reading of a later chapter on computational aspects and modeling techniques will show, there is absolutely no guarantee that the various assumptions made will lead to a satisfactory filter in practice. The first step after design of the filter is almost always to run a computer simulation.

Computer simulation results for an underwater tracking problem of much the same form as that just discussed are studied in [21]. Since the noise processes w and v have covariances which depend on the state $\bar{x}$, the filtering error covariance also depends on the actual measurements. In [21], ten different Monte Carlo runs were taken in order to achieve root-mean-square error statistics. The results confirm what one would intuitively suspect, namely that when the moving object is making sharp maneuvers, the error increases.

Several other points also arise in [21]. First, it may be possible to estimate adaptively the covariance R_k, which changes but slowly with k. Second, it is possible to cope with the case when measurements of speed are also available (typically from a Doppler shift measurement). With these additional measurements, improved performance at lower average errors are obtained. Third, it is possible to deal with the case where there is an average speed in one direction. This is also dealt with in a problem. Finally, we comment that we have described here a solution to an essentially nonlinear filtering problem that in precise terms is nonlinear, but in operational terms is essentially linear. Nonlinear solutions, in some cases exact and in other cases approximate, are known for classes of nonlinear filtering problems (see, e.g., [22]), and almost certainly a nonlinear solution could be found for the problem considered here. It would undoubtedly involve a filter of greater complexity than that suggested here.

Problem 4.1. Consider the tracking problem discussed in this section. Assume now that in addition to the range r and bearing θ measurement data available, a noisy Doppler measurement of dr/dt data is available. What would be the change to the model for this case?

Problem 4.2. Suppose that the tracking problem of the section is studied, with the additional information that the target is maintaining an average speed in a certain direction. Show that two more components m^x and m^y of the state vector can be introduced to model this effect, and that if the average speed and course are known, one obtains

$$\dot{x}_{k+1} = \sqrt{\frac{S^2 - \sigma_s^2}{S^2}}\,\dot{x}_k + \left(1 - \sqrt{\frac{S^2 - \sigma_s^2}{S^2}}\right)m_k^x + w_k^{(1)}$$

$$m_{k+1}^x = m_k^x$$

with similar equations for m_{k+1}^y and $\dot{y}_{k+1}$. Extend the model to the case when the average speed is initially not known exactly.

Problem 4.3. Suppose that $s_{k+1} = \alpha s_k + u_k$, with $\{u_k\}$ a white noise process with $E[u_k] = 0$, $E[u_k^2] = \sigma_s^2$. Show that if $E[s_k^2] = S^2$, for all k, it is necessary that $\alpha^2 = (S^2 - \sigma_s^2)/S^2$.

REFERENCES

[1] ASTROM, K. J., *Introduction to Stochastic Control Theory*, Academic Press, Inc., New York, 1970.

[2] ATHANS, M., and E. TSE, "A Direct Derivation of the Optimal Linear Filter Using the Maximum Principle," *IEEE Trans. Automatic Control*, Vol. AC-12, No. 6, December 1967, pp. 690–698.

[3] KASHYAP, R. L., "Maximum Likelihood Identification of Stochastic Linear Systems," *IEEE Trans. Automatic Control*, Vol. AC-15, No. 1, February 1970, pp. 25–33.

[4] MAYNE, D. Q., "Optimal Non-stationary Estimation of the Parameters of a Linear System with Gaussian Inputs," *J. Electronics and Control*, Vol. 14, January 1963, pp. 107–112.

[5] GOODWIN, G. C., and R. L. PAYNE, *Dynamic System Identification: Experiment Design and Data Analysis*, Academic Press, Inc., New York, 1977.

[6] SAWARAGI, Y., *et al.*, "The Prediction of Air Pollution Levels by Nonphysical Models Based on Kalman Filtering Method," *J. Dynamic Systems, Measurement and Control*, Vol. 98 No. 4, December 1976, pp. 375–386.

[7] SOLIMAN, M. A., "Application of Non-linear Filtering to the Stainless-steel Decarbonization Process," *Int. J. Control*, Vol. 20, No. 4, October 1974, pp. 641–653.

[8] PARLIS, H. J., and B. OKUMSEINDE, "Multiple Kalman Filters in a Distributed Stream Monitoring System," **15***th Proc. Joint Automatic Control Conf.*, University of Texas, Austin, June 1974, pp. 615–623.

[9] NAHI, N. E., "Role of Recursive Estimation in Statistical Image Enhancement," *Proc. IEEE*, Vol. 60, July 1972, pp. 872–877.

[10] NORTON, J. P., "Optimal Smoothing in the Identification of Linear Time-varying Systems," *Proc. IEEE*, Vol. 122, No. 6, June 1975, pp. 663–668.

[11] SNYDER, D. L., *The State Variable Approach to Continuous Estimation*, The M.I.T. Press, Cambridge, Mass., 1970.

[12] MOORE, J. B., and P. HETRAKUL, "Optimal Demodulation of PAM Signals," *IEEE Trans. Inform. Theory*, Vol. IT-19, No. 2, March 1973, pp. 188–197.

[13] TAM, P. K. S., and J. B. MOORE, "Improved Demodulation of Sampled FM Signals in High Noise," *IEEE Trans. Comm.*, Vol. COM-25, No. 9, September 1977, pp. 935–942.

[14] WOOD, M. G., J. B. MOORE, and B. D. O. ANDERSON, "Study of an Integrand Equation Arising in Detection Theory," *IEEE Trans. Inform. Theory*, Vol. IT-17, No. 6, November 1971, pp. 677–687.

[15] VAN TREES, H. L., *Detection, Estimation and Modulation Theory*, Parts I, II, III, John Wiley & Sons, Inc., New York, 1968, 1971, 1972.

[16] PEARSON, J. B., "Kalman Filter Applications in Airborne Radar Tracking," *IEEE Trans. Aerospace and Electronic Systems*, Vol. AES-10, No. 3, May 1974, pp. 319–329.

[17] KLEIBANOV, S. B., V. B. PRIVAL'SKII, and I. V. TINRE, "Kalman Filter for Equalizing of Digital Communicating Channel," *Automation and Remote Control*, Vol. 35, No. 7, Part I, July 1974, pp. 1097–1102.

[18] GODARD, D., "Channel Equalization Using a Kalman Filter for Fast Data Transmissions," *IBM J. Research and Development*, Vol. 18, No. 3, May 1974, pp. 267–273.

[19] LARSON, R. W., *et al.*, "State Estimation in Power Systems: Parts I, II, Theory and Feasibility," *IEEE Trans. Power App. Sys.*, Vol. PAS-89, March 1970, pp. 345–353.

[20] TAPLEY, B. D., "Orbit Determination in the Presence of Unmodeled Accelerations," *IEEE Trans. Automatic Control*, Vol. AC-18, No. 4, August 1973, pp. 369–373.

[21] HOLDSWORTH, J., and J. STOLZ, "Marine Applications of Kalman Filtering," in *Theory and Applications of Kalman Filtering* (ed. C. T. Leondes), Nato Advisory Group for Aerospace Research and Development, AGARDograph 139, February, 1970.

[22] JAZWINSKI, A. H., *Stochastic Processes and Filtering Theory*, Academic Press, Inc., New York, 1970.

CHAPTER 4

TIME-INVARIANT FILTERS

4.1 BACKGROUND TO TIME INVARIANCE
OF THE FILTER

We recall from the last chapter that the general form of the Kalman filter is

$$\hat{x}_{k+1/k} = [F_k - K_k H'_k]\hat{x}_{k/k-1} + K_k z_k \tag{1.1}$$

Here, $\{z_k\}$ is the measurement process and $\hat{x}_{k/k-1}$ is the conditional mean $E[x_k | Z_{k-1}]$. For the definition of other quantities, see the last chapter.

In general, F_k, H_k, and K_k depend on k; that is, (1.1) represents a time-varying filter. From the point of view of their greater ease of construction and use, time-invariant filters, or those with F_k, H_k, and K_k independent of k, are appealing. This is one reason for their study. The other reason lies in their frequency of occurrence. Some special assumptions on the system upon which the filter operates lead to the filter being time invariant; these assumptions, detailed later in the chapter, are frequently fulfilled.

Evidently, for (1.1) to represent a time-invariant filter, K_k must be constant and, unless there is some unlikely cancellation between the time-variation in F_k and $K_k H'_k$ to force $F_k - K_k H'_k$ to be constant, both F_k and H_k must be constant. This suggests that perhaps the underlying system must be time invariant, and a moment's reflection then suggests strongly that the conditions for time invariance of the filter might be:

1. Time invariance of the system being filtered.
2. Stationarity of the random processes associated with the underlying system. (This is not necessarily implied by time invariance of the system; if, for example, the system is unstable, this condition will not hold.)

As we shall show in the later sections, these two assumptions are in fact sufficient to guarantee time invariance of the filter. Actually, they are a little stronger than necessary.

A question of vital interest regarding the performance of a filter is whether or not the filter is stable. We shall leave aside consideration of the stability of time-varying filters and be concerned in this chapter with explaining when the following time-invariant filter is stable:

$$\hat{x}_{k+1/k} = [F - KH']\hat{x}_{k/k-1} + Kz_k \tag{1.2}$$

As described in Appendix C, an equivalent task is to explain when eigenvalues of $F - KH'$ lie inside $|z| < 1$. The techniques for studying this question are studied in the next section, and they also allow us, in Sec. 4.3, to expand on the notion of stationarity of the underlying random processes associated with the system being filtered. It turns out that stability of the system is normally required to guarantee stationarity of, for example, the random process $\{x_k\}$, where x_k is the system state vector.

In Sec. 4.4, we present precise conditions under which the filter is time invariant and stable. Section 4.5 discusses some important frequency domain formulas.

Problem 1.1. Assume (1.1) represents a time-invariant filter, and assume also that the filter performance, as measured by the unconditioned error covariance matrix, is independent of time. Show that the second assumption and an assumption that the input (and output) noise is stationary suggests that G_k in the signal model is independent of time.

4.2 STABILITY PROPERTIES OF LINEAR, DISCRETE-TIME SYSTEMS

In this section, we look at stability properties of the equations

$$x_{k+1} = Fx_k \tag{2.1}$$

and

$$x_{k+1} = Fx_k + Gu_k \tag{2.2}$$

As we know (see Appendix C), (2.1) is asymptotically stable—in fact, exponentially asymptotically stable—if and only if $|\lambda_i(F)| < 1$ for all i. Under this condition, (2.2) is bounded-input, bounded-output (or bounded-state) stable. Our main aim is to prove with the aid of the Lyapunov theorems of Appendix D an important result characterizing matrices F associated with systems with desirable stability properties. This result will also be of use in considering the behaviour of (2.2) when the input sequence is white noise.

Characterization of Stability via a Linear Matrix Equation

Here we want to study an equation which arises in testing a matrix F to see whether $|\lambda_i(F)| < 1$. The equation is as follows, where F and Q are known $n \times n$ matrices and P is unknown:

$$P - FPF' = Q \qquad (2.3)$$

In order to study the equation, we need a preliminary result, obvious for scalars on account of the properties of geometric series. We omit the proof.

LEMMA 2.1. Suppose F is an $n \times n$ matrix with $|\lambda_i(F)| < 1$. Let A be an arbitrary $n \times n$ matrix. Then

$$B = \sum_{k=0}^{\infty} F^k A F'^k$$

exists and is finite.

Now we return to (2.3). The reader should recall that the pair $[F, G]$ is completely reachable if $[G, FG, \ldots, F^{n-1}G]$ has rank n (see also Appendix C). An equivalent statement is that $\sum_{i=0}^{n-1} F^i GG' F'^i$. is nonsingular.

THEOREM 2.1.* Suppose Q is nonnegative definite symmetric, and let G be such that $Q = GG'$, with $[F, G]$ completely reachable. Then if $|\lambda_i(F)| < 1$, the solution P of (2.3) exists, is unique, and is positive definite. Conversely, if a positive definite solution exists, it is unique and $|\lambda_i(F)| < 1$.

Before proving the theorem, we make several comments.

1. Because Q is nonnegative definite, there exists an infinity of matrices G such that $Q = GG'$ (see Appendix A). Triangular G are readily found from Q (see [1]).
2. If G_1 and G_2 are such that $Q = G_1 G_1' = G_2 G_2'$, then complete reachability of $[F, G_1]$ is equivalent to complete reachability of $[F, G_2]$.

*The result is a composite of results due to Lyapunov, Kalman, and Stein. It is sometimes termed the discrete-time lemma of Lyapunov.

(Observe that

$$\sum_{i=0}^{n-1} F^i G_k G'_k F'^i = \sum_{i=0}^{n-1} F^i Q F'^i$$

for $k = 1, 2$.)

3. So far, we have not commented on how (2.3) might be solved. This matter will be taken up further below; suffice it to say here that solving (2.3) is equivalent to solving a linear equation of the form $Ax = b$, where A and b are a known matrix and vector, respectively, and x is an unknown vector.

4. Assuming solution of (2.3) is easy, the theorem contains an implicit procedure for testing if $|\lambda_i(F)| < 1$ for a prescribed F. One selects an arbitrary $Q \geq 0$ for which the reachability property holds—$Q = I$ is a universally possible choice. One solves (2.3) and then checks whether $P > 0$. The only possible circumstance under which (2.3) cannot be solved arises when the constraint $|\lambda_i(F)| < 1$ fails; however, failure of $|\lambda_i(F)| < 1$ does not *necessarily* imply inability to solve (2.3).

Proof. Suppose that $|\lambda_i(F)| < 1$. Define the matrix $\bar{P}$ by

$$\bar{P} = \sum_{k=0}^{\infty} F^k Q(F')^k \tag{2.4}$$

Then $\bar{P}$ exists by Lemma 2.1 and is easily seen to be positive definite. For $\bar{P} \geq \sum_{k=0}^{n-1} F^k GG'F'^k > 0$, the first inequality following from (2.4) and the second from the complete reachability assumption.

Next observe that

$$\bar{P} - F\bar{P}F' = \sum_{k=0}^{\infty} F^k Q(F')^k - \sum_{k=1}^{\infty} F^k Q(F')^k$$
$$= Q$$

So $\bar{P}$ *satisfies* (2.3). Let $\tilde{P}$ be any other solution. Then (2.3), with P replaced by first $\bar{P}$ and then $\tilde{P}$, yields

$$(\bar{P} - \tilde{P}) - F(\bar{P} - \tilde{P})F' = 0$$

from which for all k

$$F^{k-1}(\bar{P} - \tilde{P})(F')^{k-1} - F^k(\bar{P} - \tilde{P})(F')^k = 0$$

Adding such relations, we find

$$(\bar{P} - \tilde{P}) - F^k(\bar{P} - \tilde{P})(F')^k = 0$$

Now let $k \to \infty$, and use the fact that $F^k \to 0$. It follows that $\bar{P} = \tilde{P}$. *This establishes uniqueness.*

We now prove the converse result. Accordingly, suppose that (2.3) holds, with P positive definite. Associate with the homogeneous system

$$x_{k+1} = F'x_k \tag{2.5}$$

(note the prime!) the positive definite function $V(x_k) = x_k'Px_k$. Then

$$\Delta V(x_k) = V(x_{k+1}) - V(x_k)$$
$$= x_k'FPF'x_k - x_k'Px_k$$
$$= -x_k'Qx_k$$

where we have used (2.5) and (2.3) to obtain the second and third equalities. Evidently, $\Delta V \leq 0$ and so, by the Lyapunov theory, $x_{k+1} = F'x_k$ is certainly stable. We can conclude asymptotic stability if ΔV is identically zero only on the zero trajectory. Let us prove this. Thus suppose $x_k'Qx_k = 0$ for $k = 0, 1, 2, \ldots$. Then $x_0'F^kQF'^kx_0 = 0$ for all k, and so

$$x_0'\left[\sum_{k=0}^{n-1} F^kQF'^k\right]x_0 = 0$$

The complete reachability assumption implies $x_0 = 0$. Uniqueness follows as before.

Now that the proof of the theorem has been presented, the reader may understand the following additional points.

1. The same results as proved above hold, mutatis mutandis, for the equation $P - F'PF = Q$, for the eigenvalues of F' are the same as those of F.
2. The formula of (2.4) defines a solution of (2.3). Despite the fact that it represents an infinite sum, it may be an effective way to compute a solution, especially if the F^k decay quickly to zero. This will be the case if the $|\lambda_i(F)|$ are bounded by a number significantly less than 1. Further comments on the solution of (2.3) are given below.
4. Define a sequence of matrices $\{P_k\}$ by

$$P_{k+1} = FP_kF' + Q \qquad P_0 = 0 \qquad (2.6)$$

Then if $|\lambda_i(F)| < 1$, $\lim_{k \to \infty} P_k$ exists and is $\bar{P}$. To see this, observe by direct calculation that

$$P_1 = Q$$
$$P_2 = FQF' + Q$$
$$\vdots$$
$$P_{k+1} = \sum_{l=0}^{k} F^lQF''$$

The formula for P_{k+1} yields the required limit. Also, as one would expect, taking the limit in (2.6) recovers Eq. (2.3), which is satisfied by $\bar{P}$.

5. If $[F, G]$ is not completely reachable and if $|\lambda_i(F)| < 1$, then (2.4) still defines a unique solution of (2.3), as examination of the earlier proof will show. The solution, however, is not positive definite. (This point is explored in the problems.) Derivation of a converse result when $[F, G]$ is not completely reachable is, however, more difficult, and there is no simple statement of such a result.

To conclude this section, we comment on solution procedures for (2.3).

The first procedure is as follows: By equating each entry on each side of (2.3), using literals $p^{(ij)}$ for the entries of P, one obtains a set of linear equations in the $p^{(ij)}$. These may be written in the form $Ax = b$, where the entries of x are the $p^{(ij)}$, the entries of b are the $q^{(ij)}$, and the entries of A are derived from the entries of F. In principle, this equation can be solved for x. In practice, this may be difficult since A is of dimension $n^2 \times n^2$ or, if advantage is taken of the symmetry of P, of dimension $\frac{1}{2}n(n + 1) \times \frac{1}{2}n(n + 1)$. Actually one can cut down the dimension of A further, to $\frac{1}{2}n(n - 1) \times \frac{1}{2}n(n - 1)$ (see [2]). Those familiar with the Kronecker product [3] will recognize A to be the matrix $I - F \otimes F'$, which has eigenvalues $1 - \lambda_i \lambda_j$, where λ_i is an eigenvalue of F. Accordingly, $Ax = b$ is solvable if $1 - \lambda_i \lambda_j$ is nonzero for all i and j; a sufficient condition for this is $|\lambda_i| < 1$ for all i.

A second procedure for solving (2.3) is to use (2.6) and find $\lim_{k \to \infty} P_k$.

A third procedure involves a simple speed-up of the second procedure. By updating two $n \times n$ matrices, one can obtain a "doubling" algorithm:

$$M_{k+1} = (M_k)^2 \qquad\qquad M_1 = F \qquad\qquad (2.7)$$

$$N_{k+1} = M_k N_k M_k' + N_k \qquad N_1 = Q \qquad\qquad (2.8)$$

One easily verifies that $M_{k+1} = F^{2^k}$ and that $N_{k+1} = P_{2^k}$. Then $\bar{P} = \lim_{k \to \infty} N_k$, with convergence occurring faster than when (2.6) is used in its raw form.

Main Points of the Section

Given that $[F, G]$ is completely reachable, the condition $|\lambda_i(F)| < 1$ is necessary and sufficient for $P - FPF' = GG'$ to have a unique positive definite solution. The solution is definable either via an infinite series or by solving an equation of the form $Ax = b$, where A and b are a known matrix and vector, respectively. Rapid procedures for summing the series are available.

Problem 2.1. Suppose $|\lambda_i(F)| < 1$ and $Q \geq 0$. Show that the equation $P - FPF' = Q$ has a unique nonnegative definite solution and that all vectors α in the nullspace of the solution P of $P - FPF' = Q$ lie in the nullspace of $Q, QF', \ldots,$ $Q(F')^{n-1}$ and conversely.

Problem 2.2. Suppose that

$$
F = \begin{bmatrix}
0 & 0 & \cdots & 0 & -a_n \\
1 & 0 & \cdots & 0 & -a_{n-1} \\
0 & 1 & \cdots & 0 & -a_{n-2} \\
\cdot & \cdot & & \cdot & \cdot \\
\cdot & \cdot & & \cdot & \cdot \\
\cdot & \cdot & & \cdot & \cdot \\
0 & 0 & \cdots & 1 & -a_1
\end{bmatrix}, \quad
G = \begin{bmatrix}
1 - a_n^2 \\
a_1 - a_n a_{n-1} \\
a_2 - a_n a_{n-2} \\
\cdot \\
\cdot \\
\cdot \\
a_{n-1} - a_n a_1
\end{bmatrix}
$$

Show that one solution P of the equation $P - FPF' = GG'$ has ij entry

$$
p^{(ij)} = \sum_{p=1}^{\min (i,\, j)} (a_{i-p} a_{j-p} - a_{n-i+p} a_{n-j+p})
$$

The matrix P is called the Schur-Cohn matrix. Show that positive definiteness of P is necessary and sufficient for the zeros of $z^n + a_1 z^{n-1} + \cdots + a_n$ to lie inside $|z| = 1$.

Problem 2.3. Suppose that $[F, G]$ is completely reachable and the equation $P - \rho^2 FPF' = GG'$ has a positive definite solution P for some known scalar ρ. What can one infer concerning the eigenvalues of F?

Problem 2.4. Show that if the recursion $P_{k+1} = FP_k F' + Q$ is used with arbitrary initial P_0 and if $|\lambda_i(F)| < 1$, then $P_k \longrightarrow \bar{P}$, where $\bar{P} - F\bar{P}F' = Q$.

Problem 2.5. Suppose that $[F, G]$ is completely stabilizable, but not necessarily completely reachable. Show that $P - FPF' = GG'$ has a unique nonnegative definite symmetric solution if and only if $|\lambda_i(F)| < 1$. [The "if" part is easy; approach the "only if" part as follows: Let w be an eigenvector of F for which $|\lambda_i(F)| < 1$ fails. Study $w'^* GG' w$ and show that $w' F^i G = 0$ for all i. Deduce a contradiction. This technique may also be used to prove Theorem 2.1.]

Problem 2.6. Let X_k be a sequence of nonnegative definite matrices such that for some nonnegative symmetric X and for all k, $X \geq X_{k+1} \geq X_k$. Show that $\lim_{k \to \infty} X_k$ exists, as follows: Let $e^{(i)}$ be a vector comprising all zeros, save for 1 in the ith position. Consider $e^{(i)'} X_k e^{(i)}$ to conclude that $X_k^{(ii)}$ converges, and then consider $(e^{(i)'} + e^{(j)'}) X_k (e^{(i)} + e^{(j)})$ to conclude that $X_k^{(ij)}$ converges.

4.3 STATIONARY BEHAVIOUR OF LINEAR SYSTEMS

In this section, we consider the time-invariant system

$$
x_{k+1} = Fx_k + Gw_k \tag{3.1}
$$

with associated measurement process

$$
z_k = H'x_k + v_k \tag{3.2}
$$

We shall assume that v_k and w_k are independent, zero mean, stationary, white gaussian processes, with covariances given by

$$E[v_k v_l'] = R\delta_{kl} \qquad E[w_k w_l'] = Q\delta_{kl} \qquad (3.3)$$

We shall attempt to answer the question: when is the $\{x_k\}$ process, and consequently the $\{z_k\}$ process, stationary? We remind the reader that a gaussian process $\{a_k\}$ is stationary if and only if

$$E[a_k] = m \qquad E[a_k a_l'] = C_{k-l} \qquad (3.4)$$

In other words, the mean of the process is constant, and the correlation between the random variables defined by sampling the process at two time instants depends only on the difference between the time instants. Normally, it is understood that the process $\{a_k\}$ is defined on $-\infty < k < \infty$. If it is defined on $0 \leq k < \infty$, an acceptable definition of stationarity might be provided by (3.4) with $k, l \geq 0$; although not a standard convention, we shall adopt this usage of the word stationary.

As we shall see, in order to guarantee the stationarity of $\{x_k\}$, it proves convenient to introduce the stability condition we studied in the last section: $|\lambda_i(F)| < 1$. Intuitively, the reader can probably appreciate that with $|\lambda_i(F)| > 1$ for some i, the noise w_k for one k could excite the system so that the resulting x_{k+1} initiated an instability, according to $x_{k+n} = F^{n-1}x_{k+1}$. Therefore, it seems reasonable that for stationarity the inequality $|\lambda_i(F)| \leq 1$ should be satisfied. That the equality sign may rule out stationarity is less obvious; however, a simple example illustrates what can happen in case the equality sign holds.

EXAMPLE 3.1. Consider the system

$$x_{k+1} = x_k + w_k$$

where $E[w_k w_l] = \delta_{kl}$ and $x_0 = 0$; here, x_k is a scalar. Squaring the defining equation and taking expectations leads to

$$E[x_{k+1}^2] = E[x_k^2] + 1$$

or

$$E[x_k^2] = k$$

Clearly, x_k is not stationary. Rather, its variance is an unbounded function of k.

EXAMPLE 3.2. Consider now the autonomous system with scalar state x_2:

$$x_{k+1} = x_k$$

with x_0 a gaussian random variable of mean m and covariance P_0. Clearly, $E[x_k] = m$ for all $k \geq 0$ and $E[(x_k - m)(x_l - m)] = P_0$ for all $k, l \geq 0$. Therefore, stationarity is present.

Both Examples 3.1 and 3.2 deal with systems for which $\lambda_i(F) = 1$. The difference between the two systems, however, is that in the case of the nonstationary Example 3.1, the system is completely reachable from the input, while in the case of the stationary Example 3.2, this is not so.

The idea above can be generalized to the case of more general systems. However, in the interests of simplicity, we shall relegate such a generalization to the problems and restrict ourselves here to the situation when $|\lambda_i(F)| < 1$. The first result is the following.

THEOREM 3.1. Consider the arrangement of Eqs. (3.1) and (3.3), and suppose that at the initial time k_0, x_{k_0} is a gaussian random variable of mean m and covariance P_0. Suppose that $|\lambda_i(F)| < 1$. Then when $k_0 \rightarrow -\infty$, $\{x_k\}$ is a stationary process of mean zero and covariance

$$E[x_k x_l'] = F^{k-l}\bar{P} \qquad k \geq l$$
$$= \bar{P}(F')^{l-k} \qquad l \geq k \qquad (3.5)$$

where $\bar{P}$ is the unique solution of

$$\bar{P} - F\bar{P}F' = GQG' \qquad (3.6)$$

Further, if $m = 0$, k_0 is finite and fixed, and $P_0 = \bar{P}$, then $\{x_k\}$ for $k \geq k_0$ is stationary and has covariance as above.

Before proving the theorem, we offer the following comments.

1. The technique used to define $\{x_k\}$ for $-\infty < k < \infty$ should be noted —start with $k \geq k_0$, and then let $k_0 \rightarrow -\infty$.
2. The mean m and covariance P_0 of x_{k_0} are forgotten when $k_0 \rightarrow -\infty$, in the sense that m and P_0 do not affect the mean and covariance of $\{x_k\}$. This forgetting property is tied up with the fact that $|\lambda_i(F)| < 1$.
3. When $m = 0$, $P_0 = \bar{P}$, and k_0 is finite, in essence a stochastic initial state is being set which gives the initial state the same statistics it would have had if the process had been running from $-\infty$. This is a rough explanation of the resulting stationarity of $\{x_k\}$, $k \geq k_0$.
4. The question of whether or not $\bar{P}$ is singular is not taken up in the theorem statement. As we know from the previous section, $\bar{P}$ will be nonsingular if and only if $[F, GG_1]$ is completely reachable for any G_1 such that $G_1 G_1' = Q$. Lack of complete reachability would imply that the noise process w_k failed to excite some modes of the system. In turn, such nonexcitation would correspond to a zero variance of the modes, or singularity in $\bar{P}$.

Proof. For finite k_0. one can show, as outlined earlier, that

$$E[x_k] = FE[x_{k-1}] = F^{k-k_0}m$$

Since $|\lambda_i(F)| < 1$, then $\lim_{k_0 \rightarrow -\infty} F^{k-k_0} = 0$ and $E[x_k] = 0$ for all k, as required.

Next, recall that

$$E[x_k x_k'] = FE[x_{k-1} x_{k-1}']F' + GQG'$$

$$= F^{k-k_0} P_0 (F')^{k-k_0} + \sum_{m=0}^{k-k_0-1} F^m GQG'(F')^m$$

Letting $k_0 \rightarrow -\infty$, there obtains

$$E[x_k x_k'] = \sum_{m=0}^{\infty} F^m GQG'(F')^m$$

$$= \bar{P}$$

where $\bar{P}$ is the unique solution of

$$\bar{P} - F\bar{P}F' = GQG'$$

The formula for $E[x_k x_i']$ is then immediate.

Finally, suppose $m = 0$, k_0 is fixed, and $P_0 = \bar{P}$. It is trivial to see that $E[x_k] = 0$ for all k. Second, observe that for $k \geq k_0$,

$$E[x_k x_k'] = F^{k-k_0}\bar{P}(F')^{k-k_0} + \sum_{m=0}^{k-k_0-1} F^m GQG'(F')^m$$

$$= F^{k-k_0}\bar{P}(F')^{k-k_0} + \sum_{m=0}^{k-k_0-1} F^m \bar{P}(F')^m$$

$$- \sum_{m=0}^{k-k_0-1} F^m F\bar{P}F'(F')^m$$

$$= \bar{P}$$

Immediately, $E[x_k x_i']$ is as given by (3.5), and stationarity is established.

Theorem 3.1 gives a sufficient condition for stationarity, namely, $|\lambda_i(F)| < 1$. As argued with the aid of earlier examples, this condition is almost a necessary one. The following theorem, in essence, claims that, if all modes of (3.1) are excited by white noise, $|\lambda_i(F)| < 1$ is necessary for stationarity. Proof is requested in the problems.

THEOREM 3.2. Consider the arrangement of Eqs. (3.1) and (3.3). Suppose that the $\{x_k\}$ process is stationary and that $[F, GG_1]$ is completely reachable for any G_1 with $G_1 G_1' = Q$. Then $|\lambda_i(F)| < 1$.

Further results are obtainable, in case complete reachability is not present. These are explored in the problems.

Suppose now that $|\lambda_i(F)| < 1$, that the initial time k_0 is finite, and that m and P_0 are not necessarily zero and $\bar{P}$, respectively. Does one then necessarily have a nonstationary process? The answer is yes, but one does have what is termed *asymptotic stationarity*, and one can regard $\{x_k\}$ as a process consisting of a stationary part and a nonstationary part dying away as $k \rightarrow$

∞. Thus, for example,

$$E[x_k] = F^{k-k_0}m$$

and as $k \longrightarrow \infty$, this tends to zero. Also,

$$E[x_k x_k'] = F^{k-k_0}P_0(F')^{k-k_0} + \sum_{m=0}^{k-k_0-1} F^m GQG'(F')^m$$

$$= F^{k-k_0}\bar{P}(F')^{k-k_0} + \sum_{m=0}^{k-k_0-1} F^m GQG'(F')^m$$

$$+ F^{k-k_0}(P_0 - \bar{P})(F')^{k-k_0}$$

$$= \bar{P} + F^{k-k_0}(P_0 - \bar{P})(F')^{k-k_0}$$

Thus $E[x_k x_k']$ consists of a stationary part $\bar{P}$ and a nonstationary part which decays to zero as $k \longrightarrow \infty$. The nonstationary part may not be nonnegative definite. A common situation would occur with $x_0 = 0$, i.e., $m = 0$ and $P_0 = 0$. Then $E[x_k x_k'] \leq \bar{P}$ for all k and increases monotonically towards $\bar{P}$.

Evidently, after an interval $k - k_0$ large enough such that F^{k-k_0} is very small, to all intents and purposes the $\{x_k\}$ process has become stationary.

The Output Process

What now of the $\{z_k\}$ process, defined via (3.2)? The calculation of the mean and covariance of the process is straightforward; using the ideas of the last chapter, we have, for a stationary $\{x_k\}$ process,

$$E[z_k] = 0 \qquad E[z_k z_l'] = H'F^{k-l}\bar{P}H \qquad k > l$$

$$= R + H'\bar{P}H \qquad k = l \qquad (3.7)$$

$$= H'\bar{P}(F')^{l-k}H \qquad k < l$$

In essence, the only potentially difficult step in passing from (3.1), (3.2), and (3.3) to the covariance of $\{z_k\}$ is the calculation of $\bar{P}$ via (3.6), so long at least as $|\lambda_i(F)| < 1$. In the event that $[F, GG_1]$ is completely reachable for any G_1 with $G_1 G_1' = Q$, existence and positive definiteness of $\bar{P}$ provide a check on $|\lambda_i(F)| < 1$.

Now suppose that the input and output noise are correlated, with

$$E[w_k v_l'] = S\delta_{kl} \qquad (3.8)$$

for some constant matrix S of appropriate dimensions. Equations (3.7) need to be modified, and we compute the modifications in the following way. Since

$$x_k = F^{k-k_0}x_{k_0} + \sum_{m=k_0}^{k-1} F^{k-1-m}Gw_m$$

it follows that

$$E[x_k v_l'] = F^{k-1-l}GS \qquad k > l$$

$$= 0 \qquad k \leq l$$

Then, in the stationary case,

$$
\begin{aligned}
E[z_k z_l'] &= H'E[x_k x_l']H + E[v_k x_l']H + H'E[x_k v_l'] \\
&\quad + E[v_k v_l'] \\
&= H'F^{k-l}\bar{P}H + H'F^{k-l-1}GS && k > l \qquad (3.9) \\
&= R + H'\bar{P}H && k = l \\
&= H'\bar{P}(F')^{l-k}H + S'G'(F')^{l-k-1}H && k < l
\end{aligned}
$$

We sum up the result as a theorem.

THEOREM 3.3. Adopt the same hypotheses as for Theorem 3.1, except for assuming now that (3.8) holds. Then when $k_0 \rightarrow -\infty$, $\{z_k\}$ is a stationary process of mean zero and covariance as given in (3.9). Here $\bar{P}$ is the solution of (3.6).

Frequency Domain Calculations

Let $\{a_k\}$ be a stationary gaussian process with mean zero and covariance $E[a_k a_l'] = C_{k-l}$. Recall that one defines

$$
\Phi_{AA}(z) = \sum_{k=-\infty}^{+\infty} z^{-k}C_k
$$

(assuming the sum is well defined in some annulus $\rho < |z| < \rho^{-1}$) as the power spectrum associated with $\{a_k\}$ (see Appendix A).

EXAMPLE 3.3. The power spectrum of the white process $\{v_k\}$ is R, a constant independent of z.

Let us calculate now the power spectra of the $\{x_k\}$ and $\{z_k\}$ processes when these are stationary. Using (3.5), we have

$$
\Phi_{XX}(z) = \sum_{k=0}^{\infty} z^{-k}F^k\bar{P} + \sum_{k=1}^{\infty} z^k\bar{P}(F')^{-k}
$$

Now for $|z| > \max_i |\lambda_i(F)|$, one has

$$
(I - z^{-1}F)^{-1} = I + z^{-1}F + z^{-2}F^2 + \cdots
$$

and for $|z| < [\max_i |\lambda_i(F)|]^{-1}$, one has

$$
[I - z(F')^{-1}]^{-1} = I + z(F')^{-1} + z^2(F')^{-2} + \cdots
$$

Therefore,

$$
\Phi_{XX}(x) = (I - z^{-1}F)^{-1}\bar{P} + \bar{P}(I - zF')^{-1} - \bar{P}
$$

A more illuminating expression for $\Phi_{XX}(z)$ can, however, be obtained via some rearrangement. Thus

$$(I - z^{-1}F)^{-1}\bar{P} + \bar{P}(I - zF')^{-1} - \bar{P} = (I - z^{-1}F)^{-1}[\bar{P}(I - zF') + (I - z^{-1}F)\bar{P}$$
$$- (I - z^{-1}F)\bar{P}(I - zF')][I - zF']^{-1}$$
$$= (I - z^{-1}F)^{-1}(\bar{P} - F\bar{P}F)(I - zF')^{-1}$$
$$= (I - z^{-1}F)^{-1}GQG'(I - zF')^{-1}$$
$$= (zI - F)^{-1}GQG'(z^{-1}I - F')^{-1}$$

Now the transfer function matrix linking $\{w_k\}$ to $\{x_k\}$ is $M(z) = (zI - F)^{-1}G$, while $Q = \Phi_{WW}(z)$. Thus

$$\Phi_{XX}(z) = M(z)QM'(z^{-1}) \tag{3.10}$$

With $W(z) = H'(zI - F)^{-1}G$, one gets

$$\Phi_{ZZ}(z) = W(z)QW'(z^{-1}) + W(z)S + S'W'(z^{-1}) + R \tag{3.11}$$

A derivation is called for in the problems.

The formulas (3.10) and (3.11) are examples of a general theorem, noted in Appendix A. We state it here without proof.

THEOREM 3.4. Suppose $\{u_k\}$ is a zero mean, stationary process with power spectrum $\Phi_{UU}(z)$ and is the input to a time-invariant, asymptotically stable system with transfer function matrix $W(z)$. Then the output process $\{y_k\}$ is asymptotically stationary with spectrum $\Phi_{YY}(z) = W(z)\Phi_{UU}(z)W'(z^{-1})$.

By and large, the result of Theorem 3.4 is the easier to use in calculating power spectra. The approach via z-transform of the covariance should, however, be kept in mind.

Wide-sense Stationarity

A gaussian process $\{a_k\}$ is stationary when (3.4) is satisfied, If $\{a_k\}$ is not gaussian, however, (3.4) is necessary but not sufficient for stationarity; one calls $\{a_k\}$ *wide-sense stationary* when (3.4) holds, but higher order moments possibly fail to exhibit stationarity. If one drops gaussian assumptions on $\{v_k\}$, $\{w_k\}$, and x_0, then the theorems hold with stationarity replaced by wide-sense stationarity.

Main Points of the Section

Time invariance of a signal model driven by stationary white noise is necessary but not sufficient for stationarity of the state and output processes. Normally, asymptotic stability is also required, as is either a commencement at time $k_0 = -\infty$ with arbitrary initial state mean and covariance, or commencement at some finite k_0 with a special initial state mean and covariance. With k_0 finite and arbitrary initial state mean and covariance, state and output processes are asymptotically stationary. For gaussian processes, the stationarity is strict, otherwise it is wide-sense.

With signal model defined by the sextuple $\{F, G, H, Q, R, S\}$, the crucial equations are as follows in the stationary case:

$$E[x_k x_l'] = F^{k-l}\bar{P} \qquad k \geq l$$

where

$$\bar{P} - F\bar{P}F' = GQG'$$

$$E[z_k z_l'] = H'F^{k-l}\bar{P}H + H'F^{k-l-1}GS \qquad k > l$$

$$= R + H'\bar{P}H \qquad k = l$$

$$\Phi_{xx}(z) = (zI - F)^{-1}GQG'(z^{-1}I - F')^{-1}$$

and

$$\Phi_{zz}(z) = H'(zI - F)^{-1}GQG'(z^{-1}I - F')^{-1}H + H'(zI - F)^{-1}GS$$
$$+ S'G'(z^{-1}I - F')^{-1}H + R$$

Problem 3.1. Consider the system

$$x_{k+1} = \begin{bmatrix} F^{(1)} & 0 & F^{(13)} \\ 0 & F^{(2)} & F^{(23)} \\ 0 & 0 & F^{(3)} \end{bmatrix} x_k + \begin{bmatrix} G^{(1)} \\ 0 \\ 0 \end{bmatrix} w_k$$

with $E[w_k w_k'] = I$. Suppose that $E[x_0] = 0$ and $E[x_0 x_0'] = P$ with

$$P = \begin{bmatrix} P^{(1)} & 0 & 0 \\ 0 & P^{(2)} & 0 \\ 0 & 0 & 0 \end{bmatrix}$$

Show that if $|\lambda_i(F^{(1)})| < 1$ and $|\lambda_i(F^{(2)})| = 1$, then there exist nonnegative definite solutions $\bar{P}^{(1)}$ and $\bar{P}^{(2)}$ of

$$\bar{P}^{(1)} - F^{(1)}\bar{P}^{(1)}F^{(1)'} = G^{(1)}G^{(1)'}$$

$$\bar{P}^{(2)} - F^{(2)}\bar{P}^{(2)}F^{(2)'} = 0$$

with $\bar{P}^{(2)}$ not necessarily unique, and able to be nonsingular if $\bar{F}^{(2)}$ has distinct eigenvalues. Further if $P^{(1)} = \bar{P}^{(1)}$, $P^{(2)} = \bar{P}^{(2)}$, then $\{x_k\}$ is stationary for $k \geq 0$.

Problem 3.2. Prove Theorem 3.2 using $V = x_k'\bar{P}x_k$ as a Lyapunov function.

Problem 3.3. Suppose that $\Phi(z)$ is a power spectrum. Show that $\Phi(z) = \Phi'(z^{-1})$ and that $\Phi(e^{j\omega})$ is real and nonnegative definite hermitian for all real ω. (The second half of this problem is standard, but hard to prove without assuming the result of Theorem 3.4. See reference [2] of Appendix A.)

Problem 3.4. Suppose that $x_{k+1} = Fx_k + Gw_k$, $z_k = H'x_k + v_k$, with

$$E\begin{bmatrix} w_k \\ v_k \end{bmatrix} = 0 \qquad E\left\{ \begin{bmatrix} w_k \\ v_k \end{bmatrix} [w_l' \quad v_l'] \right\} = \begin{bmatrix} Q & S \\ S' & R \end{bmatrix} \delta_{kl}$$

and $|\lambda_i(F)| < 1$ for all i. Show that $\{z_k\}$ is the output sequence of a linear system driven by

$$\begin{bmatrix} w_k \\ v_k \end{bmatrix}$$

and show by two methods that

$$\Phi_{zz}(z) = H'(zI - F)^{-1}GQG'(z^{-1}I - F')^{-1}H$$
$$+ H'(zI - F)^{-1}GS + S'G'(z^{-1}I - F')^{-1}H + R$$

Problem 3.5. Show that knowledge of $W(z)W'(z^{-1})$ and the fact that $W(z)$ is exponentially stable is not sufficient to determine $W(z)$ alone. [It is enough to consider scalar, rational $W(z)$.]

Problem 3.6. For the system $x_{k+1} = Fx_k + Gw_k$, suppose that

$$F = \begin{bmatrix} 0 & 1 & 0 & \cdot & \cdot & \cdot & 0 \\ 0 & 0 & 1 & \cdot & \cdot & \cdot & 0 \\ \cdot & \cdot & \cdot & \cdot & & & \cdot \\ \cdot & \cdot & \cdot & & \cdot & & \cdot \\ \cdot & \cdot & \cdot & & & \cdot & 1 \\ -\alpha_n & \cdot & \cdot & \cdot & \cdot & -\alpha_2 & -\alpha_1 \end{bmatrix}, \quad G = \begin{bmatrix} 0 \\ 0 \\ \cdot \\ \cdot \\ 0 \\ 1 \end{bmatrix}$$

with w_k a zero mean, white process of covariance $E[w_k^2] = 1$. Show that if $|\lambda_i(F)| < 1$, the solution $\bar{P}$ of $\bar{P} - F\bar{P}F' = GG'$ is Toeplitz, i.e., $p_{ij} = p_{i-j}$ for the entries p_{ij} of $\bar{P}$. Prove also that if $\bar{P}$ is a positive definite symmetric Toeplitz matrix, the numbers β_i defined by

$$\bar{P} \begin{bmatrix} \beta_1 \\ \beta_2 \\ \cdot \\ \cdot \\ \cdot \\ \beta_n \end{bmatrix} = \begin{bmatrix} 1 \\ 0 \\ \cdot \\ \cdot \\ \cdot \\ 0 \end{bmatrix}$$

are such that all roots of $z^n + \beta_1 z^{n-1} + \beta_2 z^{n-2} + \cdots + \beta_n$ lie in $|z| < 1$. (For the first part, give a covariance interpretation to p_{ij}, and for the second part relate the β_i to the α_i.)

4.4   TIME INVARIANCE AND ASYMPTOTIC STABILITY OF THE FILTER

In this section, we are interested in pinning down conditions which guarantee simultaneously that the optimal filter is time invariant, or asymptotically time invariant, and that it is also asymptotically stable. (Conditions separately guaranteeing these properties are obtainable, but are not of great interest). Time invariance, or asymptotic time invariance, arises when there is a constant, or asymptotically constant (i.e., limiting), solution of the variance equation*

$$\Sigma_{k+1/k} = F[\Sigma_{k/k-1} - \Sigma_{k/k-1}H(H'\Sigma_{k/k-1}H + R)^{-1}H'\Sigma_{k/k-1}]F' + GQG' \tag{4.1}$$

with $\bar{\Sigma}$ a constant or limiting solution to (4.1). The associated gain is

$$K = F\bar{\Sigma}H(H'\bar{\Sigma}H + R)^{-1} \tag{4.2}$$

and the question arises as to whether the eigenvalues of $F - KH'$ have $|\lambda_i(F - KH')| < 1$, guaranteeing asymptotic stability of the filter.

It is certainly intuitively reasonable that $\Sigma_{k+1/k}$ could approach a limit when the signal process is stationary; however, it is not so clear that the associated filter should be asymptotically stable, though in fact it is (see point 1 below). Indeed, much more is true [4] (see point 2 below). The main conclusions are:

1. If the signal process model is time invariant and asymptotically stable, i.e., $|\lambda_i(F)| < 1$, then
   (a) For any nonnegative symmetric initial condition Σ_{k_0/k_0-1} one has

$$\lim_{k \to \infty} \Sigma_{k+1/k} = \bar{\Sigma} \tag{4.3}$$

   with $\bar{\Sigma}$ independent of Σ_{k_0/k_0-1} and satisfying a steady-state version of (4.1):

$$\bar{\Sigma} = F[\bar{\Sigma} - \bar{\Sigma}H(H'\bar{\Sigma}H + R)^{-1}H'\bar{\Sigma}]F' + GQG' \tag{4.4}$$

   Eq. (4.4) is sometimes termed a steady-state Riccati equation.
   (b) If k is held fixed and the value of the initial condition matrix is held fixed, but the initial time is allowed to vary, then

$$\lim_{k_0 \to -\infty} \Sigma_{k+1/k} = \bar{\Sigma} \tag{4.5}$$

   Again, the value of the initial condition matrix is immaterial, so long as it is held fixed at a nonnegative symmetric value while $k_0 \to -\infty$.
   (c) $\qquad\qquad\qquad |\lambda_i(F - KH')| < 1 \tag{4.6}$

   where k is as in (4.2)
2. If the signal process model is time invariant and not necessarily asymptotically stable, but the pair $[F, H]$ is completely detectable† and the pair $[F, GG_1]$ is completely stabilizable† for any G_1 with $G_1G_1' = Q$, then points 1(a) and 1(c) hold.

---

*Though results are obtainable for the case when $H'\Sigma_{k/k-1}H + R$ is singular, they are a good deal more complicated. We shall not discuss them here.

†See Appendix C. Note that any asymptotically stable model is automatically completely detectable and stabilizable.

Actually, under the conditions listed in point 2, (4.5) holds. But if the system is not stable it does not make sense to let $k_0 \rightarrow -\infty$, since at any finite time, the system state can be expected to have an infinite variance. Actually, it hardly makes more sense to let $k \rightarrow \infty$, since on $[k_0, \infty]$ the variance of the signal state and the output will be unbounded, though finite for all finite k. Whether or not the error variance tends to a constant is perhaps an academic point. Thus the results described under point 2 are of limited utility.

Outline of Derivations

Our discussion will break up into four parts:

1. We shall show that for an arbitrary but fixed nonnegative symmetric Σ_{k_0/k_0-1}, $\Sigma_{k/k-1}$ is bounded for all k. The detectability property is crucial, as is an appeal to optimality.
2. We shall show that with $\Sigma_{k_0/k_0-1} = 0$, $\Sigma_{k+1/k}$ is monotone increasing with k; in conjunction with the bound of 1, this establishes the existence of $\lim_{k \rightarrow \infty} \Sigma_{k+1/k} = \bar{\Sigma}$. Equation (4.4) and the limit (4.5) will also be recovered.
3. The stability property (4.6) will be obtained.
4. We shall allow arbitrary nonnegative symmetric Σ_{k_0/k_0-1} and shall obtain (4.3) [and (4.5) where appropriate].

Bound on the Error Covariance

The general strategy to exhibit a bound for $\Sigma_{k+1/k}$ with arbitrary fixed Σ_{k_0/k_0-1} is to define a suboptimal filter whose associated error covariance must overbound $\Sigma_{k+1/k}$; we also arrange for the suboptimal filter to have bounded error covariance.

In view of the complete detectability of $[F, H]$, there is a matrix K_e such that $|\lambda_i(F - K_e H')| < 1$. (This result is noted in Appendix C and also explored in the problems.) Define a suboptimal, asymptotically stable filter by

$$x_{k+1/k}^e = F x_{k/k-1}^e + K_e[z_k - H'x_{k/k-1}^e] \qquad (4.7)$$

with

$$x_{k_0/k_0-1}^e = 0 \qquad (4.8)$$

The error performance of this suboptimal filter is simply the covariance of $x_{k+1} - x_{k+1/k}^e$, which satisfies the linear equation

$$x_{k+1} - x_{k+1/k}^e = (F - K_e H')(x_k - x_{k/k-1}^e) + [G \quad -K_e]\begin{bmatrix} w_k \\ v_k \end{bmatrix}$$

Thus

$$\Sigma_{k+1/k}^e = (F - K_e H')\Sigma_{k/k-1}^e(F - K_e H')' + K_e R K_e' + GQG' \qquad (4.9)$$

If we are comparing (4.7) with an optimal filter initialized by Σ_{k_0/k_0-1}, the initial uncertainty in x_{k_0} must be Σ_{k_0/k_0-1} and, in view of (4.8), we must have $\Sigma^e_{k_0/k_0-1} = \Sigma_{k_0/k_0-1}$. However, by the suboptimality of (4.7), $\Sigma^e_{k/k-1} \geq \Sigma_{k/k-1}$ ≥ 0 in general. Because of the stability of (4.7), Eq. (4.9) has a bounded solution for any initial condition (the bound depends on the initial condition, but this is of no concern). Thus the bound on $\Sigma_{k/k-1}$ is obtained.

Note that if $|\lambda_i(F)| < 1$, we can take $K_e = 0$, and $\Sigma^e_{k+1/k}$ then agrees with the signal state covariance P_{k+1}. As we know, $P_{k+1} \geq \Sigma_{k+1/k}$, and the state covariance is bounded by virtue of the asymptotic stability of the signal process model.

Use of Zero Initial Covariance

Suppose now that $\Sigma_{k_0/k_0-1} = 0$. We shall show that $\Sigma_{k+1/k}$ is increasing with k. One of the key steps in the argument, as will be seen, is that if the same Riccati equation is solved forward for two initial conditions, one greater than the other, this "ordering property" is preserved at subsequent time instants. (This simply says that if two filtering problems are considered which are identical save that the initial state uncertainty is greater for one than the other, then the associated errors in estimating the state at an arbitrary time instant will be similarly ordered).

Consider the variance equation (4.1), with two initial conditions Σ_{k_0/k_0-1} $= 0$ and $\Sigma_{k_0-1/k_0-2} = 0$. We shall distinguish the two corresponding solutions by $\Sigma_{k+1/k,k_0-1}$ and $\Sigma_{k+1/k,k_0-2}$. We shall show that

$$\Sigma_{k+1/k,k_0-1} \leq \Sigma_{k+1/k,k_0-2} \tag{4.10}$$

for all k.

First observe that the result is trivial for $k = k_0 - 1$. For then it reads

$$0 = \Sigma_{k_0/k_0-1,k_0-1} \leq \Sigma_{k_0/k_0-1,k_0-2} \tag{4.11}$$

The inequality follows because $\Sigma_{k_0/k_0-1,k_0-2}$ is a covariance matrix, and is accordingly nonnegative. What (4.10) implies is that the inequality (4.11) propagates for all k forward in time. Figure 4.4-1 (see next page) depicts the situation in a one-dimensional case, and may convince the reader of the intuitive reasonableness of the result.

The proof of the result is straightforward on an inductive basis. We have shown that (4.10) is true for $k = k_0 - 1$. Assume it is true for $k = k_0 - 1, \ldots, i - 1$. Then an "optimal version" of (4.9) yields

$$\Sigma_{i/i-1,k_0-2} = \min_K [(F - KH')\Sigma_{i-1/i-2,k_0-2}(F - KH')' + GQG' + KRK']$$

$$= (F - K^*H')\Sigma_{i-1/i-2,k_0-2}(F - K^*H')' + GQG' + K^*RK^{*\prime}$$

(where K^* is the minimizing K)

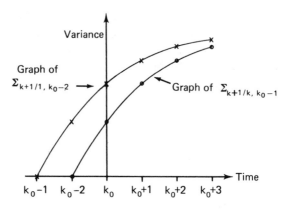

Fig. 4.4-1   Propagation of "greater than" relation for error covariances. (A smooth curve is drawn through the discrete points of the graph.)

$$\geq (F - K^*H)\Sigma_{i-1/i-2, k_0-1}(F - K^*H')' + GQG' + K^*RK^*'$$

$$\text{(by the inductive hypothesis)}$$

$$\geq \min_K [(F - KH')\Sigma_{i-1/i-2, k_0-1}(F - KH')' + GQG' + KRK']$$

$$= \Sigma_{i/i-1, k_0-1}$$

Now the underlying time-invariance of all quantities in the variance matrix equation save the variance matrix itself implies that

$$\Sigma_{k+1/k, k_0-1} = \Sigma_{k/k-1, k_0-2}$$

so that by (4.9), $\Sigma_{k+1/k, k_0-2} \geq \Sigma_{k/k-1, k_0-2}$. Since k_0 is arbitrary, this means that with zero initial condition, $\Sigma_{k/k-1}$ is monotone increasing. Because it is bounded above, as shown in the previous subsection, the limit (4.3) exists when $\Sigma_{k_0/k_0-1} = 0$. Moreover, the time-invariance shows that

$$\Sigma_{k+1/k, k_0} = \Sigma_{k-k_0+1/k-k_0, 0}$$

so that

$$\lim_{k_0 \to -\infty} \Sigma_{k+1/k, k_0} = \lim_{j \to \infty} \Sigma_{j+1/j, 0} = \bar{\Sigma}$$

verifying (4.5) in case $\Sigma_{k_0/k_0-1} = 0$. Equation (4.4) follows by simply taking limits in (4.1).

Asymptotic Stability of the Filter

Let us assume $[F, GG_1]$ is completely stabilizable. Asymptotic stability can now be shown, with argument by contradiction. Suppose asymptotic stability does not hold, and that $(F - KH')'\omega = \lambda\omega$ for some λ with $|\lambda| \geq 1$ and some nonzero ω. Then since, as may be checked,

$$\bar{\Sigma} = (F - KH')\bar{\Sigma}(F - KH')' + KRK' + GQG' \qquad (4.12)$$

we have (with a superscript asterisk denoting complex conjugation)

$$(1 - |\lambda|^2)\omega'^*\bar{\Sigma}\omega = \omega'^*KRK'\omega + \omega'^*GG_1G_1'G'\omega$$

The right side is clearly nonnegative, while the left side is nonpositive. Therefore both sides are zero, so that $K'\omega = 0$ and $G_1'G'\omega = 0$. But $K'\omega = 0$ forces $F'\omega = \lambda\omega$, and this combined with $G_1'G'\omega = 0$ implies lack of complete stabilizability.

Nonzero Initial Covariances

Our aim now is to demonstrate that the limits (4.3) and (4.5) hold for an arbitrary nonnegative symmetric Σ_{k_0/k_0-1}. The arguments are somewhat technical and might be omitted at a first reading. The reader should, however, recognize that the result is considered to be of major importance.

A transition matrix bound. Let us rewrite the variance equation (4.1) as

$$\Sigma_{k+1/k} = [F - K_kH']\Sigma_{k/k-1}[F - K_kH']' + K_kRK_k' + GQG' \quad (4.13)$$

Let $\Psi_{k,l}$ be the transition matrix associated with $F - K_kH'$, i.e.,

$$\Psi_{k,l} = [F - K_{k-1}H'][F - K_{k-2}H']\ldots[F - K_lH']$$

Then it easily follows that

$$\Sigma_{k/k-1} = \Psi_{k,k_0}\Sigma_{k_0/k_0-1}\Psi_{k,k_0}' + \text{nonnegative definite terms}$$
$$\geq \Psi_{k,k_0}\Sigma_{k_0/k_0-1}\Psi_{k,k_0}'$$

Recall that $\Sigma_{k/k-1}$ is bounded for all k for any fixed initial value Σ_{k_0/k_0-1}. Take $\Sigma_{k_0/k_0-1} = \rho I$ for some positive ρ. It follows that $\rho\Psi_{k,k_0}\Psi_{k,k_0}'$ and therefore Ψ_{k,k_0} is bounded for all k.

Note that $\Psi_{k,l}$, depending on K_k, depends on Σ_{k_0/k_0-1}. So we have not proved for *arbitrary* Σ_{k_0/k_0-1}, but only for Σ_{k_0/k_0-1} of the form ρI, that the associated Ψ_{k,k_0} is bounded. (It is however possible to extend the argument above to the case of arbitrary nonsingular Σ_{k_0/k_0-1}).

Proof of a limited convergence result. Using the steady-state variance equation (4.12) and the "transient" variance equation (4.1), one can establish

$$\Sigma_{k+1/k} - \bar{\Sigma} = (F - KH')(\Sigma_{k/k-1} - \bar{\Sigma})(F - K_kH')'$$

(calculations are requested in the problems).

Now suppose $\Sigma_{k_0/k_0-1} = \rho I$ for some ρ, we have

$$\Sigma_{k/k-1} - \bar{\Sigma} = (F - KH')^{k-k_0}(\Sigma_{k_0/k_0-1} - \bar{\Sigma})\Psi_{k,k_0}'$$

with Ψ_{k,k_0} bounded for all k. Letting $k \to \infty$ shows that $\Sigma_{k/k-1} \to \bar{\Sigma}$, since $|\lambda_i(F - KH')| < 1$. Convergence will in fact occur essentially at a rate determined by the magnitude of the maximum eigenvalue of $F - KH$.

$$\|\Sigma_{k/k-1} - \bar{\Sigma}\| = O(\max_i |\lambda_i(F - KH')|^{k-k_0})$$

Convergence for arbitrary Σ_{k_0/k_0-1}. Let Σ_{k_0/k_0-1} be an arbitrary nonnegative definite symmetric matrix. Choose ρ such that $\rho I > \Sigma_{k_0/k_0-1}$. The solution of the variance equation (4.1) initialized by ρI overbounds the solution of (4.1) initialized by Σ_{k_0/k_0-1}, for, as argued earlier, inequalities in the initial condition propagate. The solution of (4.1) initialized by the zero matrix likewise underbounds the solution initialized by Σ_{k_0/k_0-1}. The underbounding and overbounding solutions both tend to $\bar{\Sigma}$, whence the solution initialized by Σ_{k_0/k_0-1} must have this property. Thus (4.3) holds. Equation (4.5) follows by using the underlying time invariance, just as for the case when $\Sigma_{k_0/k_0-1} = 0$.

Necessity for Complete Detectability and Stabilizability

The question arises as to whether complete detectability is needed to ensure the existence of $\bar{\Sigma}$. We shall only argue heuristically. Suppose there is a mode that is not observed and is not asymptotically stable, yet is excited by the input or via a nonzero random initial condition. Since it is not observed, the best estimate of it is simply zero, and the error variance will be the variance of the mode. Since it is excited and not asymptotically stable, the variance will be unbounded and a steady-state value, therefore, cannot exist. In other words, complete detectability of all modes excited by the input or with a nonzero random initial condition is necessary for the existence of $\bar{\Sigma}$.

A more complicated argument can be used to conclude that if $\bar{\Sigma}$ exists with $F - KH'$ asymptotically stable, one must have complete stabilizability of $[F, GG_1]$.

Miscellaneous Points

Time-invariant filters. Because, in general, $\Sigma_{k+1/k}$ is only asymptotically constant, the associated filter will be asymptotically time invariant; however, if $\Sigma_{k_0/k_0-1} = \bar{\Sigma}$, then $\Sigma_{k+1/k} = \bar{\Sigma}$ for all k and the filter will be time invariant. Also if the initial time tends to minus infinity then, again under the conditions noted earlier, $\Sigma_{k+1/k} = \bar{\Sigma}$ for all k.

Filter stability. Formally, we have only argued that the steady-state filter is asymptotically stable. Actually, the time-varying (but asymptotically time-invariant) filter is also asymptotically stable, as would be imagined.

Other solutions of the steady-state equation. The steady-state equation (4.4) is a nonlinear equation. Viewed in isolation, it can in general be expected to have more than one solution. Only one, however, can be nonnegative definite symmetric. For suppose that $\hat{\Sigma} \neq \bar{\Sigma}$ is such a solution. Then with $\Sigma_{k_0/k_0-1} = \hat{\Sigma}$, (4.4) yields $\Sigma_{k+1/k} = \hat{\Sigma}$ for all k, while (4.3) yields $\lim_{k\to\infty} \Sigma_{k+1/k} = \bar{\Sigma} \neq \hat{\Sigma}$, which is a contradiction.

Limiting behaviour of $\Sigma_{k/k}$. Since

$$\Sigma_{k/k} = \Sigma_{k/k-1} - \Sigma_{k/k-1}H(H'\Sigma_{k/k-1}H + R)^{-1}H'\Sigma_{k/k-1}$$

it is clear that if $\Sigma_{k/k-1}$ approaches a limit, so must $\Sigma_{k/k}$. Also since $\Sigma_{k+1/k}$ $= F\Sigma_{k/k}F' + GQG'$, we see that if $\Sigma_{k/k}$ approaches a limit, so does $\Sigma_{k+1/k}$. Therefore, conditions for $\lim_{k\to\infty} \Sigma_{k/k}$ to exist are identical with those for $\lim_{k\to\infty} \Sigma_{k+1/k}$ to exist.

Suboptimal filter design via time-invariant filters. A number of practical situations arise where the signal model is time-invariant, input and output noises are stationary, and the signal model is started at time k_0 with a known initial state. For ease of implementation, a time-invariant filter may be used rather than the optimal time-varying filter (which is asymptotically time invariant). Serious loss of performance can occur in some situations (see [5, 6]).

Time-varying filter stability. Results on the stability of time-varying filters tend to be complex [7, 8]. They usually involve generalizations of the detectability and stabilizability notions.

EXAMPLE 4.1. Let us suppose that F, G, H, Q, R are all scalars f, g, h, q, r. Let σ_k denote $\Sigma_{k/k-1}$. Then

$$\sigma_{k+1} = f^2\left[\sigma_k - \frac{h^2\sigma_k^2}{h^2\sigma_k + r}\right] + g^2q = \frac{f^2r\sigma_k}{h^2\sigma_k + r} + g^2q$$

Suppose $h = 0$. Then $\sigma_{k+1} = f^2\sigma_k + g^2q$, and if $|f| \geq 1$, convergence cannot hold if either $gq \neq 0$ or $\sigma_0 \neq 0$. Of course, $h = 0$ and $|f| \geq 1$ corresponds to lack of complete detectability. On the other hand, with $h \neq 0$,

$$\frac{f^2r\sigma_k}{h^2\sigma_k + r} = \frac{f^2r}{h^2 + (r/\sigma_k)} \leq \frac{f^2r}{h^2}$$

for all $\sigma_k \geq 0$. Thus $\sigma_{k+1} \leq (f^2r/h^2) + g^2q$ for all $\sigma_k \geq 0$; i.e., for an arbitrary $\sigma_0 \geq 0$, σ_{k+1} will be bounded for all k. By choosing particular values of f, g, etc., and taking $\sigma_0 = 0$, we can verify the monotone nature of σ_k. (Actually, clever algebra will verify it for arbitrary f, g, etc.) The limit $\bar{\sigma}$ satisfies

$$\bar{\sigma} = \frac{f^2r\bar{\sigma}}{h^2\bar{\sigma} + r} + g^2q$$

If $g \neq 0$, this has two solutions, one only being nonnegative definite, as follows:

$$\bar{\sigma} = -\frac{(r - f^2r)}{2h^2} + \sqrt{\frac{(r - f^2r)^2}{4h^4} + \frac{g^2q}{h^2}}$$

The feedback gain $k = f\bar{\sigma}h(h^2\bar{\sigma} + r)^{-1}$ and then

$$f - kh = fr(h^2\bar{\sigma} + r)^{-1}$$

The formula for $\bar{\sigma}$ can be used to show $|f - kh| < 1$. If $g = 0$ and $\sigma_0 = 0$, then $\sigma_j = 0$ for all j, whence $\bar{\sigma} = k = 0$. Thus $f - kh$ is stable precisely when (f, g) is stabilizable, i.e., $|f| < 1$.

Main Points of the Section

Under assumptions of complete detectability and stabilizability, the filter will be asymptotically time invariant and asymptotically stable for arbitrary initial error covariance. In case the initial time is $-\infty$, requiring the signal model to be asymptotically stable, the filter is time invariant and asymptotically stable.

Problem 4.1. Complete detectability is equivalent to any unobservable state being asymptotically stable. Show that if $[F, H]$ is completely detectable, there exists K such that $|\lambda_i(F - KH')| < 1$ for all i. (Use the following two properties associated with the observability notion. The pair $[F, H]$ is completely observable if and only if there exists a K such that an arbitrary characteristic polynomial of $F - KH'$ can be obtained. If $[F, H]$ is not completely observable, there exists a coordinate basis in which

$$F = \begin{bmatrix} F_{11} & 0 \\ F_{21} & F_{22} \end{bmatrix} \qquad H' = [H_1' \quad 0]$$

with $[F_{11}, H_1]$ completely observable.)

Problem 4.2. Consider the case of scalar $F, G, H, Q, R,$ and $\Sigma_{k+1/k}$ in (4.1). Take $G = H = Q = R = 1$ and $F = a$ and verify that $\Sigma_{k+1/k}$ approaches a limit that is the positive solution of $\sigma^2 + (1 - a^2)\sigma - 1 = 0$. Verify that $|F - KH'| < 1$.

Problem 4.3. Consider $x_{k+1} = ax_k$, $z_k = x_k + v_k$, $E[v_k v_l] = \delta_{kl}$, with x_k scalar. Show that if $|a| < 1$, arbitrary positive $\Sigma_{0/-1}$ causes $\Sigma_{k+1/k}$ to approach $\bar{\Sigma}$, but that this is not so if $|a| > 1$.

Problem 4.4. (Null space of $\bar{\Sigma}$). For the system $x_{k+1} = Fx_k + Gw_k$ with w_k zero mean, gaussian and with $E[w_k w_l'] = Q\delta_{kl}$, we have noted the evolution of $P_k = E[x_k x_k']$ when $E[x_0] = 0$ according to the equation $P_{k+1} = FP_k F' + GQG'$. Observe that the evolution of the quantity $E(a'x_k)^2$ for an arbitrary row vector a can then be obtained readily from the P_k. Now suppose $P_0 = 0$ and that there are unreachable states. Therefore there are costates $a \neq 0$ such that $a'F^iGG_1 = 0$ for all i. Show that (1) $\Sigma_{k+1/k}a = 0$ for all k, through use of the variance equation; (2) $P_k a = 0$; (3) $P_k a = 0$ for all k implies on physical grounds that $\Sigma_{k+1/k}a = 0$ for all k; and (4) $\bar{\Sigma}a = 0$ if and only if $a'F^iGG_1$ for all i, so that the set of costates orthogonal to the reachable states coincides with the nullspace of $\bar{\Sigma}$.

Problem 4.5. Establish Eq. (4.14). The steady-state variance equation may be written as $\bar{\Sigma} = (F - KH')\bar{\Sigma}F' + GQG'$, and the transient equation as $\Sigma_{k+1/k} = F\Sigma_{k/k-1}(F - K_kH')' + GQG'$. By subtraction, show that

$$\Sigma_{k+1/k} - \bar{\Sigma} = (F - KH')(\Sigma_{k/k-1} - \bar{\Sigma})(F - K_kH')' \\ + KH'\Sigma_{k/k-1}(F - K_kH')' - (F - KH')\bar{\Sigma}HK_k'$$

Substitute expressions involving the error variance, H and R for K and K_k in the last two terms, and show that together they come to zero.

Problem 4.6. Suppose that conditions are fulfilled guaranteeing the existence of $\bar{\Sigma}$, that R is nonsingular, and that $F - KH'$ is known to be asymptotically stable.

Show that $[F, GG_1]$ must be completely stabilizable. [Assume there exists a scalar λ with nonnegative real part and a nonzero vector q such that $q'F = \lambda q'$ and $q'GG_1 = 0$. With the aid of the result of Problem 4.4, show that $q'K = 0$ and deduce a contradiction].

Problem 4.7. (Stability Improvement Property of the Filter). Since the closer to the origin the eigenvalues of a system matrix are, the greater is the degree of stability of a system, $|\Pi\lambda_i(F)| = |\det F|$ is a rough measure of the degree of stability of $x_{k+1} = Fx_k$. Show that when the optimal $F - KH'$ is guaranteed to be asymptotically stable, $|\det (F - KH')| \leq |\det F|$. (It may be helpful to use the facts that

$$\det [I + BA] = \det [I + AB] \quad \text{and} \quad \det [I + CD] = \det [I + D^{1/2}CD^{1/2}]$$

if $D = D' \geq 0$).

4.5 FREQUENCY DOMAIN FORMULAS

In this section, our aim is to relate the time-invariant optimal filter to the signal process model via a frequency domain formula.

Relation of the Filter and Signal Process Model

We recall that the signal process model is

$$x_{k+1} = Fx_k + Gw_k \qquad E[w_k w_l'] = Q\delta_{kl} \tag{5.1}$$

$$y_k = H'x_k + v_k \qquad E[v_k v_l'] = R\delta_{kl} \tag{5.2}$$

while the optimal filter is

$$\hat{x}_{k+1/k} = (F - KH')\hat{x}_{k/k-1} + Kz_k \tag{5.3}$$

where

$$K = F\bar{\Sigma}H(H'\bar{\Sigma}H + R)^{-1} \tag{5.4}$$

and

$$\bar{\Sigma} = F[\bar{\Sigma} - \bar{\Sigma}H(H'\bar{\Sigma}H + R)^{-1}H'\bar{\Sigma}]F' + GQG' \tag{5.5}$$

Of course, we are implicitly assuming existence of the time-invariant filter and of the inverse of $H'\bar{\Sigma}H + R$. Sufficient conditions for inverse existence are nonsingularity of R, or nonsingularity of $\bar{\Sigma}$ with H possessing full rank. We shall derive the following formula and then comment on its significance.

$$[I + H'(zI - F)^{-1}K][R + H'\bar{\Sigma}H][I + K'(z^{-1}I - F')^{-1}H]$$
$$= R + H'(zI - F)^{-1}GQG'(z^{-1}I - F')^{-1}H \tag{5.6}$$

To prove the formula, observe first that for all z one has

$$\bar{\Sigma} - F\bar{\Sigma}F' = (zI - F)\bar{\Sigma}(z^{-1}I - F') + F\bar{\Sigma}(z^{-1}I - F') + (zI - F)\bar{\Sigma}F'$$

Using this identity, one has from (5.5)

$$(zI - F)\bar{\Sigma}(z^{-1}I - F') + F\bar{\Sigma}(z^{-1}I - F') + (zI - F)\bar{\Sigma}F'$$
$$+ F\bar{\Sigma}H(H'\bar{\Sigma}H + R)^{-1}H'\bar{\Sigma}F' = GQG'$$

Next, premultiply by $H'(zI - F)^{-1}$ and postmultiply by $(z^{-1}I - F')^{-1}H$. There results

$$H'\bar{\Sigma}H + H'(zI - F)^{-1}F\bar{\Sigma}H + H'\bar{\Sigma}F'(z^{-1}I - F')^{-1}H$$
$$+ H'(zI - F)^{-1}F\bar{\Sigma}H(H'\bar{\Sigma}H + R)^{-1}H'\bar{\Sigma}F'(z^{-1}I - F')^{-1}H$$
$$= H'(zI - F)^{-1}GQG'(z^{-1}I - F')^{-1}H$$

Now use the formula for K [see (5.4)] to obtain

$$H'\bar{\Sigma}H + H'(zI - F)^{-1}K(H'\bar{\Sigma}H + R) + (H'\bar{\Sigma}H + R)K'(z^{-1}I - F')^{-1}H$$
$$+ H'(zI - F)^{-1}K(H'\bar{\Sigma}H + R)K'(z^{-1}I - F')^{-1}H$$
$$= H'(zI - F)^{-1}GQG'(z^{-1}I - F')^{-1}H$$

Equation (5.6) is then immediate.

Now let us comment on the significance of (5.6).

1. $H'(zI - F)^{-1}G = W_m(z)$ is the transfer function matrix of the signal process model; the quantity $R + H'(zI - F)^{-1}GQG'(z^{-1}I - F')^{-1}H$, which can be written as $R + W_m(z)QW'_m(z^{-1})$, was earlier shown to be the power spectrum of the output process $\{z_k\}$, at least when this process is stationary, or when $|\lambda_i(F)| < 1$ for all i. Defining the transfer function $W_K(z)$ to be $H'(zI - F)^{-1}K$, (5.6) becomes

$$[I + W_K(z)][R + H'\Sigma H][I + W'_K(z^{-1})] = R + W_m(z)QW'_m(z^{-1})$$
$$(5.7)$$

2. Equations (5.6) and (5.7) in essence define a *spectral factorization* of the power spectral density of $\{z_k\}$. A spectral factorization of a power spectrum matrix $\Phi(z)$ is a factorizing of $\Phi(z)$ into the form

$$\Phi(z) = W(z)W'(z^{-1}) \qquad (5.8)$$

The matrix $[I + H'(zI - F)^{-1}K](H'\Sigma H + R)^{1/2}$ serves the role of $W(z)$ in (5.6).

3. If the Kalman filter is asymptotically stable, the quantity

$$I + H'(zI - F)^{-1}K$$

is *minimum phase*, in the sense that $\det[I + H'(zI - F)^{-1}K]$ is never zero for $|z| \geq 1$.* To see this, observe that

$$[I + H'(zI - F)^{-1}K]^{-1} = I - H'(zI - \overline{F - KH'})^{-1}K$$

Zeros of $\det[I + H'(zI - F)^{-1}K]$ therefore correspond to eigenvalues

---

*Most definitions of minimum phase also require $\det[I + H'(zI - F)^{-1}K]$ to be analytic in $|z| \geq 1$. With $|\lambda_i(F)| < 1$, this is assured.

of $F - KH'$, which all lie inside $|z| = 1$ on account of the asymptotic stability of the filter.

4. In case the signal process model is also asymptotically stable, the zeros and poles of det $[I + H'(zI - F)^{-1}K]$ lie inside $|z| < 1$. Now there are procedures known for factoring a prescribed power spectrum matrix $\Phi(z)$ as depicted in the form of (5.8), with det $W(z)$ possessing all its zeros and poles in $|z| < 1$ and with $W(\infty)$ finite and nonsingular [9–11]. Further, the resulting $W(z)$ is known to be uniquely determined to within right multiplication by an arbitrary orthogonal matrix. Consequently, *essentially classical procedures* allow the determination of the transfer function matrix

$$[I + H'(zI - F)^{-1}K][R + H'\Sigma H]^{1/2}$$

to within right multiplication by an arbitrary orthogonal matrix, in case $|\lambda_i(F)| < 1$. Now, knowing

$$W(z) = [I + H'(zI - F)^{-1}K][R + H'\Sigma H]^{1/2}$$

to within an arbitrary orthogonal matrix, it is easy, by setting $z = \infty$, to identify $[R + H'\Sigma H]^{1/2}$ as $[W(\infty)W'(\infty)]^{1/2}$ and, thence, $I + H'(zI - F)^{-1}K$. Finally, knowing this transfer function matrix together with F and H, and assuming $[F, H]$ is completely observable, K can be found uniquely.

5. Think of the filter as a signal estimator, rather than a state estimator. The filter output thus becomes

$$\hat{y}_{k/k-1} = H'\hat{x}_{k/k-1} \qquad (5.9)$$

and the transfer function matrix of the optimal filter, regarded as having input $\{z_k\}$, is

$$W_f(z) = H'(zI - \overline{F - KH'})^{-1}K \qquad (5.10)$$

It can be shown algebraically that

$$W_f(z) = W_K(z)[I + W_K(z)]^{-1} \qquad (5.11)$$

This can also be seen by block diagram manipulation (see Fig. 4.5-1 on next page).

Main Points of the Section

The filter transfer function is related to the signal power spectrum. In particular,

$$[I + W_K(z)][R + H'\overline{\Sigma}H][I + W'_K(z^{-1})]$$
$$= R + H'(zI - F)^{-1}GQG'(z^{-1}I - F')^{-1}H$$

where $W_K(z) = I + H'(zI - F)^{-1}K$.

Problem 5.1. Show that

$$R + H'\overline{\Sigma}H = \{I - H'[zI - (F - KH')]^{-1}K\}R\{I - K'[z^{-1}I - (F - KH')']^{-1}H\}$$
$$+ H'[zI - (F - KH')]^{-1}GQG'[z^{-1}I - (F - KH')']^{-1}H$$

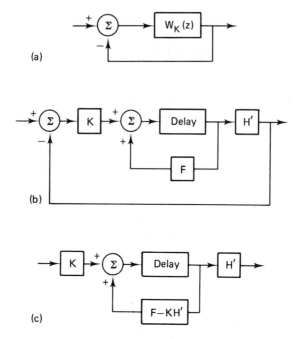

Fig. 4.5-1 Three equivalent block diagram representations of $W_f(z)$.

REFERENCES

[1] GANTMACHER, F. R. *The Theory of Matrices*, Chelsea Publishing Co., New York, 1959.

[2] BARNETT, S., "Simplification of the Lyapunov Matrix Equation $A^T P A - P = Q$," *IEEE Trans. Automatic Control*, Vol. AC-19, No. 4, August 1974, pp. 446–447.

[3] BELLMAN, R. E., *Introduction to Matrix Analysis*, 2nd ed., McGraw-Hill Book Company, New York, 1970.

[4] KUCERA, V., "The Discrete Riccati Equation of Optimal Control," *Kybernetika*, Vol. 8, No. 5, 1972, pp. 430–447.

[5] SINGER, R. A., and P. A. FROST, "On the Relative Performance of the Kalman and Wiener Filters," *IEEE Trans. Automatic Control*, Vol. AC-14, No. 4, August 1969, pp. 390–394.

[6] TITUS, H. A., and S. R. NEAL, "Filter Application to Naval Systems," *Proc. 2nd Symp. on Nonlinear Estimation Theory and its Applications*, San Diego, 1971.

[7] JAZWINSKI, A. H., *Stochastic Processes and Filtering Theory*, Academic Press, Inc., New York, 1970.

[8] ANDERSON, B. D. O., "Stability Properties of Kalman Bucy Filters," *J. Franklin Inst.*, Vol. 291, No. 2, February 1971, pp. 137–144.

[9] POPOV, V. M., "Hyperstability and Optimality of Automatic Systems with Several Control Functions," *Revue Roumaine des Sciences Techniques, Electrotechn. et Energ.*, Vol. 9, No. 4, 1964, pp. 629–690.

[10] MOTYKA, P. R., and J. A. CADZOW, "The Factorization of Discrete-process Spectral Matrices," *IEEE Trans. Automatic Control*, Vol. AC-12, No. 6, December 1967, pp. 698–706.

[11] MURTHY, D. N. P., "Factorization of Discrete-process Spectral Matrices," *IEEE Trans. Inform. Theory*, Vol. IT-19, No. 5, September 1973, pp. 693–696.

CHAPTER 5

KALMAN FILTER PROPERTIES

5.1 INTRODUCTION

In this chapter, we shall explore some properties of the Kalman filter. We shall begin our discussion in Sec. 5.2 by noting two different types of estimators, the minimum variance or conditional mean estimator encountered earlier and the linear minimum variance estimator. Although the latter estimator has not been formally encountered hitherto, linear minimum variance estimators have arisen in an indirect way since, in the gaussian case, they happen to agree with conditional mean estimators. The reason for the introduction of the linear minimum variance estimator is that it opens the way to apply simple Euclidean geometry concepts such as orthogonality and projection to estimation problems.

In Sec. 5.3, we introduce the concept of an innovations process. In rough terms, given a measurement process $\{z_k\}$ the innovations process $\{\tilde{z}_k\}$ is such that $\tilde{z}_k$ consists of that part of z_k containing new information not carried in $z_{k-1}, z_{k-2}, \ldots$. It turns out that the innovations process is white, and it is this property which opens the way in Sec. 5.4 for a new demonstration of the optimality of the Kalman filter as a conditional mean estimator in the gaussian case, and a linear minimum variance estimator otherwise. Of course, in the nongaussian case, there may well be nonlinear estimators which would outperform the Kalman filter. In the derivation, we also use a more general signal model than that encountered hitherto, permitting correlation between input and output noise, and the insertion of known inputs as well as the input noise to the signal model.

In Sec. 5.5, we turn to the development of equations for the evolution of the time-filtered estimate $\hat{x}_{k/k}$ and the associated error covariance matrix. Also in this section, we demonstrate a property of the Kalman filter with classical communication systems overtones: it improves signal-to-noise ratio!

Section 5.6 considers the question of testing an operating filter to check its optimality. By and large, a filter is optimal if that quantity which should be the innovations sequence is zero mean and white. It turns out that there are simplifications in checking these conditions.

Many of the background ideas developed in this chapter, particularly those developed in early sections, are relevant to problems other than Kalman filtering, and their assimilation may well be of general benefit to the reader.

History

A historical survey of the development of Kalman filtering can be found in [1]. As the title of [1] implies, the origins of Kalman filtering lie in the late eighteenth century usage of least squares ideas by Gauss in the study of planetary orbits [2]. More recent major ideas bearing on the Kalman filter are those of maximum likelihood estimation due to Fisher [3]; the stationary filtering theory of Wiener [4] and Kolmogorov [5], with an emphasis on linear minimum variance estimation; and in the case of [5], the innovations idea. The use of a recursive approach in estimating constant parameters to cope with new measurements essentially goes back to Gauss; however, the idea of recursion when there is dynamic evolution of the quantity being estimated, at the same time as more measurements become available, is much more recent. (Reference [1] suggests that the recursive approach is due to Follin [6].) The notion of using state-variable rather than impulse response or transfer function descriptions of linear systems is very much associated with the name of Kalman, although Swerling published in 1958 an internal report and in 1959 a journal paper [7] which many consider contain in essence the same method as that of Kalman's famous 1960 paper [8]. However, there is no doubt that subsequent work by Kalman, e.g., [9], through pursuing such matters as stationary filters and stability, went substantially beyond that of [7].

Kalman's method in [8] for deriving the filter is based on the orthogonality properties associated with linear minimum variance estimation, as discussed later in this chapter. Kalman of course recognized that in the guassian situation, the quantities computed by his recursive equations are a mean and covariance of a conditioned gaussian random variable, and thus define recursively a conditional probability density.

Various other approaches to the derivation of the filter can be used, e.g., one based on least squares theory with recursion included, as in [8], or one based on conversion to a dual optimal control problem, as in [10].

One further approach is also worth mentioning, that of [11]. All conventional formulations of the Kalman filter require full knowledge of the signal model. In [11] it is shown that the Kalman filter gain, but not the performance, can be determined using quantities appearing in the covariance of the $\{z_k\}$ process. Knowledge of the signal model allows computation of these quantities, but not vice versa, since there are actually an infinity of signal models having the same covariance for the $\{z_k\}$ process. Thus, less information is used to obtain the filter gain than normal. This will be explored in a later chapter.

5.2 MINIMUM VARIANCE AND LINEAR MINIMUM VARIANCE ESTIMATION; ORTHOGONALITY AND PROJECTION

Let X and Y be jointly distributed random vectors. We have noted earlier the significance of the quantity $E[X \mid Y = y]$ as an estimate of the value taken by X given that $Y = y$: this estimate has the property that

$$E\{\|X - E[X \mid Y = y]\|^2 \mid Y = y\} \leq E\{\|X - \hat{z}\|^2 \mid Y = y\}$$

for any other estimate $\hat{z}$, and indeed also

$$E\{\|X - E[X \mid Y]\|^2\} \leq E\{\|X - \hat{Z}(Y)\|^2\}$$

where now the expectation is over X and Y, and $E[X \mid Y]$ and $\hat{Z}(Y)$ are both functions of the random variable Y.

The functional dependence on Y of $E[X \mid Y]$ will naturally depend on the form of the joint probability density of X and Y and will not necessarily be linear. From the computing point of view, however, a linear estimator, possibly less accurate than the minimum variance estimator, may be helpful. Therefore, one defines a *linear estimator** of X given Y as an estimator of the form $AY + b$, where A is a fixed matrix and b a fixed vector, and one defines *a linear minimum variance estimator* as one in which A and b are chosen to minimize the expected mean square error, i.e.,

$$E^*[X \mid Y] = A^0 Y + b^0 \qquad (2.1)$$

with

$$E\{\|X - A^0 Y - b^0\|^2\} \leq E\{\|X - AY - b\|^2\} \text{ for all } A, b$$

---

*Strictly an affine estimator, as discussed later in the section. However, usage of the phrase "linear estimator" is entrenched.

Here, $E^*[X|Y]$, the linear minimum variance estimator, is *not* an expectation; the notation is, however, meant to suggest a parallel with the (possibly nonlinear) minimum variance estimator $E[X|Y]$.

Let us now explore some properties of the linear minimum variance estimator. The more important properties are highlighted in theorem statements.

The matrices A^0 and b^0 defining $E^*[X|Y]$ can be found in terms of the first and second order statistics (mean and covariance matrix) *of the random variable $[X'\quad Y']$*. (Note the contrast with $E[X|Y]$, the calculation of which normally requires the whole joint probability density; obviously $E^*[X|Y]$ may be much more convenient to obtain.) In fact, we have the following results:

THEOREM 2.1. Let the random variable $[X'\quad Y']'$ have mean and covariance

$$\begin{bmatrix} m_x \\ m_y \end{bmatrix} \text{ and } \begin{bmatrix} \Sigma_{xx} & \Sigma_{xy} \\ \Sigma_{yx} & \Sigma_{yy} \end{bmatrix}$$

Then

$$E^*[X|Y] = m_x + \Sigma_{xy}\Sigma_{yy}^{-1}(Y - m_y) \tag{2.2}$$

If Σ_{yy} is singular, $\Sigma_{xy}\Sigma_{yy}^{-1}$ is replaced by $\Sigma_{xy}\Sigma_{yy}^{\#} + \bar{A}$, for any $\bar{A}$ with $\bar{A}\Sigma_{yy} = 0$.

Proof: We make two preliminary observations. First for an arbitrary vector random variable Z, one has

$$E[\|Z\|^2] = E[\text{trace } ZZ'] = \text{trace cov}(Z, Z) + \text{trace }\{E[Z]E[Z']\}$$

Second, the mean and covariance of $X - AY - b$ are

$$m_x - Am_y - b \qquad \Sigma_{xx} - A\Sigma_{yx} - \Sigma_{xy}A' + A\Sigma_{yy}A'$$

Tying these observations together yields

$$\begin{aligned} E\{\|X - AY - b\|^2\} &= \text{trace }[\Sigma_{xx} - A\Sigma_{yx} - \Sigma_{xy}A' + A\Sigma_{yy}A'] \\ &\quad + \|m_x - Am_y - b\|^2 \\ &= \text{trace }\{[A - \Sigma_{xy}\Sigma_{yy}^{-1}]\Sigma_{yy}[A' - \Sigma_{yy}^{-1}\Sigma_{yx}]\} \\ &\quad + \text{trace }\{\Sigma_{xx} - \Sigma_{xy}\Sigma_{yy}^{-1}\Sigma_{yx}\} \\ &\quad + \|m_x - Am_y - b\|^2 \end{aligned}$$

All three terms are nonnegative. The second is independent of A and b, while the first and third are made zero by taking $A^0 = \Sigma_{xy}\Sigma_{yy}^{-1}$, $b^0 = m_x - Am_y$. This proves the claim of the theorem in case Σ_{yy}^{-1} exists. We shall omit proof of the case of Σ_{yy} singular.

Other minimization properties of the linear minimum variance estimator.
The above argument (with removal of the trace operators) also shows that

A^0 and b^0 serve to minimize the error covariance matrix

$$E\{[X - AY - b][X - AY - b]'\}.$$

(This is a nontrivial fact for the reason that a set of symmetric nonnegative definite matrices need not necessarily have a minimum element.) Further, A^0 and b^0 minimize $E\{[X - AY - b]'M[X - AY - b]\}$ for any positive definite M; this also may be shown by minor variation of the above argument.

Jointly gaussian X and Y. The linear minimum variance estimate is familiar in the situation of jointly gaussian X and Y:

THEOREM 2.2 If X and Y are jointly gaussian, the minimum variance and linear minimum variance estimators coincide.

The proof is a direct consequence of (2.2) and formulas of Chap. 2.

The linear minimum variance estimator is linear* in the following additional sense: if $E^*[X \mid Y]$ is a linear minimum variance estimate of X, then $FE^*[X \mid Y] + e$ is a linear minimum variance estimator of $FX + e$, where F and e are a fixed matrix and vector of appropriate dimensions, respectively. (This is easily seen by direct calculation.) This form of linearity is also possessed incidentally by the minimum variance estimator $E[X \mid Y]$; in the latter instance, it is clearly a consequence of the well-known property of linearity of the expectation operator.

Property of being unbiased. An important property often sought in estimators is lack of bias:

THEOREM 2.3. The linear minimum variance estimator is unbiased, i.e.,

$$E\{X - E^*[X \mid Y]\} = 0 \qquad (2.3)$$

The proof is immediate form (2.2). Being another property held in common with the estimator $E[X \mid Y]$, this property provides further heuristic justification for use of the notation $E^*[X \mid Y]$.

Uncorrelated conditioning quantities. The next property shows how a linear minimum variance estimator can be split up when the conditioning quantities are uncorrelated.

THEOREM 2.4. Suppose that $X, Y_1, \ldots, Y_k$ are jointly distributed, with $Y_1, \ldots, Y_k$ mutually uncorrelated, i.e., $\Sigma_{y_i y_j} = 0$ for $i \neq j$. Then

$$E^*[X \mid Y_1, Y_2, \ldots, Y_k] = E^*[X \mid Y_1] + \cdots + E^*[X \mid Y_k] - (k - 1)m_x$$
$$(2.4)$$

Proof: Think of $W = [Y_1' \quad Y_2' \quad \cdots \quad Y_k']'$ as a random vector and apply the main formula (2.2). Thus

---

*More correctly, we could say that taking linear minimum variance estimates and applying an affine transformation are commutative operations.

$$E^*[X|W] = m_x + [\Sigma_{xy_1} \quad \Sigma_{xy_2} \quad \cdots \quad \Sigma_{xy_k}] \, \text{diag} \, [\Sigma_{y_iy_i}^{-1}] \begin{bmatrix} Y_1 - m_{y_1} \\ \cdot \\ \cdot \\ \cdot \\ Y_k - m_{y_k} \end{bmatrix}$$

while

$$E^*[X|Y_i] = m_x + \Sigma_{xy_i}\Sigma_{y_iy_i}^{-1}(Y_i - m_{y_i})$$

Equation (2.4) is immediate.

Dispensing with the means. If X and Y are jointly distributed with non-zero means, $\bar{X} = X - m_x$ and $\bar{Y} = Y - m_y$ are also jointly distributed, with the same covariance, but with zero mean. One has

$$E^*[\bar{X}|\bar{Y}] = \Sigma_{xy}\Sigma_{yy}^{-1}\bar{Y}$$

Compare this formula with that for $E^*[X|Y]$. It is evident that there is no loss of generality in working with zero mean quantities in the following sense. One can subtract off the a priori mean value of all measured variables from the measurement, estimate the unknown variable less its mean with the simpler formula, and then recover an estimate of the unknown variable by addition of its a priori mean.

Change of conditioning variable. Let X and Y be jointly distributed, and let $Z = MY + n$ for some specified nonsingular matrix M and vector n. Then $E^*[X|Y] = E^*[X|Z]$; put another way, invertible linear (or better, affine) transformations of the conditioning variable leave the estimate unaltered. One can verify this by direct calculation, but the reader should be able to see intuitively that this is so in essence because any linear estimator of the form $AY + b$ is also of the form $CZ + d$, and conversely.

Orthogonality principle. With X and Y jointly distributed, X and Y are termed *orthogonal* if $E[XY'] = 0$. We then have the following most important result.

THEOREM 2.5 (Projection Theorem). Let X, Y be jointly distributed. Then the error $X - E^*[X|Y]$ associated with a linear minimum variance estimate of X by Y is orthogonal to Y:

$$E\{[X - E^*[X|Y]]Y'\} = 0 \tag{2.5}$$

Conversely, if for some A and b one has $E\{[X - AY - b]Y'\} = 0$ and $E[X - AY - b] = 0$, then $E^*[X|Y] = AY + b$.

Proof: Using the basic formula (2.2), we have

$$E\{[X - E^*[X|Y]]Y'\} = E\{[X - m_x - \Sigma_{xy}\Sigma_{yy}^{-1}(Y - m_y)]Y'\}$$
$$= E[XY'] - m_xE[Y'] - \Sigma_{xy}\Sigma_{yy}^{-1}E[YY']$$
$$+ \Sigma_{xy}\Sigma_{yy}^{-1}m_yE[Y']$$
$$= 0$$

Suppose that $AY + b \neq A^0 Y + b^0 = E^*[X \mid Y]$ is such that $E\{[X - AY - b]Y'\} = 0$. Then subtraction of this equation from (2.5) yields $E\{[A - A^0)Y + (b - b^0)]Y'\} = 0$ or

$$(A - A^0)\Sigma_{yy} + (A - A^0)m_y m_y' + (b - b^0)m_y' = 0$$

With $E[X - AY - b] = 0$ and $E[X - A^0 Y - b^0] = 0$, one has $(A - A^0)m_y + (b - b^0) = 0$. Therefore $(A - A^0)\Sigma_{yy} = 0$. The result follows by Theorem 2.1.

Theorem 2.5 is often known as the projection theorem. Why is this so? Consider first the simpler case where X, Y_i are scalar, zero mean, random variables. The unbiased property of the linear minimum variance estimate implies that $E^*[X \mid Y_1, Y_2, \ldots, Y_k]$ is of the form $\Sigma \alpha_i^* Y_i$. In analogy with geometrical reasoning, let us say that *the linear subspace spanned by a set of zero mean random variables Y_i is the set of random variables $\Sigma \alpha_i Y_i$ where α_i* ranges over the set of real numbers. Let us also say that *the projection of a random variable X on the subspace generated by the Y_i is the linear combination $\Sigma \alpha_i^* Y_i$ such that the error, $X - \Sigma \alpha_i^* Y_i$, is orthogonal to the subspace*, i.e.,

$$E\{[X - \Sigma \alpha_i^* Y_i][\Sigma \alpha_i Y_i]\} = 0 \qquad \forall \alpha_i$$

In view of our statement of the orthogonality principle, the *projection of X on the subspace generated by the Y_i is precisely $E^*[X \mid Y_1, Y_2, \ldots, Y_k]$.*

More generally, consider the situation postulated in the theorem statement. The space one projects onto is spanned by a vector Y and all constant vectors, i.e., the space is the collection $\{AY + b\}$ for all A, b. The projection of X onto this subspace is the particular element of the subspace $A^0 Y + b^0$ such that $X - (A^0 Y + b^0)$ is orthogonal to the subspace, i.e.,

$$E\{[X - A^0 Y - b^0][Y'A' + b']\} = 0 \qquad \forall A, b$$

Equivalently,

$$E\{[X - A^0 Y - b^0]Y'\} = 0 \qquad E[X - A^0 Y - b^0] = 0$$

According to the theorem, $A^0 Y + b^0 = E^*[X \mid Y]$. See Fig. 5.2-1 for an illustration.

There is great practical and theoretical utility in this idea. The practical utility rests in the fact that the equations expressing orthogonality can often be taken as a starting point for obtaining a linear minimum variance estimate. The theoretical utility lies in the fact that there is some direction offered as to how to proceed with estimation involving infinite-dimensional quantities, e.g., continuous-time random processes. By and large, as the advanced reader will know, projections coupled with minimum norm ideas are naturally viewed as taking place in a Hilbert space; it turns out that estimation problems that may involve continuous-time random processes can be structured as projection-in-Hilbert-space problems.

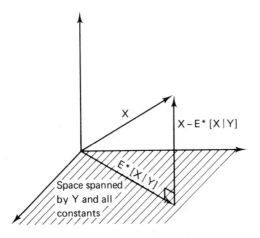

Fig. 5.2-1 Illustration of projection theorem. The error is orthogonal to the measurements.

For the purposes of this book, the reader need not, however, be concerned with the Hilbert space overtones of the projection theorem. All that is important are the notions of orthogonality, spanning of a subspace by random variables, and projection onto a subspace.

Conditional minimum variance or not? Recall that the estimator $E[X\,|\,Y]$ evaluated at $Y = y$ is a *conditional* minimum variance estimate, in the sense that

$$E\{\|\,X - E[X\,|\,Y = y]\,\|^2\,|\,Y = y\}$$

is minimized. Also, since $E[\|\,X - E[X\,|\,Y]\|^2]$ is minimized (with the expectation over X and Y), $E[X\,|\,Y]$ is a minimum variance estimator. Now $E^*[X\,|\,Y]$ has a property which parallels the second property of $E[X\,|\,Y]$, since it is a minimum variance estimator within a certain class. One might ask whether there is a parallel for $E^*[X\,|\,Y]$ evaluated at $Y = y$ of the conditional minimum variance property. The answer is, in general, no. One would be seeking the property that A^0 and b^0 should minimize

$$\int_{-\infty}^{+\infty} \|\,x - A^0 y - b^0\,\|^2 p_{X|Y}(x\,|\,y)\,dx$$

irrespective of y. It is immediately evident that the existence of such A^0 and b^0 (working for all y) cannot be guaranteed. In fact, it is necessary and sufficient for the existence of A^0 and b^0 that $E[X\,|\,Y] = CY + d$ for some C, d, or equivalently $E^*[X\,|\,Y] = E[X\,|\,Y]$. (See Prob. 2.5.)

EXAMPLE 2.1. Let $Y = X + N$, where X and N are independent, zero mean, scalar random variables. We evaluate $E^*[X\,|\,Y]$. To apply the formula,

$$E^*[X\,|\,Y] = \Sigma_{xy}\Sigma_{yy}^{-1}\,Y$$

observe that $E[XY] = E[X^2 + XN] = E[X^2]$ and $E[Y^2] = E[X^2] + E[N^2]$. Therefore,

$$E^*[X|Y] = \frac{E[X^2]}{E[X^2] + E[N^2]} Y$$

EXAMPLE 2.2. Let $x(t)$ be a zero mean, random process, let $X = \int_0^1 x(t)dt$, and let $Y' = [x(0) \quad x(\frac{1}{2}) \quad x(1)]$. We seek $E^*[X|Y]$, or an approximation to the integral in terms of the value of the integrand at particular points. Obviously, some second order statistics will be required.

From the orthogonality principle, for the optimal a_0, $a_{1/2}$, and a_1 such that $E^*[X|Y] = a_0 x(0) + a_{1/2} x(\frac{1}{2}) + a_1 x(1)$, we have

$$E\left[\left\{\int_0^1 x(t)dt - [a_0 x(0) + a_{1/2} x(\frac{1}{2}) + a_1 x(1)]\right\}\{x(0) \quad x(\frac{1}{2}) \quad x(1)\}\right] = 0$$

This yields three separate equations:

$$E\left\{\int_0^1 x(t)dt \, x(0)\right\} = a_0 E[x^2(0)] + a_{1/2} E[x(0)x(\tfrac{1}{2})] + a_1 E[x(0)x(1)]$$

$$E\left\{\int_0^1 x(t)dt \, x(\tfrac{1}{2})\right\} = a_0 E[x(\tfrac{1}{2})x(0)] + a_{1/2} E[x^2(\tfrac{1}{2})] + a_1 E[x(\tfrac{1}{2})x(1)]$$

$$E\left\{\int_0^1 x(t)dt \, x(1)\right\} = a_0 E[x(1)x(0)] + a_{1/2} E[x(1)x(\tfrac{1}{2})] + a_1 E[x^2(1)]$$

The a_i follow by solving these equations. Knowledge of $E[x(t)x(s)]$ for all t and s would be sufficient, but not actually necessary, to obtain the constants.

EXAMPLE 2.3. Let $Y = H'X + V$, where X, V are independent random variables with means m_x, m_v and covariances Σ_{xx}, Σ_{vv}. We compute $E^*[X|Y]$. Observe that $\Sigma_{xy} = E[XY'] - m_x m_y' = \Sigma_{xx}H$, while $\Sigma_{yy} = H'\Sigma_{xx}H + \Sigma_{vv}$. Accordingly,

$$E^*[X|Y] = m_x + \Sigma_{xx}H(H'\Sigma_{xx}H + \Sigma_{vv})^{-1}(Y - m_y)$$

with $m_y = H'm_x + m_v$. The associated mean square error is

$$\text{trace}\,(\Sigma_{xx} - \Sigma_{xy}\Sigma_{yy}^{-1}\Sigma_{yx}) = \text{trace}\,[\Sigma_{xx} - \Sigma_{xx}H(H'\Sigma_{xx}H + R)^{-1}H'\Sigma_{xx}]$$

A common classical problem has no information available about X at all. This is taken up in Prob. 2.7.

Ruling out affine estimators. Throughout this section, the estimators considered have really been affine, i.e., of the form $Y \rightarrow AY + b$ for some A, b rather than linear, i.e., of the form $Y \rightarrow AY$. We might ask what happens when strictly linear estimators are considered. In case all quantities have zero mean, there is no alteration to the results. If this is not the case, there is a difference. Problem 2.6 explores this issue. Crucial points are that the estimator is generally biased, but a projection theorem still holds.

Orthogonality and nonlinear estimates. If X and Y are jointly distributed, it turns out that one can characterize $E[X|Y]$ as that function of Y such that the estimation error is orthogonal to all functions of Y. We shall not use this fact, however.

Main Points of the Section

The linear minimum variance estimator is given in terms of first and second order statistics as

$$E^*[X \mid Y] = m_x + \Sigma_{xy}\Sigma_{yy}^{-1}(Y - m_y)$$

with replacement of the inverse by a pseudo-inverse if necessary. For jointly gaussian X and Y, $E^*[X \mid Y] = E[X \mid Y]$. The linear minimum variance estimator of $FX + e$ is $FE^*[X \mid Y] + e$, and if $Y = MZ + n$ for M invertible, then $E^*[X \mid Y] = E^*[X \mid Z]$.

The linear minimum variance estimator is unbiased, and the error $X - E^*[X \mid Y]$ is orthogonal to the subspace spanned by Y and all constant vectors (projection theorem or orthogonality principle).

In case $Y_1, Y_2, \ldots, Y_k$ are uncorrelated, one has the important formula

$$E^*[X \mid Y_1, \ldots, Y_k] = \sum_{j=1}^{k} E^*[X \mid Y_j] - (k-1)m_x$$

Strictly linear, nonaffine, minimum variance estimators can be defined and they agree with the usual linear estimator when all quantities have zero mean. In general, they yield a biased estimate. A form of the projection theorem still holds.

Problem 2.1. Let X and Y be jointly distributed random variables with known mean and covariance. Suppose Y is used to estimate X via $E[X \mid Y]$. For what probability density on X and Y will the mean square error be *greatest*, and why?

Problem 2.2. Suppose $\{x_k\}$ is a zero mean, scalar random sequence with known stationary covariance $E[x_k x_l] = R_{k-l}$. What is the best one-step predictor $E^*[x_k \mid x_{k-1}]$? Explain how an n-step predictor $E^*[x_{k+n-1} \mid x_{k-1}]$ could be found.

Problem 2.3. Suppose that $x_k + \sum_{i=1}^{n} a_i x_{k-i} = w_k$ defines a random sequence $\{x_k\}$ in terms of a white, zero mean sequence $\{w_k\}$. Evaluate $E^*[x_k \mid x_{k-n}, \ldots, x_{k-1}]$ and $E^*[x_k \mid x_{k-n}, \ldots, x_{k-2}]$.

Problem 2.4. Suppose $\{x_k\}$ is a zero mean, scalar random sequence with known stationary covariance $E[x_k x_l] = R_{k-l}$. Find the best interpolator $E^*[x_k \mid x_{k-1}, x_{k+1}]$.

Problem 2.5. Show that $E^*[X \mid Y]$ evaluated at $Y = y$ is a conditional minimum variance estimate for all y if and only if $E[X \mid Y] = CY + d$ for some C, d, and show that this implies $E^*[X \mid Y] = E[X \mid Y]$. (Evaluate the conditional error variance as a function of A^0, b^0 and first and second moments of $p_{X|Y}(x \mid y)$. Show that it is minimized when $A^0 y + b^0 = E[X \mid Y = y]$.)

Problem 2.6. Let X, Y be jointly distributed, and let $E^{**}[X \mid Y] = A^0 Y$ be such that $E_{X,Y}[\| X - A^0 Y \|^2] \leq E_{X,Y}[\| X - A Y \|^2]$ for all fixed A and some fixed A^0.

1. Show that second order moments $E[xy'] = R_{xy}$, $E[yy'] = R_{yy}$ determine A^0 as $R_{xy} R_{yy}^{-1}$; compute the associated mean square error.
2. Show that the estimator may be biased.
3. Show that $E\{[X - E^{**}[X \mid Y]] Y'\} = 0$ and that if $E\{[X - A Y] Y'\} = 0$ then $E^{**}[X \mid Y] = A Y$.

Problem 2.7. (Classical Least Squares). Consider the situation of Example 2.3 in which, for convenience, $m_x = 0$, $m_v = 0$, and $\Sigma_{xx} = \lim_{p \to \infty} pI$. (This corresponds to having no a priori information about X.) Suppose that HH' is nonsingular (Y has dimension at least as great as X). Show that

$$E^*[X \mid Y] = (H \Sigma_{vv}^{-1} H')^{-1} H \Sigma_{vv}^{-1} Y$$

with mean square error $\mathrm{tr}\,(H \Sigma_{vv}^{-1} H')^{-1}$. [The formula $A(BA + I)^{-1} = (AB + I)^{-1} A$ may be found useful in studying the behaviour of $pH(pH'H + \Sigma_{vv})^{-1} = pH\Sigma_{vv}^{-1}(pH'H\Sigma_{vv}^{-1} + I)^{-1}$ as $p \to \infty$.] Show that if H is square and invertible, this reduces to the intuitively obvious formula $E^*[X \mid Y] = (H')^{-1} Y$. What happens if HH' is singular?

5.3 THE INNOVATIONS SEQUENCE

We have already made mention of the innovations sequence, in the following terms: Suppose $\{z_k\}$ is a sequence of gaussian random variables, possibly vectors. Then the innovations process $\{\tilde{z}_k\}$ is such that $\tilde{z}_k$ consists of *that part of z_k containing new information not carried in* $z_{k-1}, z_{k-2}, \ldots$. What exactly does this mean?

To fix ideas, suppose $\{z_k\}$ is defined for $k \geq 0$ and the mean and covariance of the process are known. Knowing z_0 but not z_1, we would estimate the value of z_1 by $E[z_1 \mid z_0]$. The new information contained in z_1 is therefore

$$\tilde{z}_1 = z_1 - E[z_1 \mid z_0]$$

Likewise, knowing z_0 and z_1, we could estimate z_2 by $E[z_2 \mid z_0, z_1]$, and the new information contained in z_2 becomes

$$\tilde{z}_2 = z_2 - E[z_2 \mid z_0, z_1]$$

More generally,

$$\tilde{z}_k = z_k - E[z_k \mid z_0, z_1, \ldots, z_{k-1}] = z_k - E[z_k \mid Z_{k-1}] \qquad (3.1)$$

(Here, Z_{k-1} denotes the set $z_0, z_1, \ldots, z_{k-1}$. Note that Z_{k-1} is not that random variable taking particular value z_{k-1}.) We shall also make the reasonable definition

$$\tilde{z}_0 = z_0 - E[z_0] \qquad (3.2)$$

These definitions ensure that $E[\tilde{z}_k] = 0$ for all k. (Why?) What of the covariance of the process $\{\tilde{z}_k\}$? The definition of $\tilde{z}_k$ as the new information contained in z_k given the values of $z_0, \ldots, z_{k-1}$ suggests that $\tilde{z}_k$ must be independent of $z_0, \ldots, z_{k-1}$ and therefore of $\tilde{z}_l$ for $l < k$, since $\tilde{z}_l$ is determined by $z_0, \ldots, z_{k-1}$. In other words, this suggests that $E[\tilde{z}_k \tilde{z}_l'] = 0$, or that the $\tilde{z}_k$ *process is white.* Let us verify this claim, by means of the following two observations.

1. $\tilde{z}_k$ *is a linear function of* $z_0, z_1, \ldots, z_k$. This follows because the gaussian property of the sequence $\{z_k\}$ implies that $E[z_k | z_0, z_1, \ldots, z_{k-1}]$ is a linear combination of $z_0, z_1, \ldots, z_{k-1}$.
2. *By the orthogonality principle,* $\tilde{z}_k$ *is orthogonal to* $z_0, z_1, \ldots, z_{k-1}$ and all linear combinations of these random variables. This is immediate from the definition of $\tilde{z}_k$ and the orthogonality principle. (Recall that for jointly gaussian X and Y, $E[X | Y] = E^*[X | Y]$.)

These two observations show that $E[\tilde{z}_k \tilde{z}_l'] = 0$ for $k > l$ and then also $k < l$.

Let us now study a number of other properties of the innovations process.

1. Observation 1 illustrates the fact that *the sequence* $\{\tilde{z}_k\}$ *can be obtained from* $\{z_k\}$ *by a causal linear operation*; i.e., one can conceive of a black box with input the sequence $\{z_k\}$ and output the sequence $\{\tilde{z}_k\}$, operating in real time and processing the $\{z_k\}$ linearly. Strictly, the linearity, as opposed to affine, aspect of the processing is only valid in case $\{z_k\}$ is a zero mean sequence. This linearity property also implies that $\{\tilde{z}_k\}$ inherits the gaussian property of $\{z_k\}$.
2. Our next point is that *the causal linear operation producing* $\{\tilde{z}_k\}$ *from* $\{z_k\}$ *has a causal linear inverse*, i.e., for each k, z_k is constructible from $\tilde{z}_l$ for $l \leq k$ by affine transformations which are linear in case all variables have zero mean. One can argue by induction; observe first that $z_0 = \tilde{z}_0 + E[z_0]$, and suppose that for $i = 1, 2, \ldots, k-1$, z_i is expressible as an affine combination of $\tilde{z}_0, \tilde{z}_1, \ldots, \tilde{z}_i$. Now $z_k = \tilde{z}_k + E[z_k | Z_{k-1}]$ and $E[z_k | Z_{k-1}]$ is expressible as an affine combination of $z_0, z_1, \ldots, z_{k-1}$ and, by the inductive hypothesis, as an affine combination of $\tilde{z}_0, \tilde{z}_1, \ldots, \tilde{z}_{k-1}$. Hence z_k is expressible as an affine combination of $\tilde{z}_0, \tilde{z}_1, \ldots, \tilde{z}_k$.
3. Points 1 and 2 together establish that conditioning on the variables $z_0, z_1, \ldots, z_{k-1}$ is equivalent to conditioning on $\tilde{z}_0, \tilde{z}_1, \ldots, \tilde{z}_{k-1}$. In particular, we have $E[z_k | Z_{k-1}] = E[z_k | \tilde{Z}_{k-1}]$ and an alternative definition for the innovations:

$$\tilde{z}_k = z_k - E[z_k | \tilde{Z}_{k-1}] \tag{3.3}$$

More generally, for any vector w jointly distributed with $\{z_k\}$, $E[w | Z_{k-1}] = E[w | \tilde{Z}_{k-1}]$. Because the $\tilde{z}_i$ are independent random variables, conditioning on $\tilde{Z}_{k-1}$ can be more helpful than conditioning on Z_{k-1}.

4. Following on from point 3, suppose $\{x_k\}$ and $\{z_k\}$ are two jointly gaussian, zero mean, random processes and that the variables $z_0, z_1, \ldots, z_{k-1}$ are used to estimate x_k via $E[x_k \,|\, Z_{k-1}]$. Then one has

$$E[x_k \,|\, Z_{k-1}] = E[x_k \,|\, \tilde{Z}_{k-1}]$$
$$= E[x_k \,|\, \tilde{z}_0] + E[x_k \,|\, \tilde{z}_1] + \cdots + E[x_k \,|\, \tilde{z}_{k-1}]$$

Here, we have used the important result of Theorem 2.4 of the previous section. (In case the processes are not zero mean, minor adjustment is of course needed.) The device just noted will be used in the next section to obtain a derivation of the Kalman filter.

5. There is an interesting sidelight shed on a prediction problem by the innovations idea. One can think of the one-step predictor as being a system $\mathcal{P}$ with input the sequence $\{z_k\}$ and output the sequence $\{z_{k/k-1}\}$, where $z_{k/k-1} = E[z_k \,|\, Z_{k-1}]$. Then, drawing the predictor as a unity feedback system, with the forward part $\mathcal{F}$, it follows from the innovation sequence definition that $\mathcal{F}$ is driven by the innovations as illustrated in Fig. 5.3-1. The block $\mathcal{F}$ can in fact serve to define both the linear system generating the $\{\tilde{z}_k\}$ sequence from the $\{z_k\}$ sequence and the $\{z_k\}$ sequence from the $\{\tilde{z}_k\}$ sequence. For $\{\tilde{z}_k\}$ is the output of a system with unity forward gain and negative feedback $\mathcal{F}$, driven by $\{z_k\}$, while $\{z_k\}$ is the output of a system with forward part $\mathcal{F}$ paralleled with unity gain forward part and driven by $\{\tilde{z}_k\}$. (Check this!)

Let us now sum up the major points.

THEOREM 3.1 Let $\{z_k\}$ defined for $k \geq 0$ be a gaussian random sequence. Define $\tilde{z}_0 = z_0 - E[z_0]$ and $\tilde{z}_k = z_k - E[z_k \,|\, Z_{k-1}]$. Then

1. $\{\tilde{z}_k\}$ is zero mean and white.
2. $\tilde{z}_k$ is an affine combination of z_l for $l \leq k$ and z_k an affine combination of $\tilde{z}_l$ for $l \leq k$. The combinations are linear if $E[z_k] = 0$ for all k.
3. With w and $\{z_k\}$ jointly distributed,

$$E[w \,|\, Z_k] = E[w \,|\, \tilde{Z}_k] = \sum_{i=0}^{k} E[w \,|\, \tilde{z}_i]$$

provided $E[w] = 0$.
4. With a one-step predictor arranged as a unity feedback system as in Fig. 5.3-1, the forward part is driven by the innovations.

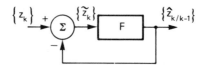

Fig. 5.3-1 One step predictor $\mathcal{P}$ drawn as unity feedback system.

Further properties of the innovations sequence will be taken up in later sections.

Pseudo-innovations

Hitherto, we have assumed that $\{z_k\}$ is a gaussian process. What can be said if this is not the case? One route we can take is to retain the definition of the innovations process and determine the properties [now normally excluding the linear relationship between $\{\tilde{z}_k\}$ and $\{z_k\}$] which are retained. Martingale theory has proved a most helpful tool for this purpose [12–18]. Another route, aimed at preserving the linear relationship, is to define a *pseudo-innovations process* by

$$\tilde{z}_k = z_k - E^*[z_k \,|\, Z_{k-1}] \tag{3.4}$$

Assuming arbitrary mean for $\{z_k\}$, it still follows that $\{\tilde{z}_k\}$ is zero mean and that $E[\tilde{z}_k \tilde{z}_l'] = 0$ for $k \neq l$; obviously $\tilde{z}_k$ depends linearly (or in an affine way) on $z_0, z_1, \ldots, z_{k-1}$. Moreover it is still true that:

1. $\tilde{z}_k$ can be obtained from z_l for $l \leq k$ by a causal affine (linear in the zero mean case) operation.
2. z_k can be obtained from $\tilde{z}_l$ for $l \leq k$ by a causal affine or linear operation.
3. $\tilde{z}_k = z_k - E^*[z_k \,|\, \tilde{Z}_{k-1}]$.
4. Conditioning on Z_{k-1} is equivalent to conditioning on $\tilde{Z}_{k-1}$ in any linear minimum variance estimator.
5. The one-step predictor idea (point 4 of Theorem 3.1) is valid.

Initial Time in the Infinitely Remote Past

Change of the initial time instant from $k = 0$ to $k = k_0$ for arbitrary finite k_0 is trivial to implement. What happens, though, if $k_0 \rightarrow -\infty$? The situation is a lot more delicate. We require that $\mathfrak{F}$ in Fig. 5.3-1 can be defined, and in order to eliminate the dependence on initial conditions, we require that the closed loop be asymptotically stable. We also require $E\{z_k^2\}$ to be finite. In case, for example, $\{z_k\}$ is the output of an asymptotically stable, linear, finite-dimensional system with independent input and output noise, all these requirements are fulfilled as noted in the last chapter.

The causal dependence of $\{z_k\}$ on $\{\tilde{z}_k\}$ can be expressed as

$$z_k = \tilde{z}_k + \sum_{i \geq 1} a_{i,k} \tilde{z}_{k-i} \tag{3.5}$$

and that of $\{\tilde{z}_k\}$ on $\{z_k\}$ by

$$\tilde{z}_k = z_k + \sum_{i \geq 1} b_{i,k} z_{k-i} \tag{3.6}$$

where zero mean processes are assumed. In the stationary case, the $a_{i,k}$ and $b_{i,k}$ are independent of k.

To discuss (3.5) and (3.6) in depth here would take us too far afield. Let us however make several remarks. For simplicity, we confine attention to the stationary case.

1. In order that finite $E[\tilde{z}_k^2]$ should produce finite $E[z_k^2]$, one needs $\sum_{i \geq 1} a_i^2 < \infty$. This condition ensures that values of $\tilde{z}_{k-i}$ well before time k are given little weighting in (3.5); i.e., there is a forgetting of initial conditions.

2. Likewise, the condition $\sum_{i \geq 1} b_i^2 < \infty$ causes forgetting of old values of z in obtaining $\tilde{z}$.

3. The quantities a_i and b_i are, naturally, related. In fact, it is easily seen that

$$
\begin{bmatrix}
1 & a_1 & a_2 & a_3 & \cdot & \cdot \\
0 & 1 & a_1 & a_2 & \cdot & \cdot \\
0 & 0 & 1 & a_1 & \cdot & \cdot \\
\cdot & \cdot & \cdot & \cdot & \cdot & \cdot
\end{bmatrix}
\begin{bmatrix}
1 & b_1 & b_2 & b_3 & \cdot & \cdot \\
0 & 1 & b_1 & b_2 & \cdot & \cdot \\
0 & 0 & 1 & b_1 & \cdot & \cdot \\
\cdot & \cdot & \cdot & \cdot & \cdot & \cdot
\end{bmatrix}
=
\begin{bmatrix}
1 & 0 & 0 & \cdot & \cdot \\
0 & 1 & 0 & \cdot & \cdot \\
0 & 0 & 1 & \cdot & \cdot \\
\cdot & \cdot & \cdot & \cdot & \cdot
\end{bmatrix}
$$

(Multiply both sides by the column vector $[z_k \quad z_{k-1} \quad z_{k-2} \quad \cdots]'$.)

4. If the quantities a_i are known, it is very easy to obtain a formula for $E[z_k | Z_{k-N}]$ for all $N \geq 1$. Because of the orthogonality of the $\tilde{z}_k$, one has

$$
\sum_{i \geq N} a_i \tilde{z}_{k-i} = E[z_k | \tilde{Z}_{k-N}] = E[z_k | Z_{k-N}]
$$

Main Points of the Section

For a gaussian random sequence $\{z_k\}$, the innovations sequence $\{\tilde{z}_k\}$ is a zero mean, white sequence obtainable by linear or affine causal transformation of the original process, and the transformation is also causally invertible. The fact that the $\tilde{z}_k$ are uncorrelated with each other may aid the evaluation of a quantity $E[w | Z_k] = E[w | \tilde{Z}_k]$. The one-step predictor generating $\hat{z}_{k/k-1}$ may be represented as a unity feedback system with forward part driven by the innovations.

In case z_k is not gaussian, one can work with linear minimum variance estimates; the main difference is that the $\tilde{z}_k$ are no longer independent, though $E[\tilde{z}_k \tilde{z}_l'] = 0$ for $k \neq l$. Alternatively, one can still work with minimum variance estimates, but thereby lose the linearity properties relating $\{z_k\}$ and $\{\tilde{z}_k\}$.

An initial time in the infinitely remote past can be assumed, provided that the system generating $\tilde{z}_{k/k-1}$ is well defined and is asymptotically stable.

Problem 3.1. Suppose that a process z_k is generated by

$$
z_k + a_1 z_{k-1} + \cdots + a_n z_{k-n} = w_k
$$

where w_k is a sequence of independent, zero mean, unit variance, gaussian random variables and the a_i are known constants. Show that for $k > n$, $\tilde{z}_k = z_k - E[z_k \,|\, z_{k-n}, z_{k-n+1}, \ldots, z_{k-1}]$.

Problem 3.2. Suppose $z_0, z_1, \ldots$ is a sequence of scalar, zero mean random variables with $E[z_k z_l] = r_{kl}$ known for all k and l. One can form the infinite matrix R with kl entry r_{kl}, and it will be nonnegative definite symmetric. Show that the coefficients used in expressing z_k in terms of $\tilde{z}_0, \tilde{z}_1, \tilde{z}_2, \ldots, \tilde{z}_k$ can be used to define a factorization of R as $R = T'ST$, where T is upper triangular and S is diagonal. Can you find recursive formulas for the entries of T and S in terms of the entries of R?

Problem 3.3. Let $\{g_k\}$ and $\{z_k\}$ be jointly gaussian, zero mean sequences with $\{\tilde{z}_k\}$ available. Show that $\hat{g}_{k/k-1} = E[g_k \,|\, Z_{k-1}]$ can be derived as the output at time k of a linear system of impulse response $\{h_{kl}\}$ driven by $\{\tilde{z}_k\}$, with $h_{kl} = E[g_k \tilde{z}_l]$ to within a scaling factor depending on l but not k.

5.4 THE KALMAN FILTER

In this section, our aim is to derive the Kalman filter equations, for the one-step prediction state estimate and associated error covariance. We shall do this for the system, defined for $k \geq 0$,

$$x_{k+1} = F_k x_k + G_k w_k + \Gamma_k u_k \tag{4.1}$$

$$z_k = H'_k x_k + v_k \tag{4.2}$$

Here, $\{u_k\}$ is a known input sequence, x_0 has mean $\bar{x}_0$ and covariance P_0, $\{v_k\}$ and $\{w_k\}$ are zero mean sequences with

$$E\left\{ \begin{bmatrix} w_k \\ v_k \end{bmatrix} \begin{bmatrix} w'_l & v'_l \end{bmatrix} \right\} = \begin{bmatrix} Q_k & S_k \\ S'_k & R_k \end{bmatrix} \delta_{kl} \tag{4.3}$$

[One can obtain $S_k \neq 0$ if, for example, the input noise to the signal process feeds through directly to the output as well as through the inherent delaying mechanism associated with (4.1). This idea is developed in the problems.] The random variable x_0 is assumed independent of $[w'_k \quad v'_k]'$.

Now we can either assume x_0, $\{v_k\}$, and $\{w_k\}$ are jointly gaussian and seek $\hat{x}_{k/k-1} = E[x_k \,|\, Z_{k-1}]$ and the associated error covariance $\Sigma_{k/k-1}$ which is both conditional and unconditional or we can drop the gaussian assumption and seek $\hat{x}_{k/k-1} = E^*[x_k \,|\, Z_{k-1}]$ and the associated unconditional error covariance $\Sigma_{k/k-1}$. *The same equations arise* (why?); it is merely the interpretation of the quantities that is different. To maintain consistency of style, we shall make the gaussian assumption, and leave to the reader the restatement of results appropriate to linear minimum variance estimation. The key equations are (4.9) through (4.12).

Note that the signal model here is a development of that used earlier in two directions, since both S_k and u_k can be nonzero. As it turns out, the increased difficulty in using an innovation-based approach to filter derivation is small; a derivation along the lines given in Chap. 3 is, however, a good deal more complicated.

Evolution of the Conditional Mean

We shall find the recursive equation for

$$\hat{x}_{k/k-1} = E[x_k | Z_{k-1}] = E[x_k | \tilde{Z}_{k-1}]$$

For convenience, we define $\tilde{x}_k = x_k - \hat{x}_{k/k-1}$; notice that $\tilde{x}_k$ is *not* an innovations sequence, because $\hat{x}_{k/k-1}$ is not $E[x_k | x_0, x_1, \ldots, x_{k-1}]$.

Our strategy will be to make use of the independence of the innovations, which as we know (see, especially, Theorem 2.4) allows us to write

$$E[x_{k+1} | \tilde{z}_0, \tilde{z}_1, \ldots, \tilde{z}_k] = E[x_{k+1} | \tilde{z}_k] + E[x_{k+1} | \tilde{z}_0, \tilde{z}_1, \ldots, \tilde{z}_{k-1}] - E[x_{k+1}]$$

$$(4.4)$$

Then we shall evaluate the first and second terms in this expression separately.

We begin by evaluating $E[x_{k+1} | \tilde{z}_k]$. In view of the jointly gaussian nature of x_{k+1} and $\tilde{z}_k$, we know from the material of Sec. 5.2 (see Theorem 2.1) that

$$E[x_{k+1} | \tilde{z}_k] = E[x_{k+1}] + \text{cov}\,(x_{k+1}, \tilde{z}_k)[\text{cov}\,(\tilde{z}_k, \tilde{z}_k)]^{-1}\tilde{z}_k \qquad (4.5)$$

To evaluate the two covariances, define the error covariance matrix

$$\Sigma_{k/k-1} = E[\tilde{x}_k \tilde{x}_k'] \qquad (4.6)$$

Then we have (reasoning follows the equalities)

$$\begin{aligned}
\text{cov}\,(x_{k+1}, \tilde{z}_k) &= \text{cov}\,(F_k x_k + G_k w_k + \Gamma_k u_k, H_k' \tilde{x}_k + v_k) \\
&= E[\{F_k x_k + G_k w_k - F_k E[x_k]\}\{\tilde{x}_k' H_k + v_k'\}] \\
&= E[F_k x_k \tilde{x}_k' H_k] + G_k S_k \\
&= F_k[E(\hat{x}_{k/k-1}\tilde{x}_k') + E(\tilde{x}_k \tilde{x}_k')]H_k + G_k S_k \\
&= F_k \Sigma_{k/k-1} H_k + G_k S_k
\end{aligned}$$

To get the first equality, observe first that from (4.2), $\hat{z}_{k/k-1} = H'\hat{x}_{k/k-1}$. Subtracting this from (4.2), we have $\tilde{z}_k = H_k' \tilde{x}_k + v_k$. We also use (4.1) to obtain the first equality. To get the second equality, we use the fact that u_k is known, that w_k and v_k have zero mean by assumption, and that $\tilde{x}_k$ has zero mean, being the error in a conditional mean estimate. The third equality follows on using again the zero mean nature of $H_k' \tilde{x}_k + v_k$, from the independence of x_k and v_k, the independence of w_k and $\tilde{x}_k$, or equivalently w_k and Z_{k-1}, and the assumed dependence of w_k and v_k. The last equality uses $E[\hat{x}_{k/k-1}\tilde{x}_k'] = 0$; this follows from the fact that $\tilde{x}_k$ is the error in projecting x_k onto the subspace generated by $z_{k-1}, z_{k-2}, \ldots$ and is, therefore, orthogonal to that subspace, while $\hat{x}_{k/k-1}$ is a member of it. The last equality also uses (4.6).

Also, straightforward calculations yield

$$\text{cov}\,(\tilde{z}_k, \tilde{z}_k) = \text{cov}\,(H_k'\tilde{x}_k + v_k, H_k'\tilde{x}_k + v_k)$$
$$= H_k'\Sigma_{k/k-1}H_k + R_k$$

(Here, we use the fact that $\tilde{x}_k$ and v_k are independent.) Consequently, (4.5) becomes

$$E[x_{k+1}\,|\,\tilde{z}_k] = E[x_{k+1}] + (F_k\Sigma_{k/k-1}H_k + G_kS_k)(H_k'\Sigma_{k/k-1}H_k + R_k)^{-1}\tilde{z}_k$$
$$(4.7)$$

We also have

$$E[x_{k+1}\,|\,\tilde{Z}_{k-1}] = E[F_kx_k + G_kw_k + \Gamma_ku_k\,|\,\tilde{Z}_{k-1}]$$
$$= F_kE[x_k\,|\,\tilde{Z}_{k-1}] + \Gamma_ku_k$$
$$= F_k\hat{x}_{k/k-1} + \Gamma_ku_k \qquad (4.8)$$

Here, we have used the independence of w_k and $\tilde{Z}_{k-1}$, and the fact that $\{u_k\}$ is known. The recursive equation for $\hat{x}_{k/k-1}$ is now immediate, using (4.7) and (4.8) in (4.4):

$$\hat{x}_{k+1/k} = F_k\hat{x}_{k/k-1} + \Gamma_ku_k + K_k(z_k - H_k'\hat{x}_{k/k-1}) \qquad (4.9)$$

with

$$K_k = (F_k\Sigma_{k/k-1}H_k + G_kS_k)(H_k'\Sigma_{k/k-1}H_k + R_k)^{-1} \qquad (4.10)$$

[In (4.10), if the inverse does not exist, one can, at least formally, have a pseudo-inverse.]

Evolution of the Covariance Matrix

We have defined the error covariance in (4.6) and have used it in (4.10), but have not yet verified that it satisfies the recursive equation given in earlier chapters. This we now do. Equations (4.1), (4.2), and (4.9) yield

$$\tilde{x}_{k+1} = (F_k - K_kH_k')\tilde{x}_k + G_kw_k - K_kv_k$$

The two vectors $\tilde{x}_k$ and $[w_k'\ \ v_k']'$ are independent and have zero mean, and so

$$E[\tilde{x}_{k+1}\tilde{x}_{k+1}'] = (F_k - K_kH_k')E(\tilde{x}_k\tilde{x}_k')(F_k - K_kH_k')'$$
$$\times [G_k \quad -K_k]\begin{bmatrix} Q_k & S_k \\ S_k' & R_k \end{bmatrix}\begin{bmatrix} G_k' \\ -K_k' \end{bmatrix}$$

or

$$\Sigma_{k+1/k} = (F_k - K_kH_k')\Sigma_{k/k-1}(F_k - K_kH_k')' + G_kQ_kG_k' + K_kR_kK_k'$$
$$- G_kS_kK_k' - K_kS_k'G_k' \qquad (4.11)$$

Using the formula (4.10) for K_k, we can also write this equation in the form

$$\Sigma_{k+1/k} = F_k\Sigma_{k/k-1}F_k' - (F_k\Sigma_{k/k-1}H_k + G_kS_k)(H_k'\Sigma_{k/k-1}H_k + R_k)^{-1}$$
$$\times (F_k\Sigma_{k/k-1}H_k + G_kS_k)' + G_kQ_kG_k' \tag{4.12}$$

Also, we clearly must take $\Sigma_{0/-1} = P_0$.

The Kalman Filter Equations and Rapprochement with Earlier Results

The equations defining the filter associated with (4.1) through (4.3) are given in (4.9) through (4.12) and are repeated below. They are applicable (1) in the nongaussian case, to give a linear minimum variance estimate and unconditional error covariance; (2) in the gaussian case, to give a minimum variance estimate and an unconditional and conditional error covariance.

$$\hat{x}_{k+1/k} = F_k\hat{x}_{k/k-1} + \Gamma_k u_k + K_k(z_k - H_k'\hat{x}_{k/k-1}) \tag{4.9}$$
$$\hat{x}_{0/-1} = \bar{x}_0$$
$$K_k = (F_k\Sigma_{k/k-1}H_k + G_kS_k)(H_k'\Sigma_{k/k-1}H_k + R_k)^{-1} \tag{4.10}$$
$$\Sigma_{k+1/k} = (F_k - K_kH_k')\Sigma_{k/k-1}(F_k - K_kH_k')' + G_kQ_kG_k' + K_kR_kK_k'$$
$$- G_kS_kK_k' - K_kS_k'G_k' \tag{4.11}$$
$$= F_k\Sigma_{k/k-1}F_k' - (F_k\Sigma_{k/k-1}H_k + G_kS_k)(H_k'\Sigma_{k/k-1}H_k + R_k)^{-1}$$
$$\times(F_k\Sigma_{k/k-1}H_k + G_kS_k)' + G_kQ_kG_k' \tag{4.12}$$
$$\Sigma_{0/-1} = P_0$$

One obtains the equations encountered earlier by setting $u_k \equiv 0$ and $S_k \equiv 0$. The interpretation in the gaussian case is as earlier. However, in the non-gaussian case, we now have a wider interpretation. Earlier, we noted that the Kalman filter was the optimum among a *restricted* set of linear filters. Now, we have shown it optimum among *all* linear processors. In case the reader feels these distinctions are trivial and therefore labored unduly, we invite him or her to consider whether or not the derivation of this chapter allows F_k, H_k, etc. to depend on Z_{k-1}, as in Sec. 3.3.

EXAMPLE 4.1. This example can be more easily analyzed without the methods of Kalman filtering. But precisely because this is the case, we are able to verify easily that the result of applying Kalman filtering ideas to at least one problem is correct.

The AR system

$$z_k + A_1 z_{k-1} + \cdots + A_n z_{k-n} = w_k$$

(where $\{w_k\}$ is a white, zero mean gaussian sequence with $E[w_kw_k'] = Q_k$) has state-variable representation

$$x_{k+1} = \begin{bmatrix} -A_1 & -A_2 & \cdot & \cdot & -A_{n-1} & -A_n \\ I & 0 & \cdot & \cdot & 0 & 0 \\ 0 & I & \cdot & \cdot & 0 & 0 \\ \cdot & \cdot & & & & \cdot \\ \cdot & \cdot & & & & \cdot \\ 0 & 0 & \cdot & \cdot & I & 0 \end{bmatrix} x_k + \begin{bmatrix} I \\ 0 \\ 0 \\ \cdot \\ \cdot \\ 0 \end{bmatrix} w_k$$

$$z_k = -[A_1 \quad A_2 \quad \cdots \quad A_n]x_k + w_k$$

From the AR equation itself it is immediately evident that

$$E[z_k | Z_{k-1}] = -A_1 z_{k-1} - \cdots - A_n z_{k-n}$$

Let us check that the Kalman filter equations give the same result. Suppose that $\Sigma_{0/-1} = 0$. Note $Q_k = S_k = R_k$. From (4.10), we obtain $K_0 = G = I$, and from (4.11), $\Sigma_{1/-0} = 0$. Continuing leads to $K'_k = G' = [I \quad 0]$, $\Sigma_{k+1/k} = 0$. From (4.9), we have

$$\hat{x}_{k+1/k} = \begin{bmatrix} 0 & 0 & \cdot & \cdot & 0 & 0 \\ I & 0 & \cdot & \cdot & 0 & 0 \\ 0 & I & \cdot & \cdot & 0 & 0 \\ \cdot & \cdot & & \cdot & & \cdot \\ \cdot & \cdot & & & \cdot & \cdot \\ 0 & 0 & \cdot & \cdot & I & 0 \end{bmatrix} \hat{x}_{k/k-1} + \begin{bmatrix} I \\ 0 \\ \cdot \\ \cdot \\ \cdot \\ 0 \end{bmatrix} z_k$$

that is,

$$\hat{x}_{k+1/k} = \begin{bmatrix} z_k \\ z_{k-1} \\ \cdot \\ \cdot \\ \cdot \\ z_{k-n+1} \end{bmatrix}$$

The signal model also implies that

$$\hat{z}_{k+1/k} = -[A_1 \quad A_2 \quad \ldots \quad A_n]\hat{x}_{k+1/k} = -A_1 z_k - A_2 z_{k-1} - \cdots - A_n z_{k-n+1}$$

as predicted.

A little reflection will show why it is that $\Sigma_{k+1/k} = 0$ for all k. Examination of the signal model equations shows that the first entry of x_{k+1}, $x_{k+1}^{(1)}$, is precisely z_k. Thus this quantity is always known with zero error. The signal model equations also show that $x_{k+1}^{(2)} = x_k^{(1)} = z_{k-1}$, $x_{k+2}^{(3)} = z_{k-2}$, etc, so that all entries of x_{k+1} are known, given past measurements. Of course, if $\Sigma_{0/-1} \neq 0$, we would expect that $\Sigma_{k+1/k}$ would be zero for $k \geq n$ but not for $k < n$. (Why?)

Alternative Covariance Equations

The question of whether (4.11) or (4.12) should be preferred in practice is largely a computational one. Because

$$G_k Q_k G'_k + K_k R_k K'_k - G_k S_k K'_k - K_k S'_k G'_k = [G_k \quad -K_k]\begin{bmatrix} Q_k & S_k \\ S'_k & R_k \end{bmatrix}\begin{bmatrix} G'_k \\ -K'_k \end{bmatrix}$$

and

$$\begin{bmatrix} Q_k & S_k \\ S_k' & R_k \end{bmatrix}$$

is a covariance and therefore nonnegative, the right side of (4.11) is of the form

$$A\Sigma_{k/k-1}A' + \mathcal{b}CB'$$

with $C = C' \geq 0$. This means that with $\Sigma_{k/k-1} \geq 0$, $\Sigma_{k+1/k}$ will tend to be nonnegative definite, even in the face of round-off errors. The same is not so true of (4.12). Therefore, (4.11) may be preferable. On the other hand, the number of operations for one update may make (4.12) preferable. (Another possibility again is explored in the problems.) The whole question of computation will be taken up in the next chapter.

Relaxation of Assumptions on $\{u_k\}$

To this point, we have assumed that $\{u_k\}$ is a known sequence. Now K_k and $\Sigma_{k+1/k}$ do not depend on $\{u_k\}$. The quantity u_k is needed to compute $\hat{x}_{k+1/k}$ but no earlier estimates. This means that if u_k is not known in advance, and is perhaps not known until time k, one can still obtain a filter. In this way, $\{u_k\}$ could become a random sequence, so long as u_k is known at time k. In particular, the $\{u_k\}$ sequence could be derived *by feedback* from the measurement sequence $\{z_k\}$ with u_k permitted to depend on z_l for $l \leq k$. Problem 4.5 considers a situation of this type.

Filter as a Feedback System

Equation (4.9) provides the easiest way to visualize the filter as a feedback system, with the forward part driven by the innovations (see Fig. 5.4-1). That such a formulation is possible in the first place is a consequence of the fact that $\hat{z}_{k/k-1}$ is derivable from $\hat{x}_{k/k-1}$.

From the philosophical point of view, the discerning reader may sense a small paradox in the arrangement; $\hat{x}_{k/k-1}$ is the output of a linear system

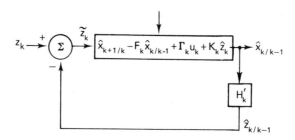

Fig. 5.4-1   Filter depicted in feedback form.

driven by the innovations, which are white noise. Since white noise is about the most chaotic of all random processes, it would seem that the filter is producing order, in the form of the estimate $\hat{x}_{k/k-1}$, out of chaos, in the form of the sequence $\{\tilde{z}_k\}$. We leave it to the reader to explain away the difficulty.

The Innovations Covariance

As we know, the innovations are a white gaussian sequence. For future reference, we can evaluate $E[\tilde{z}_k \tilde{z}_k']$. Evidently,

$$\tilde{z}_k = z_k - H_k' \hat{x}_{k/k-1} = H_k' \tilde{x}_k + v_k$$

Since $\tilde{x}_k$ and v_k are independent, we have

$$E[\tilde{z}_k \tilde{z}_k'] = H_k' E[\tilde{x}_k \tilde{x}_k'] H_k + E[v_k v_k']$$
$$= H_k' \Sigma_{k/k-1} H_k + R_k \tag{4.13}$$

Prediction

The estimate $\hat{x}_{k+1/k}$ is a one-step prediction estimate rather than a true filtered estimate. (Determination of $\hat{x}_{k/k}$ is discussed later in this chapter.) Let us now note how one may determine $\hat{x}_{k+N/k}$, where N is a fixed integer greater than 1. (The measurements z_k are, therefore, being used to predict the state variable N time units ahead of the present time.) Recall that

$$x_{k+N} = \Phi_{k+N, k+1} x_{k+1} + \sum_{i=k+1}^{k+N-1} \Phi_{k+N, i}(G_i w_i + \Gamma_i u_i)$$

where $\Phi_{k, l}$ is the usual transition matrix. Now w_i for $i = k + 1, k + 2, \ldots,$ $k + N - 1$ is independent of $z_0, z_1, \ldots, z_k$, and so, taking the conditional expectation, we obtain

$$\hat{x}_{k+N/k} = \Phi_{k+N, k+1} \hat{x}_{k+1/k} + \sum_{i=k+1}^{k+N-1} \Phi_{k+N, i} \Gamma_i u_i \tag{4.14}$$

To obtain the associated error, observe that

$$x_{k+N} - \hat{x}_{k+N/k} = \Phi_{k+N, k+1}(x_{k+1} - \hat{x}_{k+1/k}) + \sum_{i=k+1}^{k+N-1} \Phi_{k+N, i} G_i w_i$$

Taking cognizance of the independence of $x_{k+1} - \hat{x}_{k+1/k}$ and the w_i for $i = k + 1$ through $k + N - 1$, we obtain

$$\Sigma_{k+N/k} = \Phi_{k+N, k+1} \Sigma_{k+1/k} \Phi_{k+N, k+1}' + \sum_{i=k+1}^{k+N-1} \Phi_{k+N, i} G_i Q_i G_i' \Phi_{k+N, i}' \tag{4.15}$$

Observe what happens in the time-invariant situation. Then $\Phi_{k, l} = F^{k-l}$, and if $|\lambda_i(F)| < 1$, $\hat{x}_{k+N/k} \to 0$ as $N \to \infty$, irrespective of $\hat{x}_{k+1/k}$. In other words, the further one gets away from the time at which measurements are made, the less relevant those measurements become. (But $\Sigma_{k+N/k} \not\to 0$.)

Prediction of y_{k+N} and z_{k+N} is easily done. Since $y_{k+N} = H'_{k+N}x_{k+N}$, then $\hat{y}_{k+N/k} = H'_{k+N}\hat{x}_{k+N/k}$ and evidently also $\hat{z}_{k+N/k} = H'_{k+N}\hat{x}_{k+N/k}$. When z_k is a zero mean process, $z_{k+N} - \hat{z}_{k+N/k}$ can be expressed as a moving average, irrespective of the fact that z_k is the output of a finite-dimensional system. The point is explored in the problems.

Time-invariant Problems

Suppose that F_k, G_k, H_k, Q_k, S_k, and R_k are all independent of k. What can be said about the time invariance and stability of the associated filter? As before, we suppose that R_k is nonsingular.

The first question to consider is whether, with $\Sigma_{0/-1} = 0$, we might have $\Sigma_{k+1/k} \to \bar{\Sigma}$ as $k \to \infty$. The fact that u_k may be nonzero is irrelevant, since u_k does not enter the variance equation. On the other hand, it is not clear whether or not the nonzero nature of S_k is important. Actually, it is not. The same argument as given earlier can be used to conclude that $\Sigma_{k+1/k}$ is monotone increasing with k and, provided that $[F, H]$ is detectable, is bounded above. Accordingly, *with $[F, H]$ detectable, $\bar{\Sigma}$ exists.*

To obtain a stability result, we shall observe, using a little hindsight, that the stationary version of (4.10) through (4.12) can also arise for a problem with no correlation between input and output noise. Since stability conditions are available for this kind of problem, having been derived in the last chapter, we can apply them here. Either one can proceed via the result of Prob. 4.5 or as follows. Define

$$\bar{K} = K - GSR^{-1}$$

Then

$$\bar{K} = (F - GSR^{-1}H')\bar{\Sigma}H(H'\bar{\Sigma}H + R)^{-1}$$

[For $K - GSR^{-1} = (F\bar{\Sigma}H + GS)(H'\bar{\Sigma}H + R)^{-1} - GSR^{-1}(H'\bar{\Sigma}H + R)$ $\times (H'\bar{\Sigma}H + R)^{-1} = (F\bar{\Sigma}H - GSR^{-1}H'\bar{\Sigma}H)(H'\bar{\Sigma}H + R)^{-1} = \bar{K}$.] Also define

$$\bar{F} = F - GSR^{-1}H'$$

so that

$$\bar{K} = \bar{F}\bar{\Sigma}H(H'\bar{\Sigma}H + R)^{-1} \tag{4.16}$$

and

$$F - KH' = \bar{F} - \bar{K}H' \tag{4.17}$$

Then one can check that the steady-state version of (4.11) yields

$$\bar{\Sigma} = (\bar{F} - \bar{K}H')\bar{\Sigma}(\bar{F} - \bar{K}H')' + \bar{K}R\bar{K}' + G(Q - SR^{-1}S')G' \tag{4.18}$$

We recognize from (4.16) and (4.18) that $\bar{\Sigma}$ is the limiting error covariance associated with a filtering problem defined by $\{\bar{F}, G, H, \bar{Q} = Q - SR^{-1}S', R\}$ and with zero correlation between input and output noise. It follows that $|\lambda_i(\bar{F} - \bar{K}H')| < 1$ for all i if $[\bar{F}, G\bar{G}_1]$ is completely stabilizable for any $\bar{G}_1$ with $\bar{G}_1\bar{G}'_1 = \bar{Q}$. Since $F - KH' = \bar{F} - \bar{K}H'$, we conclude that the *time-*

invariant filter for the original problem is asymptotically stable provided that $[F - GSR^{-1}H', G\bar{G}_1]$ *is completely stabilizable for any* $\bar{G}_1$ *with* $\bar{G}_1\bar{G}_1' = Q - SR^{-1}S'$. Given satisfaction of this constraint, the time-varying but asymptotically time-invariant filter will be asymptotically stable, and for arbitrary nonnegative $\Sigma_{0/-1}$, we shall have $\Sigma_{k+1/k} \longrightarrow \bar{\Sigma}$. If all unstable modes are observed (which will be the case if $[F, H]$ is completely detectable), the complete stabilizability is also necessary for filter stability.

EXAMPLE 4.2. The MA equation $z_k = w_k + c_1 w_{k-1} + \cdots + c_n w_{k-n}$ (where $\{w_k\}$ is a zero mean, white gaussian process with $E[w_k^2] = 1$) gives rise to a signal model of the form

$$x_{k+1} = \begin{bmatrix} 0 & 1 & 0 & \cdots & 0 \\ 0 & 0 & 1 & \cdots & 0 \\ \cdot & \cdot & \cdot & & \cdot \\ \cdot & \cdot & \cdot & & \cdot \\ \cdot & \cdot & \cdot & & 1 \\ 0 & 0 & 0 & \cdots & 0 \end{bmatrix} x_k + \begin{bmatrix} 0 \\ 0 \\ \cdot \\ \cdot \\ \cdot \\ 1 \end{bmatrix} w_k$$

$$z_k = [c_n \quad c_{n-1} \quad \cdots \quad c_1] x_k + w_k$$

(Thus $Q = S = R = 1$.) Since F has all eigenvalues at zero, $[F, H]$ is certainly completely detectable. We also see that $F - GSR^{-1}H'$ becomes

$$\bar{F} = \begin{bmatrix} 0 & 1 & 0 & \cdots & 0 \\ 0 & 0 & 1 & \cdots & 0 \\ \cdot & \cdot & \cdot & & \cdot \\ \cdot & \cdot & \cdot & & \cdot \\ \cdot & \cdot & \cdot & & 1 \\ -c_n & -c_{n-1} & \cdots & & -c_1 \end{bmatrix}$$

while $Q - SR^{-1}S' = 0$. Consequently, while $\bar{\Sigma}$ always exists, the time-invariant filter is asymptotically stable only if $z^n + c_1 z^{n-1} + \cdots + c_n$ has all its zeros inside $|z| < 1$. It is easy in fact to compute $\bar{\Sigma}$ and $F - KH'$ to check these conclusions. Using (4.12), and setting $\Sigma_{0/-1} = 0$, we have

$$\Sigma_{1/0} = -GSR^{-1}S'G' + GQG' = 0$$

More generally, $\Sigma_{k+1/k} = 0$ and so $\bar{\Sigma} = 0$. The steady-state value of the filter gain matrix is, from (4.10), $\bar{K} = 0$, $K = G$, and accordingly $F - KH' = \bar{F}$.

Main Points of the Section

Not only is the Kalman filter the best filter among a subset of all linear filters, but it is the best filter among the set of all filters when the noise processes are gaussian and the best linear filter among the set of all linear filters otherwise. The filter can be visualized as a feedback system, with the forward part driven by the innovations, which are a white noise sequence.

The signal model equations are

$$x_{k+1} = F_k x_k + G_k w_k + \Gamma_k u_k \qquad z_k = H_k' x_k + v_k$$

$$E\left\{ \begin{bmatrix} w_k \\ v_k \end{bmatrix} [w_l' \quad v_l'] \right\} = \begin{bmatrix} Q_k & S_k \\ S_k' & R_k \end{bmatrix} \delta_{kl}$$

with the usual other assumptions, while the Kalman filter equations are

$$\hat{x}_{k+1/k} = F_k \hat{x}_{k/k-1} + \Gamma_k u_k + (F_k \Sigma_{k/k-1} H_k + G_k S_k)(H_k' \Sigma_{k/k-1} H_k + R_k)^{-1}$$
$$\times (z_k - H_k' \hat{x}_{k/k-1})$$

$$\Sigma_{k+1/k} = F_k \Sigma_{k/k-1} F_k' - (F_k \Sigma_{k/k-1} H_k + G_k S_k)(H_k' \Sigma_{k/k-1} H_k + R_k)^{-1}$$
$$\times (F_k \Sigma_{k/k-1} H_k + G_k S_k)' + G_k Q_k G_k'$$

Problem 4.1. Show that

$$\Sigma_{k+1/k} = (F_k - K_k H_k') \Sigma_{k/k-1} F_k' + (G_k Q_k - K_k S_k') G_k'$$

Problem 4.2. Derive the equations obtained in this section via the technique used in Chap. 3 in studying filtering in the gaussian situation; do not make use of the independence property of the innovations sequence.

Problem 4.3. Consider the following signal model with direct feedthrough:

$$x_{k+1} = F_k x_k + G_k w_k$$
$$z_k = H_k' x_k + v_k + J_k w_k$$

where v_k and w_k are independent, zero mean, white gaussian processes of covariances $R_k \delta_{kl}$ and $Q_k \delta_{kl}$, respectively. Show that this arrangement is equivalent to one in which $z_k = H_k' x_k + \bar{v}_k$ and $\bar{v}_k$ is a zero mean, white gaussian process of covariance $(R_k + J_k Q_k J_k')\delta_{kl}$ with $E[w_k \bar{v}_l'] = Q_k J_k' \delta_{kl}$.

Problem 4.4. Consider the arrangement in Prob. 4.3, save that $R_k = 0$ for all k, and x_0 is known to be zero. Show that if J_k is nonsingular also, $\Sigma_{k/k-1} = 0$ for all k. Explain in direct terms why this should be so.

Problem 4.5. Consider the signal model described in this section and suppose for convenience that R_k is nonsingular.
(a) Show that $\bar{w}_k = w_k - S_k R_k^{-1} v_k$ is independent of v_k.
(b) Show that $x_{k+1} = F_k x_k + G_k w_k + G_k u_k$ becomes, with the choice $u_k = -S_k R_k^{-1} z_k$,

$$x_{k+1} = (F_k - G_k S_k R_k^{-1} H_k') x_k + G_k \bar{w}_k$$

precisely.
(c) Find the filter for this arrangement, and recover from it the filter for (4.1) and (4.2).

Problem 4.6. In [19], it is pointed out that models of the form of (4.1) and (4.2) arise when discretizing a continuous-time system; in this case one has

$$E[v_k v_l'] = R_k \delta_{kl} \qquad E[w_k w_l'] = Q_k \delta_{kl} \quad \text{and} \quad E[w_k v_l'] = 0$$

for $k \neq l - 1$, $E[w_k v_l'] = S_k$ for $k = l - 1$. Find a recursive algorithm for $\hat{x}_{k/k-1}$.

Problem 4.7. Is $\{\tilde{x}_k\}$ a white sequence?

Problem 4.8. Let $\{z_k\}$ be any gaussian process. Show that

$$z_{k+N} - \hat{z}_{k+N/k} = \tilde{z}_{k+N} - E[z_{k+N}] + E[z_{k+N}\,|\,\tilde{z}_{k+1}, \ldots, \tilde{z}_{k+N-1}]$$

Conclude that if $E[z_k] = 0$ for all k, the error $z_{k+N} - \hat{z}_{k+N/k}$ is a moving average process. (The formula

$$E[z_{k+N}\,|\,\tilde{Z}_{k+N-1}] = E[z_{k+N}\,|\,\tilde{Z}_k] + E[z_{k+N}\,|\,\tilde{z}_{k+1}, \ldots, \tilde{z}_{k+N-1}] - E[z_{k+N}]$$

may be helpful).

5.5 TRUE FILTERED ESTIMATES AND THE SIGNAL-TO-NOISE RATIO IMPROVEMENT PROPERTY

In this section, we set up equations for calculating the true filtered quantities, and we make an interesting connection with classical ideas by demonstrating a signal-to-noise ratio (SNR) improvement achieved with the true filter. As in the last section, we work with the model

$$x_{k+1} = F_k x_k + G_k w_k + \Gamma_k u_k \tag{5.1}$$

$$z_k = y_k + v_k = H'_k x_k + v_k \tag{5.2}$$

with $\{u_k\}$, $\{v_k\}$, $\{w_k\}$, and x_0 as before. In particular, we suppose that

$$E\left\{ \begin{bmatrix} w_k \\ v_k \end{bmatrix} [w'_l \;\; v'_l] \right\} = \begin{bmatrix} Q_k & S_k \\ S'_k & R_k \end{bmatrix} \delta_{kl} \tag{5.3}$$

with S_k not necessarily zero.

Evolution of $\hat{x}_{k/k}$

It is easy to check that with $\bar{w}_k = w_k - S_k R_k^{-1} v_k$, one has

$$E\left\{ \begin{bmatrix} \bar{w}_k \\ v_k \end{bmatrix} [\bar{w}'_l \;\; v'_l] \right\} = \begin{bmatrix} Q_k - S_k R_k^{-1} S'_k & 0 \\ 0 & R_k \end{bmatrix} \delta_{kl}$$

and

$$\begin{aligned}
x_{k+1} &= F_k x_k + G_k \bar{w}_k + G_k S_k R_k^{-1} v_k + \Gamma_k u_k \\
&= (F_k - G_k S_k R_k^{-1} H'_k) x_k + G_k \bar{w}_k + \Gamma_k u_k + G_k S_k R_k^{-1} z_k \quad (5.4)
\end{aligned}$$

At time k, z_k is known just as u_k is known. So (5.4) and (5.2) describe a system equivalent in a sense to (5.1) and (5.2), and such that the input noise sequence $\{\bar{w}_k\}$ is uncorrelated with the measurement noise sequence $\{v_k\}$. From (5.4) we have immediately the *time-update* equation

$$\hat{x}_{k+1/k} = (F_k - G_k S_k R_k^{-1} H'_k)\hat{x}_{k/k} + \Gamma_k u_k + G_k S_k R_k^{-1} z_k \tag{5.5}$$

It is also straightforward to obtain an equation expressing $\hat{x}_{k+1/k+1}$ in terms of $\hat{x}_{k+1/k}$ and z_k. Let us proceed formally, in the same manner as was used in deriving an equation for the evolution of $\hat{x}_{k+1/k}$ from $\hat{x}_{k/k-1}$. Evidently,

$$\hat{x}_{k+1/k+1} = E[x_{k+1} \mid \tilde{Z}_{k+1}]$$

$$= E[x_{k+1} \mid \tilde{z}_{k+1}] + E[x_{k+1} \mid \tilde{Z}_k] - E[x_{k+1}]$$

$$= E[x_{k+1}] + \text{cov}\,(x_{k+1}, \tilde{z}_{k+1})[\text{cov}\,(\tilde{z}_{k+1}, \tilde{z}_{k+1})]^{-1}\tilde{z}_{k+1}$$

$$+ \hat{x}_{k+1/k} - E[x_{k+1}]$$

Now

$$\text{cov}\,(x_{k+1}, \tilde{z}_{k+1}) = E\{[\tilde{x}_{k+1} + \hat{x}_{k+1/k} - E(x_{k+1})][\tilde{x}_{k+1}H_{k+1} + v_{k+1}]\}$$

$$= E[\tilde{x}_{k+1}\tilde{x}'_{k+1}]H_{k+1}$$

$$= \Sigma_{k+1/k}H_{k+1}$$

on using arguments like those of the previous section in evaluating $\text{cov}\,(x_k, \tilde{z}_{k-1})$. We also know that

$$\text{cov}\,(\tilde{z}_{k+1}, \tilde{z}_{k+1}) = H'_{k+1}\Sigma_{k+1/k}H_{k+1} + R_{k+1}$$

from the previous section. Consequently, we have the *measurement-update* equation

$$\hat{x}_{k+1/k+1} = \hat{x}_{k+1/k} + \Sigma_{k+1/k}H_{k+1}(H'_{k+1}\Sigma_{k+1/k}H_{k+1} + R_{k+1})^{-1}$$

$$\times (z_{k+1} - H'_{k+1}\hat{x}_{k+1/k}) \tag{5.6}$$

Together, (5.5) and (5.6) provide a recursive procedure for updating $\hat{x}_{k/k}$. Let us pause to make several observations.

1. In case $S_k = 0$, (5.5) simplifies to the equation:

$$\hat{x}_{k+1/k} = F_k\hat{x}_{k/k} + \Gamma_k u_k \tag{5.7}$$

With $u_k = 0$, this equation was encountered in Chap. 3. When $S_k \neq 0$, z_k contains information about w_k (via v_k), and this information has to be used in estimating x_{k+1}; this is the reason for the additional complication in (5.5).

2. One can combine measurement- and time-update equations to get a single recursion for $\hat{x}_{k/k-1}$, as in the last section. One can also obtain a single recursion for $\hat{x}_{k/k}$. Combination of (5.5) and (5.6) leads to

$$\hat{x}_{k+1/k+1} = [I - \Sigma_{k+1/k}H_{k+1}(H'_{k+1}\Sigma_{k+1/k}H_{k+1} + R_{k+1})^{-1}H'_{k+1}]$$

$$\times [F_k - G_kS_kR_k^{-1}H'_k]\hat{x}_{k/k}$$

$$+ A_ku_k + B_kz_k + C_kz_{k+1} \tag{5.8}$$

for appropriate A_k, B_k, C_k. The important special case of $S_k = 0$ yields

$$\hat{x}_{k+1/k+1} = F_k\hat{x}_{k/k} + L_{k+1}(z_{k+1} - H'_{k+1}F_k\hat{x}_{k/k} - H'_{k+1}\Gamma_k u_k) \tag{5.9}$$

with

$$L_{k+1} = \Sigma_{k+1/k}H_{k+1}(H'_{k+1}\Sigma_{k+1/k}H_{k+1} + R_{k+1})^{-1} \tag{5.10}$$

Figure 5.5-1 illustrates this arrangement in case $u_k \equiv 0$, and also illustrates how $\hat{x}_{k/k-1}$ and the innovations $\tilde{z}_k$ can be obtained.

3. Recursive equations for $\hat{x}_{k/k}$ as in (5.9) require initialization. This is straightforward. With $\hat{x}_{0/-1} = \bar{x}_0$, one takes, on the basis of (5.6),

$$\hat{x}_{0/0} = \bar{x}_0 + P_0 H_0 (H_0' P_0 H_0 + R_0)^{-1}(z_0 - H_0' \bar{x}_0)$$

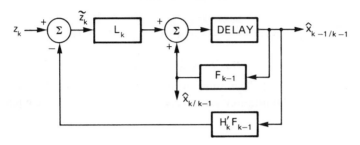

Fig. 5.5-1 Recursive filter for $\hat{x}_{k/k}$, illustrating innovations. Zero u_k is assumed.

Evolution of the Error Covariance

We now seek an expression for the error covariance

$$E\{[x_k - \hat{x}_{k/k}][x_k - \hat{x}_{k/k}]'\} = \Sigma_{k/k}$$

From (5.6), we have

$$(x_{k+1} - \hat{x}_{k+1/k+1}) + \Sigma_{k+1/k} H_{k+1}(H_{k+1}' \Sigma_{k+1/k} H_{k+1} + R_{k+1})^{-1} \tilde{z}_{k+1}$$
$$= x_{k+1} - \hat{x}_{k+1/k}$$

By the orthogonality principle, $x_{k+1} - \hat{x}_{k+1/k+1}$ is orthogonal to $\tilde{z}_{k+1}$. The covariance matrix of the whole left side is therefore the sum of the covariance matrices of the two summands, thus

$$\Sigma_{k+1/k+1} + \Sigma_{k+1/k} H_{k+1}(H_{k+1}' \Sigma_{k+1/k} H_{k+1} + R_{k+1})^{-1}(H_{k+1}' \Sigma_{k+1/k} H_{k+1} + R_{k+1})$$
$$\times (H_{k+1}' \Sigma_{k+1/k} H_{k+1} + R_{k+1})^{-1} H_{k+1}' \Sigma_{k+1/k} = \Sigma_{k+1/k}$$

that is,

$$\Sigma_{k+1/k+1} = \Sigma_{k+1/k} - \Sigma_{k+1/k} H_{k+1}(H_{k+1}' \Sigma_{k+1/k} H_{k+1} + R_{k+1})^{-1} H_{k+1}' \Sigma_{k+1/k} \tag{5.11}$$

This equation is a *measurement-update* equation. To complete the picture, let us obtain a *time-update* equation. Subtracting (5.5) from (5.1) and using the definition of $\bar{w}_k$ yields

$$x_{k+1} - \hat{x}_{k+1/k} = (F_k - G_k S_k R_k^{-1} H_k')(x_k - \hat{x}_{k/k}) + G_k \bar{w}_k$$

Using the easily checked orthogonality of $\bar{w}_k$ and $x_k - \hat{x}_{k/k}$, we obtain

$$\Sigma_{k+1/k} = (F_k - G_k S_k R_k^{-1} H_k')\Sigma_{k/k}(F_k - G_k S_k R_k^{-1} H_k')'$$
$$+ G_k(Q_k - S_k R_k^{-1} S_k')G_k' \tag{5.12}$$

This equation is the time-update equation. Combination of (5.11) and (5.12) provides a recursion for $\Sigma_{k+1/k}$, as obtained earlier, or a recursion for $\Sigma_{k/k}$. When $S_k = 0$, (5.11) is unchanged while (5.12) simplifies:

$$\Sigma_{k+1/k} = F_k \Sigma_{k/k} F_k' + G_k Q_k G_k' \tag{5.13}$$

We remark that two rewritings of the measurement-update equation are possible:

$$\Sigma_{k+1/k+1} = \Sigma_{k+1/k}[I - H_{k+1} L_{k+1}'] = [I - L_{k+1} H_{k+1}']\Sigma_{k+1/k}$$

and

$$\Sigma_{k+1/k+1} = [I - L_{k+1} H_{k+1}']\Sigma_{k+1/k}[I - H_{k+1} L_{k+1}'] + L_{k+1} R_{k+1} L_{k+1}'$$

with L_{k+1} as in (5.10). The number of operations required to compute $\Sigma_{k+1/k+1}$ and the computational accuracy differ among the three equations.

Stability

In the last section, we obtained the one-step predictor recursive equation

$$\hat{x}_{k+1/k} = F_k \hat{x}_{k/k-1} + \Gamma_k u_k + K_k(z_k - H_k' \hat{x}_{k/k-1})$$

with

$$K_k = (F_k \Sigma_{k/k-1} H_k + G_k S_k)(H_k' \Sigma_{k/k-1} H_k + R_k)^{-1}$$

Suppose the filter is time invariant and R is nonsingular. The system matrix associated with the homogeneous equation is

$$F_1 = F - KH' = F - F\bar{\Sigma}H(H'\bar{\Sigma}H + R)^{-1}H' - GS(H'\bar{\Sigma}H + R)^{-1}H'$$

From Eq. (5.8), the system matrix associated with the homogeneous version of the $\hat{x}_{k/k}$ equation is

$$F_2 = [I - \bar{\Sigma}H(H'\bar{\Sigma}H + R)^{-1}H'](F - GSR^{-1}H')$$

The matrices F_1 and F_2 can be shown to have the same eigenvalues, as developed in the problems. Accordingly the stability properties for the true filter are the same as those for the one-step predictor. Actually, this idea can be extended to the time-varying case.

A Signal-to-Noise Ratio Improvement Property

We can make an interesting connection with more classical approaches to signal processing by demonstrating a property of the true filter. Suppose that the signal process $\{y_k\}$ is scalar and stationary. (Both these assumptions may be relaxed if desired.) Suppose also that $u_k = 0$ for all k. The measurements z_k comprise signal y_k and noise n_k, with signal power $E[y_k^2] = H'PH$ and noise power $E[n_k^2] = R$. Here, P is the unique nonnegative definite solution of $P - FPF' = GQG'$. We define

$$\text{Filter input signal-to-noise ratio} = \frac{H'PH}{R} \tag{5.14}$$

The output of the true filter is $\hat{x}_{k/k}$, from which we derive $\hat{y}_{k/k} = H'\hat{x}_{k/k}$. Now suppose

$$E\{[x_k - \hat{x}_{k/k}][x_k - \hat{x}_{k/k}]'\} = \bar{\Sigma}$$

so that $E[(y_k - \hat{y}_{k/k})^2] = H'\bar{\Sigma}H$. By the orthogonality principle, $y_k - \hat{y}_{k/k}$ is orthogonal to $\hat{y}_{k/k}$. Therefore

$$E[y_k^2] = E[(y_k - \hat{y}_{k/k})^2] + E[\hat{y}_{k/k}^2] \quad \text{or} \quad E[\hat{y}_{k/k}^2] = H'(P - \bar{\Sigma})H$$

which is the total filter output power.

There are various possible definitions for the filter output signal-to-noise ratio. We shall take this ratio to be the total filter output power, viz., $H'(P - \bar{\Sigma})H$, divided by the mean square error $E[(y_k - \hat{y}_{k/k})^2]$ between the filter output and true signal. This error is in part due to the noise v_k and in part to the distortion in y_k produced by the filter. Thus

$$\text{Filter output signal-to-noise ratio} = \frac{H'(P - \bar{\Sigma})H}{H'\bar{\Sigma}H} \qquad (5.15)$$

We claim there is a signal-to-noise ratio improvement property:

$$\frac{H'(P - \bar{\Sigma})H}{H'\bar{\Sigma}H} \geq \frac{H'PH}{R} \qquad (5.16)$$

and that strict inequality normally holds.

An alternative definition of output signal-to-noise ratio which could be used is as follows. Regard the output of the filter $\hat{y}_{k/k}$ as the sum of a signal y_k and error $(\hat{y}_{k/k} - y_k)$. Notice that in this case the signal and error are correlated, which is unconventional. The signal power is $H'PH$ and noise power is $H'\bar{\Sigma}H$, so that the output signal-to-noise-ratio is $H'PH/H'\bar{\Sigma}H$. Comparing this with (5.15), we see that if (5.16) holds, we shall also have signal-to-noise ratio improvement with this alternative definition.

Yet another possibility would be to define output noise as the result of passing $\{v_k\}$ through the filter. The output noise power is then less than $H'\bar{\Sigma}H$, and so again (5.16) would imply signal-to-noise ratio improvement.

To establish (5.16), we proceed as follows. Suppose that y_k is estimated using the value of z_k but no other z_l for $l \neq k$. Since

$$E\left\{\begin{bmatrix} y_k \\ z_k \end{bmatrix}[y_k \quad z_k]\right\} = \begin{bmatrix} H'PH & H'PH \\ H'PH & H'PH + R \end{bmatrix}$$

this means that $\hat{y}_{k/k}^e = H'PH(H'PH + R)^{-1}z_k$. The associated estimation error is

$$E[(y_k - \hat{y}_{k/k}^e)^2] = H'PH - \frac{(H'PH)^2}{H'PH + R}$$

This error is underbounded by $H'\bar{\Sigma}H$ in view of the optimality of $\hat{y}_{k/k}$:

$$H'\bar{\Sigma}H \leq H'PH - \frac{(H'PH)^2}{H'PH + R} = \frac{(H'PH)R}{H'PH + R}$$

Strict inequality will normally hold. Equation (5.16) follows by simple manipulation.

Main Points of the Section

One can find a recursive equation for $\hat{x}_{k/k}$ and a filter producing $\hat{x}_{k/k}$ which, when drawn in a feedback arrangement, has the forward part driven by the innovations. One can also find recursive equations for obtaining $\hat{x}_{k/k-1}$ followed by $\hat{x}_{k/k}$, followed by $\hat{x}_{k+1/k}$, etc. The error covariance matrix $\Sigma_{k/k}$ can be expressed in terms of $\Sigma_{k/k-1}$, and $\Sigma_{k+1/k}$ can be expressed in terms of $\Sigma_{k/k}$. Stability properties of the $\hat{x}_{k/k}$ filter are the same as those for the $\hat{x}_{k/k-1}$ filter.

There is a signal-to-noise ratio improvement obtained in using the $\hat{y}_{k/k}$ filter.

The most important of the particular equations are as follows:

MEASUREMENT UPDATE:

$$\hat{x}_{k+1/k+1} = \hat{x}_{k+1/k} + \Sigma_{k+1/k}H'_{k+1}(H'_{k+1}\Sigma_{k+1/k}H_{k+1} + R_{k+1})^{-1}$$
$$\times (z_{k+1} - H'_{k+1}\hat{x}_{k+1/k})$$
$$\Sigma_{k+1/k+1} = \Sigma_{k+1/k} - \Sigma_{k+1/k}H_{k+1}(H'_{k+1}\Sigma_{k+1/k}H_{k+1} + R_{k+1})^{-1}H'_{k+1}\Sigma_{k+1/k}$$

TIME UPDATE:

$$\hat{x}_{k+1/k} = (F_k - G_kS_kR_k^{-1}H'_k)\hat{x}_{k/k} + \Gamma_ku_k + G_kS_kR_k^{-1}z_k$$
$$\Sigma_{k+1/k} = (F_k - G_kS_kR_k^{-1}H'_k)\Sigma_{k/k}(F_k - G_kS_kR_k^{-1}H'_k)' + G_k(Q_k - S_kR_k^{-1}S'_k)G'_k$$

TIME UPDATE WITH $S_k = 0$:

$$\hat{x}_{k+1/k} = F_k\hat{x}_{k/k} + \Gamma_ku_k$$
$$\Sigma_{k+1/k} = F_k\Sigma_{k/k}F'_k + G_kQ_kG'_k$$

COMBINED UPDATE WITH $S_k = 0$ FOR FILTERED STATE:

$$\hat{x}_{k+1/k+1} = F_k\hat{x}_{k/k} + \Gamma_ku_k + L_{k+1}(z_{k+1} - H'_{k+1}F_k\hat{x}_{k/k} - H'_{k+1}\Gamma_ku_k)$$
$$L_{k+1} = \Sigma_{k+1/k}H_{k+1}(H'_{k+1}\Sigma_{k+1/k}H_{k+1} + R_{k+1})^{-1}$$

Problem 5.1. Consider the one-dimensional problem in which $F_k = a, |a| < 1$, $G_k = H_k = R_k = Q_k = 1$, $S_k = 0$. Find the limiting value of $\Sigma_{k/k}$.

Problem 5.2. Carry out an alternative derivation of the material of this section using known results for the evolution of $\bar{x}_{k/k-1}$ and $\Sigma_{k/k-1}$ along the following lines. Change the time scale by halving the time interval, and with superscript hat denoting quantities measured after changing the time scale, arrange that

$$\bar{x}_{2k} = \bar{x}_{2k+1} = x_k$$

$$\bar{w}_{2k} = w_k \qquad \bar{w}_{2k+1} = 0$$

$$\bar{z}_{2k} = z_k$$

$$\bar{z}_{2k+1} \equiv 0$$

This means that $\bar{F}_{2k} = F_k$, $\bar{G}_{2k} = G_k$, $\bar{F}_{2k+1} = I$, $\bar{G}_{2k+1} = 0$, $\bar{H}_{2k} = H_k$, $\bar{H}_{2k+1} = 0$, $\bar{Q}_{2k} = Q_k$, $\bar{R}_{2k} = R_k$, $\bar{S}_{2k} = S_k$, $\bar{Q}_{2k+1} = 0$, $\bar{S}_{2k+1} = 0$ and $\bar{R}_{2k+1} = 0$. Also, one has

$$E[\bar{x}_{2k} | \bar{z}_0, \ldots, \bar{z}_{2k-1}] = E[x_k | \bar{z}_0, \bar{z}_1, \ldots, \bar{z}_{k-1}]$$

and

$$E[\bar{x}_{2k+1} | \bar{z}_0, \ldots, \bar{z}_{2k}] = E[x_k | \bar{z}_0, \bar{z}_1, \ldots, \bar{z}_k]$$

with relations also being available for the covariances. The procedure converts the task of finding $\bar{x}_{k/k}$ and $\Sigma_{k/k}$ to one of finding one-step prediction quantities.

Problem 5.3. With notation as in the section, the system matrices associated with the one-step predictor and true filter are

$$F_1 = F - F\bar{\Sigma}H(H'\bar{\Sigma}H + R)^{-1}H' - GS(H'\bar{\Sigma}H + R)^{-1}H'$$

and

$$F_2 = [I - \bar{\Sigma}H(H'\bar{\Sigma}H + R)^{-1}H'](F - GSR^{-1}H').$$

Using the fact that $\{\lambda_i(AB)\} = \{\lambda_i(BA)\}$ for square A, B show that $\{\lambda_i(F_1)\} = \{\lambda_i(F_2)\}$.

Problem 5.4. (Signal-to-Noise Ratio Improvement in the Vector Case). Take $R = I$. Show that

$$H'PH \leq (H'PH)^{1/2}(H'\bar{\Sigma}H)^{-1}(H'PH)^{1/2} - I$$

Generalize to the case of arbitrary positive definite R. Interpret the inequalities obtained. (Note that scalar inequalities can be obtained by taking the trace; it can then be helpful to use the fact that trace AB = trace BA.)

Problem 5.5. (Smoothing Formula). The true filtered estimate is related to the one-step prediction estimate by

$$\hat{x}_{k+1/k+1} = \hat{x}_{k+1/k} + \Sigma_{k+1/k}H_{k+1}(H'_{k+1}\Sigma_{k+1/k}H_{k+1} + R_{k+1})^{-1}\tilde{z}_k$$

Establish the smoothing generalization:

$$\hat{x}_{k+1/k+N} = \hat{x}_{k+1/k} + \Sigma_{k+1/k}\left[\sum_{i=k+1}^{k+N} \bar{\Phi}'(i, k+1)H_i(H'_i\Sigma_{i/i-1}H_i + R_i)^{-1}\tilde{z}_i\right]$$

where $\bar{\Phi}$ is the transition matrix such that $\bar{\Phi}(i+1, i) = F_i - K_iH'_i$. Do this by first establishing that

$$\hat{x}_{k+1/k+N} = \sum_{i=k+1}^{k+N} \text{cov}\,(x_{k+1}, \tilde{z}_i)[\text{cov}\,(\tilde{z}_i, \tilde{z}_i)]^{-1}\tilde{z}_i + \hat{x}_{k+1/k}$$

Observe then that cov $(x_{k+1}, \tilde{z}_i) = E[\tilde{x}_{k+1}\tilde{x}'_i]H_i$, and that $E(\tilde{x}_{k+1}\tilde{x}'_i)$ can be obtained from

$$\tilde{x}_{k+1} = (F_k - K_kH'_k)\tilde{x}_k + G_kw_k - K_kv_k$$

5.6 INVERSE PROBLEMS; WHEN IS A FILTER OPTIMAL?

Suppose a discrete time process z_k is being filtered by some linear system. We consider in this section what tests can be executed on the input and output of the filter to check whether or not the filter is optimal. Such tests are valuable in checking whether a filter design lives up to expectations when it is actually implemented.

A general feature of all the tests is that they involve obtaining certain first and second order statistics. To do this in practice generally requires that the various processes be stationary and ergodic; since time averages then equal ensemble averages, the statistics can readily be obtained, at least approximately. Throughout this section, therefore, we shall assume for the most part that *processes are stationary and ergodic* and the various linear systems are time invariant. We shall also assume that *all processes are zero mean*.

As a general rule, we shall present results as if all processes are gaussian. Should this not be the case, the results hold if the estimates are interpreted as being constrained to be linear.

The results fall into three categories. The first group contains those involving signal estimation and makes no use of finite dimensionality. The second group still involves signal estimation, but imposes a finite-dimensionality constraint, while the third group relates to state estimation.

Ergodicity is usually easily checked in the case of gaussian processes associated with finite-dimensional signal models (see Prob. 6.3).

Signal Estimation

We suppose that

$$z_k = y_k + v_k \tag{6.1}$$

with $\{y_k\}$ and $\{v_k\}$ jointly gaussian signal and noise processes. We suppose that $E[y_k v_l'] = 0$ for $l \geq k$. (This will be the case if a model of the form considered in the last two sections is applicable. Often of course, one has $E[y_k v_l'] = 0$ for all k, l.) Also, v_k is assumed white.

It is then clear that $\hat{y}_{k/k-1} = \hat{z}_{k/k-1}$ and, as we know, $\{\tilde{z}_k = z_k - \hat{z}_{k/k-1}\}$ is a white process. The converse question then arises: Suppose we have a process $\{q_k\}$ such that q_k is Z_{k-1}-measurable, i.e., q_k is some function of z_l for $l \leq k - 1$. Suppose also that $\{z_k - q_k\}$ is zero mean and white. Must $q_k = \hat{z}_{k/k-1}$?

Figure 5.6-1 represents the situation in mind. One can conceive of a finite initial time k_0, with $\mathfrak{F}$ containing a unit delay, and $q_0 = 0$. Then $q_1 = \mathfrak{F}(z_0)$, $q_2 = \mathfrak{F}(z_0, q_1 - z_1) = \mathfrak{F}(z_0, \mathfrak{F}(z_0) - z_1)$, etc., and evidently q_k is

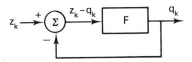

Fig. 5.6-1   An arbitrary arrangement, with $\mathfrak{F}$ injecting a unit delay.

Z_{k-1}-measurable. Alternatively, one can conceive of an initial time in the infinitely remote past provided the closed-loop arrangement has the right stability properties. Now an important aspect of this arrangement is that not only is q_k a Z_{k-1}-measurable quantity, but also $\{z_k\}$ is causally recoverable from the sequence $\{z_k - q_k\}$. This is because q_k is recoverable from $z_l - q_l$ for $l < k$, and $z_k = (z_k - q_k) + q_k$, provided that $\mathfrak{F}$ has the right stability properties in case the initial time is in the infinitely remote past. Put another away, the arrangement of Fig. 5.6.2 recovers $\{z_k\}$ causally from $\{z_k - q_k\}$. All this leads to:

THEOREM 6.1. Consider the situation depicted in Fig. 5.6-1, where $\{z_k\}$ is a zero mean, gaussian sequence, $\{q_k\}$ is Z_{k-1}-measurable, and z_k is recoverable from $\{z_l - q_l, l \le k\}$. Then $q_k = \hat{z}_{k/k-1}$ if and only if $\{z_k - q_k\}$ is zero mean and white.

Proof: The "only if" part was established earlier. To prove the "if" part, proceed as follows. Because q_k is Z_{k-1}-measurable, the sequence $\{z_k - q_k\}$ is causally dependent on $\{z_k\}$. Conversely, the sequence $\{z_k\}$ is causally dependent on $\{z_k - q_k\}$. Hence for an arbitrary random variable w,

$$E[w \,|\, Z_{k-1}] = E[w \,|\, z_l - q_l, l < k]$$

Now take $w = z_k - q_k$. Then

$$0 = E[z_k - q_k \,|\, z_l - q_l, l < k] = E[z_k - q_k \,|\, Z_{k-1}]$$

the first equality stemming from the assumptions on the $\{z_k - q_k\}$. Thus

$$E[z_k \,|\, Z_{k-1}] = E[q_k \,|\, Z_{k-1}] = q_k$$

in view of the fact that q_k is Z_{k-1}-measurable.

Note that the linearity of $\mathfrak{F}$ has not been used above, though of course one cannot have $q_k = \hat{z}_{k/k-1}$ for gaussian $\{z_k\}$ without linear $\mathfrak{F}$. As remarked in the introduction, for nongaussian $\{z_k\}$, one can constrain $\mathfrak{F}$ to be linear

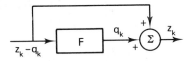

Fig. 5.6-2   The sequence $\{z_k\}$ is causally obtainable from $\{z_k - q_k\}$.

and work with linear minimum variance estimates. One can also obtain a result with nonlinear $\mathfrak{F}$ (see Prob. 6.1).

Checking the whiteness of a stationary ergodic sequence $\{p_k\}$ in theory requires the evaluation of $E[p_{k+l}p_k]$ for all l; the situation is even worse for nonstationary sequences on two counts: ensemble rather than time averaging is needed, and one must evaluate $E[p_k p_l]$ for all k and l.

By assuming a finite dimensionality property, we can get a major simplification.

Signal Estimation Using Finite-dimensional Signal Model and Filter

Let us suppose that in (6.1), $\{y_k\}$ is the output of a linear finite-dimensional signal model, and that $\mathfrak{F}$ in Fig. 5.6-1 is linear and finite dimensional. We shall also assume that the signal model and closed-loop system in Fig. 5.6-1 are time invariant and asymptotically stable, and that $\{z_k\}$ and $\{q_k\}$ are zero mean and stationary. Some relaxation of these conditions is possible (see Prob. 6.2). The first main result is as follows:

THEOREM 6.2. Let the signal model and filter be of state dimension m and n, respectively, and other assumptions be as above. Then $\{z_k - q_k\}$ is white if

$$E\{[z_{k+l} - q_{k+l}][z_k - q_k]'\} = 0$$

for $0 < l \leq m + n$.

*Proof.** Let us first establish that $\{z_k - q_k\}$ can be regarded as the output of an $(m + n)$-dimensional linear system excited by white noise. This is intuitively obvious. The details are as follows. Suppose the signal model is

$$x_{k+1} = Fx_k + Gw_k \qquad z_k = y_k + v_k = H'x_k + v_k \qquad (6.2)$$

with

$$E\left\{ \begin{bmatrix} w_k \\ v_k \end{bmatrix} [w_l \quad v_l'] \right\} = \begin{bmatrix} Q & S \\ S' & R \end{bmatrix} \delta_{kl} \qquad (6.3)$$

Suppose the filter is

$$p_{k+1} = F_1 p_k + G_1 z_k \qquad q_k = H_1' p_k \qquad (6.4)$$

for some F_1, G_1, H_1 and state vector p_k.

Together, we then have

$$\begin{bmatrix} x_{k+1} \\ p_{k+1} \end{bmatrix} = \mathbf{F} \begin{bmatrix} x_k \\ p_k \end{bmatrix} + \begin{bmatrix} G & 0 \\ 0 & I \end{bmatrix} \begin{bmatrix} w_k \\ v_k \end{bmatrix}$$

$$z_k - q_k = \mathbf{H}' \begin{bmatrix} x_k \\ p_k \end{bmatrix} + [0 \quad I] \begin{bmatrix} w_k \\ v_k \end{bmatrix}$$

*The proof may be omitted at a first reading.

where

$$\mathbf{F} = \begin{bmatrix} G & 0 \\ G_1 H' & F_1 \end{bmatrix}, \qquad \mathbf{H}' = [H' \quad -H'_1]$$

Now via the procedures of Chap. 4, one can compute $E\{[z_{k+l} - q_{k+l}][z_k - q_k]'\}$ for all k and l. The general form of this quantity, and this is all we need, is

$$\mathbf{H}'\mathbf{F}^{l-1}\mathbf{K} = E\{[z_{k+l} - q_{k+l}][z_k - q_k]'\} \qquad l > 0$$

with the particular form of $\mathbf{K}$ irrelevant. By assumption

$$0 = \mathbf{H}'\mathbf{F}^{l-1}\mathbf{K} \qquad 0 < l \leq m + n$$

Now $\mathbf{F}$ has dimension $m + n$, so by the Cayley-Hamilton theorem $\mathbf{F}^l$ for arbitrary $l \geq m + n$ is a linear combination of $I, \mathbf{F}, \ldots, \mathbf{F}^{m+n-1}$. Accordingly, $0 = \mathbf{H}'\mathbf{F}^{l-1}\mathbf{K}$ for all $l > 0$, i.e.,

$$E\{[z_{k+l} - q_{k+l}][z_k - q_l]'\} = 0$$

for all $l > 0$. Equivalently, $\{z_k - q_k\}$ is white.

When z_k is a vector, certain minor improvements to the above result are possible. This is because rank $[\mathbf{H}' \quad \mathbf{H}'\mathbf{F} \quad \ldots \quad \mathbf{H}'\mathbf{F}^{m+n-1}]$ can sometimes equal rank $[\mathbf{H}' \quad \mathbf{H}'\mathbf{F} \quad \ldots \quad \mathbf{H}'\mathbf{F}^{j-1}]$ for $j < m + n$. One significant improvement, corresponding to having $j = n$, arises if we demand that the filter have the form

$$p_{k+1} = (F - G_1 H')p_k + G_1 z_k \qquad q_k = H'p_k \tag{6.5}$$

Thus the input gain of the filter may be incorrect, but nothing else. In this case we have the following result:

THEOREM 6.3. Under the same assumptions as Theorem 6.2, and with the signal model and filter as given in (6.2) and (6.5), $\{z_k - q_k\}$ is white if

$$E\{[z_{k+l} - q_{k+l}][z_k - q_k]'\} = 0$$

for $0 < l \leq n$.

Proof.* One can work with the quantities $\mathbf{H}$ and $\mathbf{F}$ used above, or proceed possibly more quickly as follows. The signal model and filter equations give

$$x_{k+1} - p_{k+1} = (F - G_1 H')(x_k - p_k) + [G \quad -G_1]\begin{bmatrix} w_k \\ v_k \end{bmatrix}$$

$$z_k - q_k = H'(x_k - p_k) + [0 \quad I]\begin{bmatrix} w_k \\ v_k \end{bmatrix}$$

so that for some $\mathbf{K}$,

$$H'(F - G_1 H')^{l-1}\mathbf{K} = E\{[z_{k+l} - q_{k+l}][z_k - q_k]'\} \qquad l > 0$$

---

*The proof may be omitted at a first reading.

The argument now proceeds as for Theorem 6.2, noting that $F - G_1 H'$ has dimension n.

State Estimation

Let us retain the assumption that $\{z_k\}$ is the measurement process associated with a finite-dimensional system like (6.2). Suppose also that we have a linear filter $\mathcal{G}$ with input $\{z_k\}$ and output a sequence $\{p_k\}$ purporting to be the sequence $\{\hat{x}_{k/k-1}\}$. We consider now how one might check this property.

First, we must pin down a coordinate basis—otherwise the state estimation problem is ill defined. One way to do this is to assume that F and H are known. A second assumption, which is almost essential to make, is that $[F, H]$ is completely observable. (Roughly, this is because this condition is required for $\{\tilde{z}_k\}$ and $\{\hat{z}_{k/k-1}\}$ to determine uniquely the state trajectory $\{\hat{x}_{k/k-1}\}$ of the filter; the point will be explored below.)

Now since H is assumed known, we can form $\{q_k\} = \{H'p_k\}$ and check whether $\{z_k - q_k\}$ is white. This is clearly necessary for $\{p_k\}$ to be identical with $\{\hat{x}_{k/k-1}\}$. Is it sufficient? With the complete observability assumption, the answer is yes. The transfer function matrix of the system which when driven by $\{\tilde{z}_k\}$ produces $\{\hat{z}_{k/k-1}\}$ at its output is $H'[zI - F]^{-1}K$, with K the usual Kalman gain. If there were two matrices K_i such that

$$H'[zI - F]^{-1}K_1 = H'[zI - F]^{-1}K_2$$

we would not know whether the state estimate would evolve as

$$\hat{x}_{k+1/k} = F\hat{x}_{k/k-1} + K_1\tilde{z}_k \quad \text{or} \quad \hat{x}_{k+1/k} = F\hat{x}_{k/k-1} + K_2\tilde{z}_k$$

On the other hand, if K is uniquely specified by $H'[zI - F]^{-1}K$, then the state estimate equation is uniquely determined. That K is in fact uniquely specified follows by the complete observability assumption; $H'F^jK_1 = H'F^jK_2$ for all i implies $H'F^j(K_1 - K_2) = 0$ for all j and, thus, $K_1 = K_2$. We have thus shown the following:

THEOREM 6.4. Suppose the signal model is of the form (6.2) and a filter is of the form (6.5). If $[F, H]$ is completely observable and $\{q_k\} = \{\hat{z}_{k/k-1}\}$, then $\{p_k\} = \{\hat{x}_{k/k-1}\}$.

Of course, one can check that $\{q_k\} = \{\hat{z}_{k/k-1}\}$ by using the test implicit in Theorem 6.3.

Main Points of the Section

A sufficient, as well as necessary, condition for q_k in Fig. 5.6-1 to be $\hat{z}_{k/k-1}$ is that the sequence $z_k - q_k$ be zero mean and white. In case $\{z_k\}$ and $\{q_k\}$ are stationary and the closed-loop arrangement in Fig. 5.6-1 is asymptotically stable, this can be checked via time averaging. If $\{z_k\}$ is the output of a finite-dimensional system and $\mathfrak{F}$ is finite dimensional, the testing is greatly eased: only a limited number of cross correlations have to be checked to be zero. In case $[F, H]$ is fixed and completely observable and $\mathfrak{F}$ has a standard structure, $q_k = \hat{z}_{k/k-1}$ implies the state of $\mathfrak{F}$ is $\hat{x}_{k/k-1}$.

Problem 6.1. Show that the "if" part of Theorem 6.1 holds for nongaussian $\{z_k\}$ provided one assumes that $E[z_k - q_k | z_l - q_l, l < k] = 0$.

Problem 6.2. The innovations sequence $\{\tilde{z}_k\}$ and one-step prediction estimate $\hat{z}_{k/k-1}$ satisfy $E\{\hat{z}_{k/k-1} \tilde{z}_l'\} = 0$ for $l \geq k$. Establish this. This property suggests the following conjecture. Let $\{q_k\}$ be such that q_k is Z_{k-1}-measurable, with z_k recoverable from $\{z_l - q_l, l \leq k\}$. Suppose that $E\{q_k(z_l - q_l)'\} = 0$ for $l \geq k$. Then $q_k = \hat{z}_{k/k-1}$. Show that this conjecture is false.

Problem 6.3. Let $\{a_k\}$ be a stationary scalar gaussian process with covariance R_k. Then $\{a_k\}$ is ergodic if $\sum_{k=-\infty}^{+\infty} |R_k| < \infty$ (see Appendix A). Show that in case $\{a_k\}$ is the output of a linear, time-invariant, finite-dimensional system that is asymptotically stable and is excited by white noise from time $k_0 = -\infty$, then $\{a_k\}$ is ergodic.

REFERENCES

[1] SORENSON, H. W., "Least-squares Estimation: from Gauss to Kalman," *IEEE Spectrum*, Vol. 7, No. 7, July 1970, pp. 63–68.

[2] GAUSS, K. F., *Theory of the Motion of the Heavenly Bodies Moving about the Sun in Conic Sections*, Dover Publications, Inc., New York, 1963.

[3] FISHER, R. A., "On an Absolute Criterion for Fitting Frequency Curves," *Messenger of Math.*, Vol. 41, 1912, p. 155.

[4] WIENER, N., *Extrapolation, Interpolation and Smoothing of Stationary Time Series*, John Wiley & Sons, Inc., New York, 1949.

[5] KOLMOGOROV, A. N., "Interpolation and Extrapolation of Stationary Random Sequences," *Bull. de l'académie des sciences de U.S.S.R.*, Ser. Math., Vol. 5, 1941, pp. 3–14.

[6] CARLTON, A. G., and J. W. FOLLIN, *Recent Developments in Fixed and Adaptive Filtering*, Nato Advisory Group for Aerospace Research and Development, AGARDograph, No. 21, 1956.

[7] SWERLING, P., "A Proposed Stagewise Differential Correction Procedure for Satellite Tracking and Prediction," *J. Astronaut. Sci.*, Vol. 6, 1959, pp. 46–59.

[8] KALMAN, R. E., "A New Approach to Linear Filtering and Prediction Problems," *J. Basic Eng., Trans. ASME*, Series D, Vol. 82, No. 1, March 1960, pp. 35–45.

[9] KALMAN, R. E., "New Methods in Wiener Filtering Theory," in *Proc. 1st Symp. of Engineering Applications of Random Function Theory and Probability*, John Wiley & Sons, Inc., New York, 1963, pp. 270–388.

[10] ASTRÖM, K. J., *Introduction to Stochastic Control Theory*, Academic Press, Inc., New York, 1970.

[11] SON, L. H., and B. D. O. ANDERSON, "Design of Kalman Filters Using Signal Model Output Statistics," *Proc. IEE*, Vol. 120, No. 2, February 1973, pp. 312–318.

[12] WONG, E., *Stochastic Processes in Information and Dynamical Systems*, McGraw-Hill Book Company, New York, 1971.

[13] WONG, E., "Recent Progress in Stochastic Processes—A Survey," *IEEE Trans. Inform. Theory*, Vol. IT-19, No. 3, May 1973, pp. 262–274.

[14] KAILATH, T., "Sigma-Fields, Stopping Times, Martingales and all That (Almost Surely)," *Proc. IEEE*, to appear.

[15] FROST, P. A., and T. KAILATH, "An Innovations Approach to Least-squares Estimation: Part III, Nonlinear Estimation in Gaussian White Noise," *IEEE Trans. Automatic Control*, Vol. AC-16, No. 3, June 1971, pp. 217–226.

[16] BALAKRISHNAN, A. V., "A Martingale Approach to Linear Recursive State Estimation," *SIAM J. Control*, Vol. 10, No. 4, November 1972, pp. 754–766.

[17] MEYER, P. A., *Martingales and Stochastic Integrals I, Lecture Notes in Mathematics*, Vol. 284, Springer-Verlag OHG, Berlin, 1972.

[18] NEVEU, J., *Discrete Parameter Martingales*, North-Holland American Elsevier, Oxford and Amsterdam, 1975.

[19] STEINWAY, W. J., and J. L. MELSA, "Discrete Linear Estimation for Previous Stage Noise Correlation," *Automatica*, Vol. 7, No. 3, May 1971, pp. 389–391.

CHAPTER 6

COMPUTATIONAL ASPECTS

6.1 SIGNAL MODEL ERRORS, FILTER DIVERGENCE, AND DATA SATURATION

In designing Kalman filters, two important types of computational questions arise. First, what is the nature of the errors which can be encountered, and what is their effect on the performance of the filter? Secondly, how may one minimize the computational burden of design? Of course, the two questions are not entirely independent, since, for example, procedures involving a small number of computations may be procedures which offer poor error performance. In this chapter, we sketch some of the ideas that are useful for dealing with these questions—a complete study would probably run to hundreds of pages, and is therefore out of the question.

The most obvious types of error are those in which incorrect values are assumed for the system matrices or noise covariance matrices. However, many others can be envisaged. Linearization, neglect of certain system modes, neglect of a colored component in the noise, and neglect of biases, whether deliberate or unwitting, will all give rise to errors. Modeling errors aside, round-off errors in computation can also create problems.

In the remainder of this section, we discuss how one may analyze the effect of certain errors, and discuss one consequence, that of divergence, of some types of error. Brief mention is made of some techniques for eliminating these problems. Then in the next section, one of these techniques, exponential data weighting, is discussed in a little greater detail.

In later sections, methods for streamlining computations and also for avoiding the effects of computational errors are considered. In particular, a derivation is given of the information filter, and the concept of sequential data processing is developed. Square root filtering algorithms are presented, and simplified suboptimal filters for the high measurement noise case are studied. Finally, alternative algorithms for the time-invariant signal model case are derived.

Error Analysis

In this subsection, we illustrate how one can analyze, in a particular instance, the effects of a particular class of modeling errors.

We shall assume that the only errors which occur are in the values of the system and covariance matrices $\{F_k, G_k, H_k, Q_k, R_k, P_0\}$, in the mean of the initial state $\bar{x}_0$, and in the system input. In some cases, it can be useful to regard the system input as a time-varying bias term, inserted as a device to take account of linearization, neglect of modes, and the like; in this case, almost certainly the actual value and design value will be different.

Let us adopt the notation F_k^a, G_k^a, etc., to denote actual quantities and F_k^d, G_k^d, etc., to denote quantities used for design. The indices attaching to the error covariance, $\Sigma_{k/k-1}$ need, however, to be *three* in number; $\Sigma_{k/k-1}^a$ denotes the error covariance which would be obtained were the actual quantities to be used in the design equations; $\Sigma_{k/k-1}^d$ denotes the error covariance predicted by the design equations. Finally, $\Sigma_{k/k-1}^p$ denotes the performance obtained by using the filter computed by the *design* equations on the *actual* signal model, defined by F_k^a, G_k^a, etc, so that

$$\Sigma_{k/k-1}^p = E\{[x_k^a - \hat{x}_{k/k-1}^d][x_k^a - \hat{x}_{k/k-1}^d]'\}$$

Notice that $\Sigma_{k/k-1}^p$ is not the same as $\Sigma_{k/k-1}^d$.

The first matter of interest is to show how to compute $\Sigma_{k/k-1}^p$. The calculation is valid for all types of errors. First, the design quantities are used to evaluate $\Sigma_{k/k-1}^d$ according to the usual equations. (We assume for convenience that input and measurement noise are independent.)

$$\Sigma_{k+1/k}^d = F_k^d \Sigma_{k/k}^d F_k^{d\,'} + G_k^d Q_k^d G_k^{d\,'} \tag{1.1}$$

$$\Sigma_{k/k}^d = \Sigma_{k/k-1}^d - \Sigma_{k/k-1}^d H_k^d (H_k^{d\,'} \Sigma_{k/k-1}^d H_k^d + R_k^d)^{-1} H_k^{d\,'} \Sigma_{k/k-1}^d \tag{1.2}$$

Then the filter equations are

$$\hat{x}_{k+1/k}^d = F_k^d \hat{x}_{k/k-1}^d + K_k^d (z_k - H_k^{d\,'} \hat{x}_{k/k-1}^d) + u_k^d \tag{1.3}$$

Here, z_k is the measurement sequence, u_k^d is the design value of the system input at time k (multiplied by the coupling matrix), and K_k^d is the gain matrix

$$K_k^d = F_k^d \Sigma_{k/k-1}^d H_k^d (H_k^{d'} \Sigma_{k/k-1}^d H_k^d + R_k^d)^{-1} \tag{1.4}$$

The associated error when the signal model is defined by the actual quantities is $(x_k^a - \hat{x}_{k/k-1}^d)$. The evaluation of the mean-square-error performance $\Sigma_{k/k-1}^p$ is best achieved by considering the signal model augmented with the filter (1.3) as

$$\begin{bmatrix} x_{k+1}^a \\ \hat{x}_{k+1/k}^d \end{bmatrix} = \begin{bmatrix} F_k^a & 0 \\ K_k^d H_k^{a'} & F_k^d - K_k^d H_k^{d'} \end{bmatrix} \begin{bmatrix} x_k^a \\ \hat{x}_{k/k-1}^d \end{bmatrix} + \begin{bmatrix} G_k^a & 0 \\ 0 & K_k^d \end{bmatrix} \begin{bmatrix} w_k \\ v_k \end{bmatrix} + \begin{bmatrix} u_k^a \\ u_k^d \end{bmatrix}$$

This equation has the form

$$x_{k+1} = \mathscr{F}_k x_k + \mathscr{G}_k w_k + u_k \tag{1.5}$$

where w_k is a white, zero mean, gaussian sequence of known covariance, u_k is a time-varying input, and x_0 has known statistics.

From (1.5), the evolution of the mean and covariance of x_k can be obtained, and thence the mean and covariance of

$$x_k^a - \hat{x}_{k/k-1}^d = [I \quad -I] x_k$$

In general, $x_k^a - \hat{x}_{k/k-1}^d$ will not have zero mean, so the correlation matrix $\Sigma_{k/k-1}^p$ will not be the same as the covariance matrix!

At this stage, several points should be made.

1. The important thing here is the procedure for obtaining a result. The notion of tying together an actual signal model and a designed filter in a single equation set may apply to many situations other than that considered.

2. One major use of the above type of analysis is in sensitivity studies. For example, it may be known that a given system parameter fluctuates slowly 10% around its nominal value. One can then compute the effect on filter performance of this variation, when the filter is designed using the nominal value.

3. A second major use lies in drawing useful qualitative conclusions, applicable to situations in which errors are described qualitatively but not quantitatively. Examples are given below.

The analysis presented in outline form above is given more fully in [1]. Among other work on errors arising from incorrect modeling, we note [2–6], some of which contain results of simulations; reference [6] also includes equations for sensitivity coefficients.

Qualitative Conclusions from Error Analysis

The most obvious conclusion is that $\Sigma_{k+1/k}^p \geq \Sigma_{k+1/k}^a$, the inequality holding for all classes of errors. (Why is this obvious?)

Next are results generally associated with the names of Heffes and Nishimura. Suppose that the only errors are in Q_k, R_k, and P_0, with the following inequalities holding for all k.

$$Q_k^d \geq Q_k^a, \qquad R_k^d \geq R_k^a, \qquad P_0^d \geq P_0^a \qquad (1.6)$$

Equation (1.6) implies that we are assuming more input noise, more measurement noise, and more initial state uncertainty than is actually there. One might then imagine that this would lead to a conservative filter design in some sense. This is indeed what we find: the design error covariance $\Sigma_{k/k-1}^d$ and the error covariance $\Sigma_{k/k-1}^p$ resulting from using the filter designed with Q^d, etc., on the actual signal model, stand in the relation

$$\Sigma_{k/k-1}^d \geq \Sigma_{k/k-1}^p \qquad (1.7)$$

(Why is $\Sigma_{k/k-1}^p$ a covariance matrix here?) The usefulness of this result (a proof of which is called for in the problems) is as follows. Suppose one simply does not know accurately the noise covariance of the input or output, but one does know an upper bound. Then one can design assuming the noise covariance is at its upper bound, with the result that the performance of the resulting filter as measured by $\Sigma_{k/k-1}^p$ will be upper-bounded by the design performance $\Sigma_{k/k-1}^d$. In some sense a worst case design results. If the various side conditions are fulfilled which ensure that $\Sigma_{k/k-1}^d$ is bounded for all k, then $\Sigma_{k/k-1}^p$ will also be bounded for all k.

A third qualitative result (see [1]) follows from assuming that errors are possible in P_0, Q_k, R_k, and the bias term u_k, but in no other terms. The conclusion is that if the side conditions are fulfilled which ensure exponential asymptotic stability of the filter *and* if the error $u_k^a - u_k^d$ is bounded, then $\Sigma_{k/k-1}^p$ is bounded. Note that if u_k^a is known to be bounded, taking $u_k^d \equiv 0$ ensures the difference is bounded. However, a difficulty arises if u_k^a is unbounded.

A fourth qualitative result extends the above [1]; if errors in any parameter are possible, then to guarantee a bound on $\Sigma_{k/k-1}^p$, one almost always needs exponential asymptotic stability of the actual system, i.e., $\|\Phi^a(k + l, k)\| \leq \rho^l$ for some $\rho < 1$ and all k, l. (See also Prob. 1.3.) This is a severe constraint (ruling out as it does systems with stable, but not asymptotically stable, modes), and at the same time it is one that is most important to bear in mind in view of its wide applicability.

Divergence

Divergence is the name given to a class of error phenomena. Filter divergence is said to exist when the design error covariance $\Sigma_{k/k-1}^d$ remains bounded while the error performance matrix $\Sigma_{k/k-1}^p$ becomes very large relative to $\Sigma_{k/k-1}^d$ or in fact is unbounded.

Evidently, divergence is a qualitative concept, and for this reason, it is hard to pin down precise conditions which will cause it, although a number have been found (see, e.g., [7]). Divergence is typically, but not always, associated with one or more of: low or zero input noise, signal models which are not asymptotically stable, and bias errors. High input noise, signal models with a high degree of stability, and absence of bias errors will tend to eliminate divergence. Again, divergence seems to arise more from modeling error than computational errors. We illustrate some of these points in the example below.

EXAMPLE 1.1. Suppose that the design equations used are

$$x_{k+1}^d = x_k^d$$

$$z_k = x_k^d + v_k^d$$

with $E[v_k^d v_l^d] = \delta(k - l)$, $E\{[x^d(0)]^2\} = 1$, $E[x^d(0)] = 0$, and $x^d(0)$ and v_k^d are independent. One obtains $\Sigma_{k+1/k}^d = (k + 1)^{-1}$. Suppose that the actual system state equation is

$$x_{k+1}^a = x_k^a + w_k^a$$

with $E[w_k^a w_l^a] = \epsilon\delta(k - l)$, $E[w_k^a] = 0$, and the usual independence assumptions holding. The measurement equation agrees with the design equation. One can verify that the actual error variance diverges as fast as k. Similarly, in case $x_{k+1}^a = x_k^a + u_k^a$, with $u_k^a = \epsilon$ for all k, divergence occurs.

Two questions at once present themselves. How, in an operating filter, can one check whether divergence is occurring, and how may it be eliminated? The prime indicator of the presence of divergence is the inconsistency of the design statistics of the innovations sequence [zero mean, whiteness, and a certain covariance, $(R_k^d + H_k^{d\prime}\Sigma_{k/k-1}^d H_k^d)$] and the actual statistics encountered in operation. (In the event that all processes are stationary, this inconsistency will be easier to see.) A second pointer to divergence—not always encountered, and not a guaranteed indicator—is a situation in which the filter gain matrix (or, what is more or less equivalent, the design error covariance $\Sigma_{k/k-1}^d$) tends to zero as k approaches infinity. Less and less weighting is given to new measurements as time evolves (the old data is said to "saturate" the filter), and the filter state may tend asymptotically to some value. This value may be quite different from the state of the signal model; in this case the filter is said to have learned the wrong state.

Advance warning of the likelihood of divergence occurs under the conditions stated prior to Example 1.1. The reader should ponder why divergence is likely under the conditions stated.

A number of approaches to cope with the divergence problem have been suggested. In fact, they tend to be useful on any occasion when there is a significant discrepancy between design calculations and performance. Among the approaches, we note the following.

1. The input noise used in design is increased. The general thinking is that the increased noise may somehow make up for all the modeling errors and tend to promote exponential stability of the filter. A discussion of attempts at defining the amount of increase systematically can be found in [1]; the success of these attempts is viewed by [1] pessimistically.

2. As a significant refinement on an a priori adjustment of the input noise variance, one can adjust it on line, using measured innovations as the tool for the adaption of the noise variance. This is discussed in [1].

3. One can overweight the most recent data relative to the old data. One common approach involves using a finite memory, basing estimation at any instant of time on measurement data extending over an interval of fixed finite length into the past [1]. A second approach involves exponential weighting [4, 8, 9] of the measurement data. Exponential weighting is discussed in greater detail in the next section. The general thinking is that this will prevent old data from saturating the filter; the filter gain will not tend to zero. Further insight will be provided in the next section.

4. A somewhat crude technique is simply to put an ad hoc lower bound on the size of the gain matrix. Any design value smaller than the bound is not used, the bounding value being used in its place.

The easiest techniques would seem to be increase of the noise variance and use of exponential data weighting.

Main Points of the Section

Various types of modeling and computational errors can cause the performance of a Kalman filter to differ from the design value. Given sufficient data, the performance degradation can be computed. Qualitative conclusions can also be derived from an error analysis. Asymptotic stability of the signal model is almost always needed to guarantee satisfactory performance in the presence of modeling errors. A particular error phenomenon is that of divergence, for which there are available various indicators of its existence and techniques for its removal.

Problem 1.1. Suppose the conditions of (1.6) are in force, and that the only difference between actual and design parameters lies in Q_k, R_k, and P_0. Show that

$$\Sigma^p_{k+1/k} = (F_k - K^d_k H'_k)\Sigma^p_{k/k-1}(F_k - K^d_k H'_k)' + G_k Q^a_k G'_k + K^d_k R^a_k K^{d'}_k$$

With the aid of a similar equation for $\Sigma^d_{k+1/k}$, deduce that for all k, $\Sigma^p_{k+1/k} \leq \Sigma^d_{k+1/k}$.

Problem 1.2. Given a signal model with state vector $[x'_1 \quad x'_2]$, where it is required to estimate only x_1, what assumptions are involved in obtaining an approximate reduced order signal model in the standard form of dimension equal to that of x_1? How would one check whether or not a Kalman filter designed for the reduced order approximate model would perform satisfactorily when applied to the actual signal model? Illustrate by setting up the various equations that one would need to solve.

Problem 1.3. Suppose that $u_k^a, F_k^a, G_k^a, \ldots$ and $u_k^d, F_k^d, G_k^d, \ldots$ are all bounded, that F_k^a is exponentially stable, and that the closed-loop filter matrix $F_k^d - K_k^d H_k^{d'}$ is exponentially stable. Show that $\Sigma_{k+1/k}^p$ is bounded.

Problem 1.4. Compute the quantities $\Sigma_{k+1/k}^p$ in Example 1.1. Check that the filter gain $K_k^d \longrightarrow 0$ as $k \longrightarrow \infty$. Show that if input noise is added in the design equations, divergence will not occur.

Problem 1.5. Extend the error analysis calculations to compute the value of $E[v_k v_k']$, where $v_k = z_k - H_k^d \hat{x}_{k/k-1}$. Specialize first to $H_k^d = H_k^a$, and then to $u_k^a = u_k^d$, $E[x_0^a] = x_0^d$.

Problem 1.6. Situations in which divergence is likely to occur are noted prior to Example 1.1. Justify the claims made.

6.2 EXPONENTIAL DATA WEIGHTING—A FILTER WITH PRESCRIBED DEGREE OF STABILITY

Background Thinking

With usual notation, it is evident that a classical least squares approach to the estimation of an entire system trajectory $x_0, x_1, \ldots, x_N$ given measurements $z_0, z_1, \ldots, z_{N-1}$ would involve the minimization of a function of the type

$$J_N = \tfrac{1}{2}(x_0 - \bar{x}_0)'\tilde{P}_0^{-1}(x_0 - \bar{x}_0)$$
$$+ \tfrac{1}{2}\sum_{k=0}^{N-1}(z_k - H_k'x_k)'\tilde{R}_k^{-1}(z_k - H_k'x_k)$$
$$+ \tfrac{1}{2}\sum_{k=0}^{N-1} w_k'\tilde{Q}_k^{-1}w_k \tag{2.1}$$

subject to the constraints $x_{k+1} = F_k x_k + G_k w_k$. In (2.1), the matrices $\tilde{P}_0^{-1}$, $\tilde{R}_k^{-1}$, and $\tilde{Q}_k^{-1}$ are simply positive definite weighting matrices. Many authors have observed (see [1]) that if $\tilde{P}_0$, $\tilde{R}_k$, and $\tilde{Q}_k$ are identified with the quantities P_0, R_k, and Q_k of the usual signal model, and if the inverses in (2.1) are replaced by pseudo-inverses, then minimization of (2.1) is equivalent to finding the trajectory $x_0, x_1, \ldots, x_N$ maximizing the a posteriori probability density

$$p(x_0, x_1, \ldots, x_N | z_0, z_1, \ldots, z_{N-1})$$

In view of the gaussian nature of this density, it happens to be true that if $x_0^*, x_1^*, \ldots, x_N^*$ is the maximizing trajectory, then $x_i^* = E[x_i \mid z_0, \ldots, z_{N-1}]$ and, in particular, $x_N^* = \hat{x}_{N/N-1}$.

These observations can be used to develop the Kalman filter equations, but this is of little interest to us here. Rather, we use the observations to pin down how one might give greater emphasis to recent data. Since giving greater emphasis to recent data is equivalent to penalizing recent estimation errors more than old ones, classical least squares ideas suggest the way to do this is to increase the weighting matrices in (2.1) for large values of k.

This leads us to our replacing of (2.1) by

$$J_N = \tfrac{1}{2}(x_0 - \bar{x}_0)' P_0^{-1}(x_0 - \bar{x}_0)$$
$$+ \tfrac{1}{2} \sum_{k=0}^{N-1} (z_k - H_k' x_k)' \alpha^{2k} R_k^{-1}(z_k - H_k' x_k)$$
$$+ \tfrac{1}{2} \sum_{k=0}^{N-1} w_k' \alpha^{2k+2} Q_k^{-1} w_k \tag{2.2}$$

where α is some constant greater than 1. (Naturally, other methods of increasing the weighting matrices could be used. However, the resulting filters may not be so convenient, for example, being time varying rather than time invariant, nor may their properties be as easily understood.) In view of the remarks connecting the loss function (2.1) with maximum a posteriori estimation, we see that this idea is equivalent to replacing actual noise variances Q_k^a, R_k^a by design values

$$Q_k^d = \alpha^{-2k-2} Q_k^a \qquad R_k^d = \alpha^{-2k} R_k^a \tag{2.3}$$

for the purposes of obtaining a filter, and leaving other signal model quantities unaltered for design purposes, i.e., $F_k^d = F_k^a$, etc.

Filter Design Equations

It is of course a simple matter to write down filter design equations, giving in particular the values of $\Sigma_{k/k-1}^d$, $\Sigma_{k/k}^d$, and K_k^d. However, it is more convenient to work with quantities

$$\Sigma_{k/k-1}^\alpha = \alpha^{2k} \Sigma_{k/k-1}^d \qquad \Sigma_{k/k}^\alpha = \alpha^{2k} \Sigma_{k/k}^d \tag{2.4}$$

to which physical significance will be given below. From the equations for $\Sigma_{k/k-1}^d$, etc., the following readily follow.

$$\Sigma_{k+1/k}^\alpha = \alpha^2 F_k^a \Sigma_{k/k}^\alpha F_k^{a\prime} + G_k^a Q_k^a G_k^{a\prime} \tag{2.5}$$

$$\Sigma_{k/k}^\alpha = \Sigma_{k/k-1}^\alpha - \Sigma_{k/k-1}^\alpha H_k^a (H_k^{a\prime} \Sigma_{k/k-1}^\alpha H_k^a + R_k^a)^{-1} H_k^{a\prime} \Sigma_{k/k-1}^\alpha \tag{2.6}$$

$$\hat{x}_{k+1/k}^d = F_k^a \hat{x}_{k/k-1}^d + K_k^d (z_k - H_k^{a\prime} \hat{x}_{k/k-1}^d) \tag{2.7}$$

$$K_k^d = F_k^a \Sigma_{k/k-1}^\alpha H_k^a (H_k^{a\prime} \Sigma_{k/k-1}^\alpha H_k^a + R_k^a)^{-1} \tag{2.8}$$

Let us now draw some conclusions. The most important ones relate to the fact that the design procedure is equivalent to one in which input noise is increased, and to the fact that the procedure promotes exponential stability of the filter.

1. It is not difficult to see that the quantities $\Sigma^\alpha_{k/k-1}$, $\Sigma^\alpha_{k/k}$ and the filter (2.7) are the error covariances and the Kalman filter, respectively, for the signal model with $F^\alpha_k = F^a_k$, etc., but with R^α_k and $G^\alpha_k Q^\alpha_k G^{\alpha'}_k$ defined as follows:

$$R^\alpha_k = R^a_k \quad \text{and} \quad G^\alpha_k Q^\alpha_k G^{\alpha'}_k = G^a_k Q^a_k G^{a'}_k + (\alpha^2 - 1) F^a_k \Sigma^\alpha_{k/k} F^{a'}_k$$

Of course $G^\alpha_k Q^\alpha_k G^{\alpha'}_k$ can only be calculated along with the calculations for $\Sigma^\alpha_{k/k}$. There are two consequences. The first is that the filter can be viewed as flowing, not from a technique of weighting more heavily the more recent data, but simply from assuming an increased input noise variance. The amount of the increase is not computable in advance. The second consequence is that, in view of the remarks in the last section flowing from Eq. (1.6), we shall evidently have

$$\Sigma^a_{k/k-1} \leq \Sigma^p_{k/k-1} \leq \Sigma^\alpha_{k/k-1} \tag{2.9}$$

For the case of time-invariant quantities R^a, Q^a, F^a, etc., it is interesting that time-varying quantities R^d_k and Q^d_k [which do not satisfy (1.6)] produce the same effect as time-invariant R^α and $G^\alpha Q^\alpha G^{\alpha'}$ which do satisfy an appropriate modification of (1.6).

2. Again the quantities $\Sigma^\alpha_{k/k-1}$ and $\Sigma^\alpha_{k/k}$ could be derived directly from the alternative design relationships $R^\alpha = R^a$, $Q^\alpha = Q^a$, $H^\alpha = H^a$, $G^\alpha = G^a$, and $F^\alpha = \alpha F^a$. However, for this case, the filter equations are different from (2.7). The homogeneous filter equations are

$$\hat{x}_{k+1} = [F^\alpha_k - \alpha K^d_k H^{a'}_k]\hat{x}_k = \alpha[F^a_k - K^d_k H^{a'}_k]\hat{x}_k$$

where K^d_k is given by (2.8). The point of interest to us here is that asymptotic stability of $\hat{x}_{k+1} = \alpha[F^a_k - K^a_k H^{a'}_k]\hat{x}_k$ guarantees asymptotic stability of $\hat{x}_{k+1} = [F^a_k - K^d_k H^{a'}_k]\hat{x}_k$ with a degree of stability α, or equivalently guarantees the *asymptotic stability of our filter* (2.7) *with a degree of stability* α. An alternative derivation of this result can be achieved by noting that the sufficient conditions which are usually listed to ensure asymptotic stability of the optimal filter for the actual signal model (see an earlier chapter), upon a small amount of manipulation not displayed here, also ensure the asymptotic stability of $\hat{x}_{k+1} = [F^\alpha_k - \alpha K^d_k H^{a'}_k]\hat{x}_k$, and this in turn also ensures that the filter (2.7) achieves a prescribed degree of stability α.

3. The quantity $\Sigma^d_{k/k-1}$ will, in general, be unbounded, whereas the quantity $\Sigma^\alpha_{k/k-1}$ will usually be bounded. The equations for $\Sigma^\alpha_{k/k-1}$

are thus clearly better for calculation purposes than the equations for $\Sigma^d_{k/k-1}$.

4. With time-invariant F^a, G^a, H^a, Q^a, and R^a and constant α, (2.5) and (2.6) under a detectability constraint will yield an asymptotically time-invariant filter.

An examination of [1] suggests that exponential data weighting should be a more straightforward tool to use than limited memory filtering. Both techniques can cure many error problems, though possibly the latter technique is more powerful, imposing as it does a hard rather than soft limit on memory.

Main Points of the Section

Exponential data weighting has the same effect as increasing the input noise, and normally causes the filter to have a prescribed degree of stability. A performance bound is provided in the course of filter design.

Problem 2.1. Establish the claim made in the introductory subsection linking a quadratic minimization problem possessing linear constraints with a maximum a posteriori estimation problem.

Problem 2.2. Suppose that the system model is time invariant and that Q^a_k, R^a_k are constant; suppose that the associated optimal filter is asymptotically time invariant. Show that the only form of increased weighting of more recent data still yielding an asymptotically time-invariant filter is exponential.

6.3 THE MATRIX INVERSION LEMMA AND THE INFORMATION FILTER

The matrix inversion lemma is actually the name given to a number of closely related but different matrix equalities which are quite useful in obtaining various forms for estimation algorithms. In this section, the various matrix equalities are stated and derived. They are applied to the usual Kalman filter equations (which are expressed in terms of covariances $\Sigma_{k/k-1}$ and $\Sigma_{k/k}$) to yield new filter equations which are expressed in terms of the inverses of these quantities, viz., $\Sigma^{-1}_{k/k-1}$ and $\Sigma^{-1}_{k/k}$. These inverses are termed *information matrices*.

The Matrix Inversion Lemma

In terms of an $n \times n$ matrix Σ, a $p \times p$ matrix R, and an $n \times p$ matrix H, the following equalities hold on the assumption that the various inverses exist:

$$(I + \Sigma H R^{-1} H')^{-1} \Sigma = (\Sigma^{-1} + H R^{-1} H')^{-1} = \Sigma - \Sigma H (H' \Sigma H + R)^{-1} H' \Sigma \tag{3.1}$$

Multiplication on the right by $H R^{-1}$ and application of the identity

$$H' \Sigma H R^{-1} = [(H' \Sigma H + R) R^{-1} - I]$$

yields a variation of (3.1) as

$$(I + \Sigma H R^{-1} H')^{-1} \Sigma H R^{-1} = (\Sigma^{-1} + H R^{-1} H')^{-1} H R^{-1} = \Sigma H (H' \Sigma H + R)^{-1} \tag{3.2}$$

(Alternative formulations are frequently written using the quantity $C = H R^{-1}$.)

That the first equality of (3.1) holds is immediate. That

$$(I + \Sigma H R^{-1} H')[I - \Sigma H (H' \Sigma H + R)^{-1} H'] = I$$

holds can be verified in one line by direct verification. These two results together yield the remaining equality in (3.1).

The Information Filter

Application of the above matrix inversion lemma to the Kalman filter equations of earlier chapters yields an alternative filter algorithm known as the information filter [10]. This filter is now derived for the case when the input noise and output noise are independent (i.e., $S = 0$). Problem 3.4 suggests a procedure for coping with dependent noises.

Application of the matrix inversion lemma to the identity

$$\Sigma_{k/k} = \Sigma_{k/k-1} - \Sigma_{k/k-1} H_k (H'_k \Sigma_{k/k-1} H_k + R_k)^{-1} H'_k \Sigma_{k/k-1}$$

yields immediately an expression for the information matrix $\Sigma_{k/k}^{-1}$ as

$$\Sigma_{k/k}^{-1} = \Sigma_{k/k-1}^{-1} + H_k R_k^{-1} H'_k \tag{3.3}$$

A further application to the identity

$$\Sigma_{k+1/k} = F_k \Sigma_{k/k} F'_k + G_k Q_k G'_k$$

with

$$A_k = [F_k^{-1}]' \Sigma_{k/k}^{-1} F_k^{-1} \tag{3.4}$$

identified with Σ of (3.1) and $G_k Q_k G'_k$ identified with $H R^{-1} H'$ of (3.1), yields an expression for the information matrix $\Sigma_{k+1/k}^{-1}$ as

$$\Sigma_{k+1/k}^{-1} = [A_k^{-1} + G_k Q_k G'_k]^{-1}$$
$$= [I - A_k G_k [G'_k A_k G_k + Q_k^{-1}]^{-1} G'_k] A_k$$

or equivalently

$$\Sigma_{k+1/k}^{-1} = [I - B_k G'_k] A_k \tag{3.5}$$

where

$$B_k = A_k G_k [G'_k A_k G_k + Q_k^{-1}]^{-1} \tag{3.6}$$

Equations (3.3) through (3.6) provide recurrence relationships for $\Sigma_{k/k-1}^{-1}$ and $\Sigma_{k/k}^{-1}$.

Next, we observe that the filter gain matrix can be expressed in terms of $\Sigma_{k/k-1}^{-1}$. In fact, it is immediate from the expression

$$K_k = F_k \Sigma_{k/k-1} H_k (H_k' \Sigma_{k/k-1} H_k + R_k)^{-1}$$

and from (3.2) that

$$K_k = F_k \Sigma_{k/k} H_k R_k^{-1} = F_k \Sigma_{k/k}^{-1} H_k R_k^{-1} \tag{3.7}$$

Now we turn to the filter equations themselves. It proves awkward to find appropriate recursions for $\hat{x}_{k/k}$ or $\hat{x}_{k/k-1}$, and instead recursions are found for the following quantities

$$\hat{a}_{k/k-1} = \Sigma_{k/k-1}^{-1} \hat{x}_{k/k-1}$$
$$\hat{a}_{k/k} = \Sigma_{k/k}^{-1} \hat{x}_{k/k} \tag{3.8}$$

from which the state estimates may be recovered by the solution of algebraic equations without the need for actually taking inverses of $\Sigma_{k/k}^{-1}$ and $\Sigma_{k/k-1}^{-1}$ to obtain $\Sigma_{k/k}$ and $\Sigma_{k/k-1}$ explicitly. Application of (3.4) and (3.5) to (3.8) yields

$$\hat{a}_{k+1/k} = [I - B_k G_k'] A_k F_k \hat{x}_{k/k}$$
$$= [I - B_k G_k'] (F_k^{-1})' \Sigma_{k/k}^{-1} \hat{x}_{k/k}$$

or equivalently

$$\hat{a}_{k+1/k} = [I - B_k G_k'][F_k^{-1}]' \hat{a}_{k/k} \tag{3.9}$$

The measurement-update equation,

$$\hat{x}_{k/k} = \hat{x}_{k/k-1} + \Sigma_{k/k-1} H_k (H_k' \Sigma_{k/k-1} H_k + R_k)^{-1} (z_k - H_k' \hat{x}_{k/k-1})$$

leads to

$$\hat{a}_{k/k} = \hat{a}_{k/k-1} + H_k R_k^{-1} z_k \tag{3.10}$$

(A derivation is called for in the problems.)

Equations (3.3) through (3.10) constitute the information filter equations. They are, of course, algebraically equivalent to the usual Kalman filter equations, so that it is computational simplicity and error propagation properties that govern a choice between the two sets of equations. Some comparisons can be found in [11, 12].

Some of the points which should be borne in mind are the following:

1. In some situations, no information concerning the initial state is available, i.e., the situation is as if $P_0 = \Sigma_{0/-1}$ were infinite. In this case, though $\Sigma_{0/-1}$ does not exist, it is perfectly legitimate to take $\Sigma_{0/-1}^{-1} = 0$; and this, together with $\hat{a}_{0/-1} = 0$, is easily coped with in the information filter equations. (In contrast, an infinite initial condition in the covariance equations is, quite obviously, difficult to cope with.) The matrices $\Sigma_{k/k-1}^{-1}$ and $\Sigma_{k/k}^{-1}$ are still evaluated recursively;

while singular, they lack the interpretation of being inverses of finite error covariance matrices, though they can be regarded as inverses of error covariance matrices associated with some infinite errors. The case of poor a priori information (i.e., $\Sigma_{0/-1}^{-1}$ is very small, but possibly non-zero) can also be easily dealt with using the information filter.

2. The inverses of $H_k'\Sigma_{k/k-1}H_k + R_k$ and $G_k'A_kG_k + Q_{\bar{k}}^{-1}$ must be computed in the covariance and information filter recursions, respectively. If the output dimension is significantly different from the input dimension, one inverse will be easier to compute, and this will constitute an argument favouring one set of equations over the other.

3. Notwithstanding the above point, the fact that $F_{\bar{k}}^{-1}$ and $Q_{\bar{k}}^{-1}$ have to be computed to implement the information filter could make it less attractive.

4. The information filter formulation seems a more efficient vehicle for handling measurement updates than the covariance filter, but not so efficient at handling time updates.

5. In a later section, we discuss "square root" filtering. It is probably true that the square root information filter equations are much more valuable than the equations of this section.

6. A duality exists between the update equations for $\Sigma_{k+1/k}$ and $\Sigma_{k/k}$ and the update equations for the inverses of these quantities, or, more precisely, for $\Sigma_{k+1/k}^{-1}$ and $A_k = [F_{\bar{k}}^{-1}]'\Sigma_{k/k}^{-1}F_{\bar{k}}^{-1}$. (See Prob. 3.3.) The duality shows that the latter two quantities also can arise in the covariance equations associated with a certain filtering problem; the signal model for the dual problem is exponentially unstable when that for the initially given problem is exponentially stable. This suggests that there could on occasions be numerical difficulties with the use of information filter equations, in the light of the conclusions of Sec. 6.1.

7. Use of (3.5) and (3.6) can, through computational error, lead to lack of symmetry or even nonnegative definiteness in $\Sigma_{k+1/k}^{-1}$. [Equation (3.3), on the other hand, is much less likely to cause problems.] The prime method for avoiding these difficulties is to use a square root filter, but Prob. 3.5 considers other avenues.

8. It is a straightforward matter to combine the measurement-and time-update equations and give update equations taking $\Sigma_{k-1/k-1}^{-1}$ into $\Sigma_{k/k}^{-1}$ and $\hat{a}_{k-1/k-1}$ into $\hat{a}_{k/k}$, and similarly for the one-step prediction quantities.

Main Points of the Section

Information filter equations are an alternative to covariance filter equations, and on occasions may be more efficient.

Problem 3.1. (Matrix Inversion Lemma). Assuming all inverses exist, show that $[A - BC^{-1}D]^{-1} = A^{-1} + A^{-1}B[C - DA^{-1}B]^{-1}DA^{-1}$. Verify that (3.1) constitutes a special case.

Problem 3.2. Derive Eq. (3.10).

Problem 3.3. Consider one filtering problem defined by quantities $F_k, G_k, H_k,$ Q_k, and R_k (with the usual significance) and a second filtering problem defined by $\bar{F}_k = F_{k+1}^{-1\prime}, \bar{G}_k = F_{k+1}^{-1\prime}H_{k+1}, \bar{H}_k = G_k, \bar{Q}_k = R_{k+1}^{-1}, \bar{R}_k = Q_k^{-1}$, assuming of course that all inverses exist. Relate the covariance filter equations for $\bar{\Sigma}_{k+1/k}$ and $\bar{\Sigma}_{k/k}$ to the information filter equations for $A_k = [F_k^{-1}]'\Sigma_{k/k}^{-1}F_k^{-1}$ and $\Sigma_{k+1/k}^{-1}$.

Problem 3.4. Can you derive information filter equations when there is correlation between input and output noise? What happens to the duality ideas of Prob. 3.3? [*Hint:* Consider writing the state transition equation as

$$x_{k+1} = (F_k - G_k S_k R_k^{-1} H_k')x_k + G_k \bar{w}_k + G_k S_k R_k^{-1} z_k$$

with $\bar{w}_k = w_k - S_k R_k^{-1} v_k$.]

Problem 3.5. A potential difficulty with using equations like (3.5) and (3.6) is that, because of computational errors, loss of symmetry or even nonnegative definiteness of $\Sigma_{k+1/k}^{-1}$ can occur. In the covariance filter, one device used to partly eliminate this problem is to update $\Sigma_{k+1/k}$ by computing

$$(F_k - K_k H_k')\Sigma_{k/k-1}(F_k - K_k H_k')' + K_k R_k K_k' + G_k Q_k G_k'$$

rather than by

$$F_k[\Sigma_{k/k-1} - \Sigma_{k/k-1}H_k(H_k'\Sigma_{k/k-1}H_k + R_k)H_k'\Sigma_{k/k-1}]F_k' + G_k Q_k G_k'$$

or even

$$(F_k - K_k H_k')\Sigma_{k/k-1}F_k' + G_k Q_k G_k'$$

Similar alternatives exist in passing from $\Sigma_{k/k-1}$ to $\Sigma_{k/k}$. Discuss the corresponding alternatives for the information filter.

6.4 SEQUENTIAL PROCESSING

Sequential processing is the name given to the procedure in which the measurement vector is processed one component at a time. There are at least two reasons why there is sometimes an advantage to be gained from sequential processing. The first is that when the output noise covariance R_k is block diagonal, there is a reduction in processing time which can range up to fifty percent depending on the signal model and selection of data vector components. The second reason is that should there not be adequate time available to complete the processing of the data vector (as when a priority interrupt occurs, for example), then there is an effective loss of only some components of the data in sequential processing rather than a loss of the entire data vector as in simultaneous processing.

Sequential processing results can also be helpful for generalizing certain theoretical results for scalar measurement processes to vector measurement processes. In addition, sequential processing proves useful in the implementation of adaptive estimators discussed in a later chapter.

Sequential processing can be used with either the normal (covariance) Kalman filter equations or the information filter equations, or with the square root formulations of these equations discussed later in the chapter. It is very hard to pin down precisely when it should be used: the choice is governed by tradeoffs in computer time and computer storage requirements, by the relative dimensions of input, state and output vectors, by the diagonal or nondiagonal nature of the noise covariances, and so on. The most complete comparative results can be found in [12, 13]. With qualifications described in these references, it does seem that sequential processing is preferable for covariance equations, including square root formulations, but not for information filter equations, again including square root formulations. However, a modification of the sequential processing idea applied to the time-update equations does prove advantegeous for the information filter.

We shall begin by supposing that the output noise covariance matrix is block diagonal encompassing thereby the case of a strictly diagonal matrix. Of course, block diagonal or strictly diagonal covariance matrices occur commonly. Thus with signal model

$$x_{k+1} = Fx_k + Gw_k \tag{4.1}$$

$$z_k = H'x_k + v_k \tag{4.2}$$

we have $E[v_k v_k'] = R = \text{diag}(R^1, R^2, \ldots, R^r)$. (For clarity, we are omitting the time subscripting on matrices where possible.) The R^i have dimension $p^i \times p^i$ with $\sum_{i=1}^{r} p^i = p$, where p is the dimension of v_k and the measurement vector z_k. It now makes sense to partition v_k into components $v_k^1, v_k^2, \ldots, v_k^r$ and z_k into components $z_k^1, z_k^2, \ldots, z_k^r$, where v_k^i and z_k^i are of dimension p^i. A partitioning of the measurement matrix as $H = [H^1 \quad H^2 \quad \ldots H^r]$, where H^i is $n \times p^i$, allows us to rewrite the measurement equation (4.2) as

$$z_k^i = (H^i)'x_k + v_k^i \tag{4.3}$$

with $E[v_k^i v_k^j] = R^i \delta_{ij}$ for $i, j = 1, 2, \ldots, r$ Clearly, for each i the sequence $\{v_k^i\}$ is a white noise sequence.

We are now in a position to define more precisely the notion of *sequential processing* of vector data. Instead of processing z_k as a single data vector as in the simultaneous processing of earlier sections and chapters, the components $z_k^1, z_k^2, \ldots, z_k^r$ are processed one at a time, or sequentially. Thus instead of calculating $\hat{x}_{k/k} = E[x_k | Z_{k-1}, z_k]$ in terms of $\hat{x}_{k/k-1} = E[x_k | Z_{k-1}]$ and z_k, first the quantity $\hat{x}_k^1 = E[x_k | Z_{k-1}, z_k^1]$ is calculated in terms of $\hat{x}_{k/k-1}$ and z_k^1, then $\hat{x}_k^2 = E[x_k | Z_{k-1}, z_k^1, z_k^2]$ is calculated in terms of $\hat{x}_k^1$ and z_k^2; and so on until finally

$$\hat{x}_{k/k} = \hat{x}_k^r = E[x_k | Z_{k-1}, z_k^1, z_k^2, \ldots, z_k^r]$$

is obtained. How then do we achieve the intermediate estimates $\hat{x}_k^i$ for $i = 1, 2, \ldots, r$?

The estimates x_k^i are achieved by a direct application of the Kalman filter equations to the measurement process (4.3), regarding i as the running variable. (Of course, the filter equations are specialized to the case where the state is a constant vector and thus the one-step prediction estimates are identical to the true filtered estimates.) Consequently, we have (assuming independent measurement and input noise):

Measurement-update Equations (r updates, i = 1, 2, . . . , r)

$$L^i = \Sigma^{i-1} H^i [(H^i)' \Sigma^{i-1} H^i + R^i]^{-1} \tag{4.4}$$

$$\hat{x}_k^i = \hat{x}_k^{i-1} + L^i [z_k^i - (H^i)' \hat{x}_k^{i-1}] \tag{4.5}$$

$$\Sigma^i = [I - L^i (H^i)'] \Sigma^{i-1} \tag{4.6}$$

Here the measurement-update equations are initialized by $\hat{x}_k^0 = \hat{x}_{k/k-1}$ and $\Sigma^0 = \Sigma_{k/k-1}$ and terminated by the identifications $\hat{x}_{k/k} = \hat{x}_k^r$ and $\Sigma_{k/k} = \Sigma_k^r$. More symmetric forms of (4.6) can naturally be used.

Of course, $\hat{x}_{k+1/k}$ and $\Sigma_{k+1/k}$ are obtained from $\hat{x}_{k/k}$ and $\Sigma_{k/k}$ via the usual time-update equations.

One evident advantage of sequential processing, as opposed to simultaneous processing, is that instead of requiring the inversion of a $p \times p$ matrix $[H'\Sigma H + R]$ as in simultaneous processing, the inversion of $p^i \times p^i$ matrices $[(H^i)'\Sigma H^i + R^i]$ is required for $i = 1, 2, \ldots, r$. Since the latter task requires less computational effort, sequential processing may lead to considerable computational savings. Actually, it is argued in [14] that a further saving is possible when $p_i > 1$. To compute $\Sigma H[H'\Sigma H + R]^{-1}$, one first computes (ΣH) and then $[H'(\Sigma H) + R]$. But then one computes $X = \Sigma H[H'\Sigma H + R]^{-1}$ by solving the equation $X[H'\Sigma H + R] = \Sigma H$ using Gaussian elimination, rather than by explicitly evaluating the matrix inverse.

We now make some miscellaneous observations.

1. Suppose that R is diagonal. There are then a number of partitions of R into block diagonal form other than the obvious one. More generally, if R is block diagonal with $r > 2$, there is more than one block diagonal decomposition of R. Reference [13] discusses the optimal choice of decomposition. Interestingly, if $p \leq n/2$, n being the state-vector dimension, it seems that simultaneous processing is often optimal.

2. In case R is not block diagonal, suppose a Cholesky decomposition [15] of R_k is given as

$$R_k = \mathcal{S}_k \mathcal{S}'_k \tag{4.7}$$

where $\mathcal{S}_k$ is lower block triangular. Note that $\mathcal{S}_k$ is easily found from R_k (see Prob. 4.1). Then one works with measurements $\bar{z}_k = \mathcal{S}_k^{-1} z_k$ and output noise covariance matrix $\bar{R}_k = I$. The components of $\bar{z}_k$ are processed one at a time; the ith component of $\bar{z}_k$ is, incidentally a linear combination of the first i components of z_k.

3. One can use Eqs. (4.4) and (4.6) to pass from $\Sigma_{k/k-1}$ to $\Sigma_{k/k}$, benefiting from the computational simplicity, but at the same time one can dispense with the calculation of $\hat{x}_k^1, \hat{x}_k^2, \ldots$ and simply use the fact that

$$\hat{x}_{k/k} = \Sigma_{k/k} H_k R_k^{-1}[z_k - H_k' \hat{x}_{k/k-1}] + \hat{x}_{k/k-1} \tag{4.8}$$

to pass from $\hat{x}_{k/k-1}$ to $\hat{x}_{k/k}$. This gets away from the notion of sequential processing of the measurements. Much of the development above is then seen to be descriptive of a clever way of passing from $\Sigma_{k/k-1}$ to $\Sigma_{k/k}$ and nothing else. Of course, as a tool for minimizing the consequences of priority interrupts, sequential processing is justified.

4. The measurement-update equations for the information filter are given in the last section; we repeat them as

$$\Sigma_{k/k}^{-1} = \Sigma_{k/k-1}^{-1} + H_k R_k^{-1} H_k' \tag{4.9}$$

$$\hat{a}_{k/k} = \hat{a}_{k/k-1} + H_k R_k^{-1} z_k \tag{4.10}$$

A brief examination of these equations shows that there is little or no advantage to be gained from sequential processing.

5. Despite the above point, some of the ideas of this section are relevant to the information filter. In the last section, a certain duality was established between the covariance filter and information filter; an examination of this duality shows that the measurement-update equations of one filter are related to the time-update equations for the other. Consequently, we can construct alternative schemes for the time-update equation of the information filter, when Q_k is block diagonal, which may be computationally attractive. This is done below.

Suppose that $Q = \text{diag}[Q^1, Q^2, \ldots, Q^s]$ with Q^j of dimension $m_j \times m_j$, $\sum_{k=1}^{s} m_j = m$, with m the dimension of w_k. Partition the input matrix G as $[G^1 \quad G^2 \quad \ldots \quad G^s]$. We replace the following equations (omitting most time subscripts):

$$A = (F^{-1})' \Sigma_{k/k}^{-1} F^{-1} \tag{4.11}$$

$$\Sigma_{k+1/k}^{-1} = [A^{-1} + GQG']^{-1} \tag{4.12a}$$

$$= A - AG[G'AG + Q^{-1}]^{-1}G'A \tag{4.12b}$$

$$= [I - BG']A \tag{4.12c}$$

where

$$B = AG(G'AG + Q^{-1})^{-1} \tag{4.13}$$

by the following, for $j = 1, 2, \ldots, s$:

$$A^1 = (F^{-1})'(\Sigma_{k/k})^{-1}F^{-1} \qquad A^j = (\Sigma^{j-1})^{-1} \quad j > 1 \tag{4.14}$$

$$(\Sigma^j)^{-1} = [(A^j)^{-1} + (G^j)'Q^jG^j]^{-1} \tag{4.15a}$$

$$= A^j - A^jG^j[(G^j)'A^jG^j + (Q^j)^{-1}]^{-1}(G^j)'A^j \tag{4.15b}$$

$$= [I - B^j(G^j)']A^j \tag{4.15c}$$

with

$$B^j = A^jG^j[(G^j)'A^jG^j + (Q^j)^{-1}]^{-1} \tag{4.16}$$

where $\Sigma^s = \Sigma_{k+1/k}$. Note that (4.15c) is normally used in preference to (4.15a) or (4.15b). The easiest way of deriving these equations is to observe that with

$$\Sigma^1 = F\Sigma^0F' + G^1Q^1(G^1)' \tag{4.17a}$$

$$\Sigma^j = \Sigma^{j-1} + G^jQ^j(G^j)' \qquad j = 2, \ldots, s \tag{4.17b}$$

and with $\Sigma^0 = \Sigma_{k/k}$, there results

$$\Sigma^s = F\Sigma_{k/k}F' + \sum_j G^jQ^j(G^j)' = \Sigma_{k+1/k}$$

Equations (4.15a) constitute restatements of (4.17).

We can also replace the equation

$$\hat{a}_{k+1/k} = [I - BG'](F^{-1})'\hat{a}_{k/k} \tag{4.18}$$

by setting

$$\hat{a}^1 = [I - B^1(G^1)'](F^{-1})'\hat{a}_{k/k} \tag{4.19a}$$

$$\hat{a}^j = [I - B^j(G^j)']\hat{a}^{j-1} \quad j > 1 \tag{4.19b}$$

There results $a^s = \hat{a}_{k+1/k}$. Reference [12], incidentally, suggests that when $m > n/2$, it is preferable not to use (4.19), even though (4.14) through (4.16) are still used. In lieu of (4.19) one uses

$$\hat{a}_{k+1/k} = \Sigma_{k+1/k}^{-1}F\hat{x}_{k/k}$$

which is readily derived from (4.11), (4.12a), and (4.18).

Block Processing

The reverse process to sequential processing is block processing. For systems with large state dimension (say ≥ 30), there may be computational advantages in block processing the measurements and applying Fast Fourier Transform (FFT) techniques, [34]. Such techniques are outside the scope of this text.

With a diagonal R matrix (which may always be secured by a Cholesky decomposition) covariance filter formulas for the measurement-update equations exist which amount to using a sequence of scalar updates. With a diagonal Q matrix, analagous information filter formulas can be found for the time-update step.

Problem 4.1. (Cholesky Decomposition). Suppose that $R = \mathbb{S}\mathbb{S}'$ with R an $n \times n$ nonnegative definite symmetric matrix and $\mathbb{S}$ lower triangular. Show that entries of $\mathbb{S}$ can be calculated recursively; for $i = 1, 2, \ldots,$

$$\mathbb{S}_{ii} = [R_{ii} - \sum_{j=1}^{i-1} \mathbb{S}_{ij}^2]^{1/2}$$

$$\mathbb{S}_{ji} = 0 \quad j < i$$

$$= \mathbb{S}_{ii}^{-1}[R_{ji} - \sum_{k=1}^{i-1} \mathbb{S}_{jk}\mathbb{S}_{ik}] \quad j = i+1, i+2, \ldots, n$$

Problem 4.2. With $L_k = \Sigma_{k/k-1}H(H'\Sigma_{k/k-1}H + R)^{-1}$, show that

$$I - LH' = \prod_{j=1}^{r} [I - L^j(H^j)']$$

where the notation is as used in this section.

Problem 4.3. Verify the claims associated with Eqs. (4.14) through (4.18).

6.5 SQUARE ROOT FILTERING

Use of the normal Kalman filter equations for calculation of the error covariance can result in a matrix which fails to be nonnegative definite. This can happen particularly if at least some of the measurements are very accurate, since then numerical computation using ill-conditioned quantities is involved. As a technique for coping with this difficulty, Potter [3, pp. 338–340] suggested that the error covariance matrix be propagated in square root form; his ideas were restricted to the case of zero input noise and scalar measurements. Potter's ideas were later extended to cope with the presence of input noise and vector measurements [11, 12, 16–24]. Update equations for the square root of an inverse covariance matrix were also demonstrated.

Let M be a nonnegative definite symmetric matrix. A square root of M is a matrix N, normally square, but not necessarily nonnegative definite symmetric, such that $M = NN'$. Sometimes, the notation $M^{1/2}$ is used to denote an arbitrary square root of M. Let $\mathbb{S}_{k/k}$ and $\mathbb{S}_{k+1/k}$ denote square roots of $\Sigma_{k/k}$ and $\Sigma_{k+1/k}$. We shall shortly present update equations for the square roots in lieu of those for the covariances.

There are two crucial advantages to these equations, the first having been alluded to above:

1. Since the product $\underset{\sim}{S}\underset{\sim}{S}'$ is always nonnegative definite, the calculation of Σ as $\underset{\sim}{S}\underset{\sim}{S}'$ cannot lead to a matrix which fails to be nonnegative definite as a result of computational errors in the update equations.
2. The numerical conditioning of $\underset{\sim}{S}$ is generally much better than that of Σ, since the condition number of $\underset{\sim}{S}$ is the square root of the condition number of Σ. This means that only half as many significant digits are required for square root filter computation as compared with covariance filter computation, if numerical difficulties are to be avoided.

For certain applications with restrictions on processing devices, square root filtering may be essential to retain accuracy.

Square root algorithms are not always without disadvantage, for the algorithms usually, but not always, require larger computer programs for implementation; the computational burden can vary from $\frac{1}{2}$ to $1\frac{1}{2}$ times that for the standard algorithms, depending on the state, input, and output vector dimensions n, m, and r. For small r, the square root covariance filter is more efficient than the square root information filter, but for moderate or large r, the reverse is true.

The reader may recall one other technique for partially accommodating the first difficulty remedied by square root equations. When using the usual covariance filter equation, it is possible to write the update equation as

$$\Sigma_{k+1/k} = (F_k - K_k H_k')\Sigma_{k/k-1}(F_k - K_k H_k')' + K_k R_k K_k' + G_k Q_k G_k'$$

where K_k is given in terms of $\Sigma_{k/k-1}$, etc. This form of update equation tends to promote nonnegativity of $\Sigma_{k+1/k}$. Note, however, that if $\Sigma_{k/k-1}$ fails to be nonnegative for some reason, $\Sigma_{k+1/k}$ may not be. Thus nonnegativity is not as automatic as with the square root approach.

Covariance Square Root Filter

The Potter algorithm [3] was first extended by Bellantoni and Dodge [16] to handle vector measurements, and subsequently by Andrews [17] to handle process noise as well. Schmidt [18] gave another procedure for handling process noise. Vector measurements can be treated either simultaneously or sequentially; in the latter case a diagonal R matrix simplifies the calculations. Until the work of Morf and Kailath [24], time and measurement updates had been regarded as separate exercises; their work combined the two steps. In this subsection, we shall indicate several of the possible equations covering these ideas for the case of models with uncorrelated input and output noise.

Time update.

$$\hat{x}_{k+1/k} = F_k \hat{x}_{k/k} \tag{5.1}$$

$$\begin{matrix} n\{ \\ m\{ \end{matrix} \begin{bmatrix} \mathcal{S}'_{k+1/k} \\ 0 \end{bmatrix} = T \begin{bmatrix} \mathcal{S}'_{k/k}F'_k \\ Q_k^{1/2'}G'_k \end{bmatrix} \tag{5.2}$$

$$\underbrace{\phantom{\begin{bmatrix} \mathcal{S}'_{k+1/k} \\ 0 \end{bmatrix}}}_{n}$$

In (5.2), the matrix T is orthogonal, but otherwise is any matrix making $\mathcal{S}'_{k+1/k}$ in (5.2) upper triangular. It is readily checked that (5.2) implies that $\Sigma_{k+1/k} = F_k \Sigma_{k/k} F'_k + G_k Q_k G'_k$. The construction of T is a task to which much attention has been given: the main methods suggested are use of the Householder and modified Gram-Schmidt methods, while [29] suggests that a Givens transformation could be used. These are standard algorithms of numerical analysis [15]; an outline of their use in this connection can be found in, e.g., [11]. We remark that the algorithms frequently find triangular square roots—this may be advantageous if an inverse of the square root is needed.

Measurement update via simultaneous processing. Here

$$\hat{x}_{k/k} = \hat{x}_{k/k-1} + \tilde{K}_k(R_k + H'_k\Sigma_{k/k-1}H_k)^{-1/2}(z_k - H'_k\hat{x}_{k/k-1}) \tag{5.3}$$

$$\begin{bmatrix} (R_k + H'_k\Sigma_{k/k-1}H_k)^{1/2'} & \tilde{K}'_k \\ 0 & \mathcal{S}'_{k/k} \end{bmatrix} = \bar{T} \begin{bmatrix} R_k^{1/2'} & 0 \\ \mathcal{S}'_{k/k-1}H_k & \mathcal{S}'_{k/k-1} \end{bmatrix} \tag{5.4}$$

where $\bar{T}$ is orthogonal. Finding $\bar{T}$ is of course the same task as finding T in (5.2). Verification of these equations is requested in the problems. Notice that, in view of the inverse in (5.3), it is helpful to have $(R_k + H'_k\Sigma_{k/k-1}H_k)^{1/2}$ triangular.

Measurement update via sequential processing. We assume that the R matrix is diagonal (not just block diagonal), viz., $R = \text{diag}[R^1, R^2, \ldots, R^r]$. (If this is not the case, the Cholesky algorithm is used to determine a transformation producing a problem in which R is diagonal, as outlined in the last section.) Let H_k^i denote the ith column of H_k and z_k^i the ith entry of z_k. With $\hat{x}^0 = \hat{x}_{k/k-1}$, $\hat{x}^r = \hat{x}_{k/k}$, $\mathcal{S}^0 = \mathcal{S}_{k/k-1}$, $\mathcal{S}^r = \mathcal{S}_{k/k}$, one obtains for $i = 1, 2, \ldots, r$:

$$D^i = (\mathcal{S}^{i-1})'H^i \tag{5.5a}$$

$$\alpha^i = [(D^i)'D^i + R^i]^{-1} \tag{5.5b}$$

$$\gamma^i = (1 + \sqrt{\alpha^i R^i})^{-1} \tag{5.5c}$$

$$\mathcal{S}^i = \mathcal{S}^{i-1} - \gamma^i \alpha^i \mathcal{S}^{i-1}D^iD^{i'} \tag{5.6}$$

$$\hat{x}^i = \hat{x}^{i-1} + \alpha^i \mathcal{S}^{i-1}D^i[z_k^i - (H^i)'\hat{x}^{i-1}] \tag{5.7}$$

Of course, these equations define a sequence of updates corresponding to a sequence of scalar measurements. The equations agree with the original algorithm (a derivation of which is requested in the problems), and also follow by specializing (5.4) to the scalar case, in which it is always possible to specify a $\bar{T}$ explicitly, as

$$\bar{T} = I - \frac{2}{\alpha'\alpha}\alpha\alpha' \tag{5.8}$$

where α is the vector

$$\alpha = \begin{bmatrix} (R_k + H_k'\Sigma_{k/k-1}H_k)^{1/2} - R_k^{1/2} \\ -S_{k/k-1}'H_k \end{bmatrix} \tag{5.9}$$

Morf-Kailath combined update equations. With $\bar{T}$, T as defined earlier, one can readily verify the effect of following one transformation by another; using (5.2) and (5.4), one has

$$\begin{bmatrix} I & 0 \\ 0 & T \end{bmatrix}\begin{bmatrix} \bar{T} & 0 \\ 0 & I \end{bmatrix}\begin{bmatrix} R_k^{1/2'} & 0 \\ S_{k/k-1}'H_k & S_{k/k-1}'F_k' \\ 0 & Q_k^{1/2'}G_k' \end{bmatrix}$$

$$= \begin{bmatrix} I & 0 \\ 0 & T \end{bmatrix}\begin{bmatrix} (R_k + H_k'\Sigma_{k/k-1}H_k)^{1/2} & \tilde{K}_k'F_k' \\ 0 & S_{k/k}'F_k' \\ 0 & Q_k^{1/2'}G_k' \end{bmatrix}$$

$$= \begin{bmatrix} (R_k + H_k'\Sigma_{k/k-1}H_k)^{1/2'} & \tilde{K}_k'F_k' \\ 0 & S_{k+1/k}' \\ 0 & 0 \end{bmatrix} \tag{5.10}$$

This suggests that *any* orthogonal matrix $\hat{T}$ such that

$$\hat{T}\begin{bmatrix} R_k^{1/2'} & 0 \\ S_{k/k-1}'H_k & S_{k/k-1}'F_k' \\ 0 & Q_k^{1/2'}G_k' \end{bmatrix}$$

is lower triangular or simply block lower triangular generates the matrix $\tilde{K}_k$ with the relevance defined in (5.3) and square roots of $R_k + H_k'\Sigma_{k/k-1}H_k$ and $\Sigma_{k+1/k}$. This is easily verified. In this way one obtains update equations, which are as follows:

$$\hat{x}_{k+1/k} = F_k\hat{x}_{k/k-1} + F_k\tilde{K}_k(R_k + H_k'\Sigma_{k/k-1}H_k)^{-1/2}(z_k - H_k'\hat{x}_{k/k-1}) \tag{5.11}$$

$$\hat{T}\begin{bmatrix} R_k^{1/2'} & 0 \\ S_{k/k-1}'H_k & S_{k/k-1}'F_k' \\ 0 & Q_k^{1/2'}G_k' \end{bmatrix} = \begin{bmatrix} (R_k + H_k'\Sigma_{k/k-1}H_k)^{1/2'} & \tilde{K}_k'F_k' \\ 0 & S_{k+1/k}' \\ 0 & 0 \end{bmatrix} \tag{5.12}$$

Evidently, there is no need to compute $\tilde{K}_k'$ by itself; from (5.12) $\tilde{K}_k'F_k'$ is seen to be computed directly and used as a single entity in (5.11).

Information Square Root Filters

Information square root filters can be developed by using the duality between information and covariance filters, mentioned in earlier sections, or by direct calculation. Both approaches can be found in the literature; references [11, 12, 19–24] are relevant. It is interesting to note that while no one set of filtering equations is always superior to all other sets, [12] argues that the square root information filter equations are most commonly the best to use.

As one might expect from examination of the covariance square root equations: there is a straightforward measurement-update equation; there are two possible time-update equations, one using a series of scalar updates and the other a single vector update; and there is an update equation covering both measurement and time update. These equations are set out below. In the various equations, $S_{k/k}^{-1}$ is such that $(S_{k/k}^{-1})'S_{k/k}^{-1} = \Sigma_{k/k}^{-1}$ and $S_{k/k-1}^{-1}$ has a similar property. Just as update equations for $\Sigma_{k/k}^{-1}$ and $\Sigma_{k/k-1}^{-1}$ can be used when these matrices are singular (for example, when no information concerning the initial state is available) so it turns out that update equations for $S_{k/k}^{-1}$ and $S_{k/k-1}^{-1}$ can be found even if these quantities are singular.

Measurement update.

$$\begin{bmatrix} S_{k/k}^{-1} \\ 0 \end{bmatrix} = T \begin{bmatrix} S_{k/k-1}^{-1} \\ R_k^{-1/2}H_k' \end{bmatrix} \tag{5.13}$$

$$\begin{bmatrix} \hat{b}_{k/k} \\ * \end{bmatrix} = T \begin{bmatrix} \hat{b}_{k/k-1} \\ R_k^{-1/2}z_k \end{bmatrix} \tag{5.14}$$

Here T is orthogonal, $\hat{b}_{k/k} = S_{k/k}^{-1}\hat{x}_{k/k}$ and $\hat{b}_{k/k-1} = S_{k/k-1}^{-1}\hat{x}_{k/k-1}$. The general idea is to find T such that the right side of (5.13) is upper triangular. Then the various quantities can be defined.

Scalar time-update equations. We assume that $Q = \text{diag}\{Q^1, Q^2, \ldots, Q^m\}$, performing a preliminary Cholesky decomposition if necessary.

$$E^1 = (S^0)^{-1}F^{-1}G^1 \qquad E^i = (S^{i-1})^{-1}G^i \qquad i > 1 \tag{5.15a}$$

$$\alpha^i = [(E^i)'E^i + (Q^i)^{-1}]^{-1} \qquad i \geq 1 \tag{5.15b}$$

$$\gamma^i = \left(1 + \sqrt{\frac{\alpha^i}{Q^i}}\right)^{-1} \qquad i \geq 1 \tag{5.15c}$$

$$(S^1)^{-1} = (S^0)^{-1}F^{-1} - \gamma^1\alpha^1E^1(E^1)'(S^0)^{-1}F^{-1} \tag{5.15d}$$

$$(S^i)^{-1} = (S^{i-1})^{-1} - \gamma^i\alpha^iE^i(E^i)'(S^{i-1})^{-1} \qquad i \geq 1 \tag{5.15e}$$

One has $(S^0)^{-1} = (S_{k/k})^{-1}$ and $(S^m)^{-1} = (S_{k+1/k})^{-1}$. The reader will perceive an obvious parallel between these equations and those of (5.5) and (5.6); he should also consider why the F matrix appears only at the first step of the iteration. (A similar phenomenon was observed in the normal information filter equations.) The update equations for $\hat{b}_{k/k}$ are given, with $b^0 = \hat{b}_{k/k}$ and $b^m = \hat{b}_{k+1/k}$, by

$$b^i = b^{i-1} - \alpha^i\gamma^iE^i(E^i)'b^{i-1} \tag{5.16}$$

Vector time-update equations.

$$\begin{bmatrix} (Q_k^{-1} + G_k'A_kG_k)^{1/2\,\prime} & \tilde{B}_k' \\ 0 & S_{k+1/k}^{-1} \end{bmatrix} = \bar{T}\begin{bmatrix} (Q_k^{1/2})^{-1} & 0 \\ S_{k/k}^{-1}F_k^{-1}G_k & S_{k/k}^{-1}F_k^{-1} \end{bmatrix} \quad (5.17)$$

where $\bar{T}$ is orthogonal and produces the upper triangular form in (5.17). Also,

$$\begin{bmatrix} * \\ \hat{b}_{k+1/k} \end{bmatrix} = \bar{T}\begin{bmatrix} 0 \\ \hat{b}_{k/k} \end{bmatrix} \quad (5.18)$$

The combined update equations can be found easily by combining the measurement- and time-update equations, as in the covariance square root filter. A derivation is called for in the problems. Again, as with the square root filter, specialization of (5.17) to the scalar input case and selection of a suitable $\bar{T}$ will generate (5.15).

Review Remarks

The last three sections have illustrated choices which can be made in the implementation of filter algorithms: covariance or information filter; square root or not square root; sequential or simultaneous processing of state and covariance data, or covariance data only; symmetry promoting or standard form of covariance and information matrix update. Yet another choice is available for stationary problems, to be outlined in Sec. 6.7. There are also further choices available within the square root framework. Recent references [25, 26] suggest that factorizations of the form $\Sigma = MDM'$ should be used (rather than $\Sigma = SS'$), where D is diagonal and M is triangular with 1's on the diagonal; update equations are found for M and D in covariance and information filter frameworks.

Main Points of the Section

Square root filtering ensures nonnegativity of covariance and information matrices and lowers requirements for computational accuracy, generally at the expense of requiring further calculations. Information and covariance forms are available, with and without sequential processing, and with and without combination of time and measurement update. Sometimes, it is essential to use square root filtering.

Problem 5.1. The condition number of a square matrix A is $[\lambda_{\max}(A'A)/\lambda_{\min}(A'A)]^{1/2}$. Show that the condition number of S is the square root of the condition number of (SS').

Problem 5.2. Verify the measurement-update equation (5.4).

Problem 5.3. (Potter Algorithm). In the scalar measurement situation, as we know,

$$\Sigma_{k/k} = \Sigma_{k/k-1} - \Sigma_{k/k-1}H_k(R_k + H'_k\Sigma_{k/k-1}H_k)^{-1}H'_k\Sigma_{k/k-1}$$

Show that this yields

$$\Sigma_{k/k} = \mathcal{S}_{k/k-1}[I - \alpha DD']\mathcal{S}'_{k/k-1}$$

where

$$D = \mathcal{S}'_{k/k-1}H_k, \qquad \mathcal{S}_{k/k-1}\mathcal{S}'_{k/k-1} = \Sigma_{k/k-1} \quad \text{and} \quad \alpha^{-1} = D'D + R_k$$

Then show that by proper choice of the constant γ one has

$$I - \alpha DD' = (I - \alpha\gamma DD')(I - \alpha\gamma DD')$$

Deduce the square root update equation, and relate these calculations to Eqs. (5.5) through (5.7).

Problem 5.4. Verify the claims concerning the matrix $\bar{T}$ of Eqs. (5.8) and (5.9).

Problem 5.5. The matrices other than $\hat{T}$ on both sides of (5.12) define transposes of square roots of the matrix

$$M = \begin{bmatrix} R_k + H'_k\Sigma_{k/k-1}H_k & H_k\Sigma_{k/k-1}F'_k \\ F_k\Sigma_{k/k-1}H_k & F_k\Sigma_{k/k-1}F'_k + G_kQ_kG'_k \end{bmatrix}$$

Show that if $E[w_i v_j] = C_i\delta_{ij}$ with $C_i \neq 0$, square root equations for the case of dependent input and output noise can be obtained by studying square roots (one block upper triangular, the other block lower triangular) of

$$\bar{M} = \left[\begin{array}{cc|c} & M & C_k \\ & & G_kQ^k \\ \hline C'_k & Q_kG'_k & Q_k \end{array} \right]$$

Assume that Q^{-1} exists. See [24].

Problem 5.6. Derive the following combined measurement- and time-update equations for the square root information filter. With symbols possessing the usual meaning,

$$\hat{T}\begin{bmatrix} (Q_k^{1/2})^{-1} & 0 \\ \mathcal{S}_{k/k}^{-1}F_k^{-1}G_k & \mathcal{S}_{k/k}^{-1}F_k^{-1} \\ 0 & R_{k+1}^{-1/2}H'_{k+1} \end{bmatrix} = \begin{bmatrix} (Q_k^{-1} + G'_kA_kG_k)^{1/2'} & \tilde{B}'_k \\ 0 & \mathcal{S}_{k+1/k+1}^{-1} \\ 0 & 0 \end{bmatrix}$$

$$\begin{bmatrix} * \\ \hat{b}_{k+1/k+1} \\ * \end{bmatrix} = \hat{T}\begin{bmatrix} 0 \\ \hat{b}_{k/k} \\ R_{k+1}^{-1/2}z_{k+1} \end{bmatrix}$$

6.6 THE HIGH MEASUREMENT NOISE CASE

We have seen in an earlier section that for some low measurement noise filtering problems, a square root filtering algorithm may be necessary if numerical problems are to be avoided. In contrast to this, a study of the high measurement noise case shows that considerable simplification to the algorithm for calculating the error covariance is possible using approximations.

For the moment, we shall work with time-invariant, asymptotically stable signal models and filters. Later, we shall note how the ideas extend to the time-varying case.

The idea in high measurement noise filtering is the following. Instead of determining the solution of the steady-state Riccati equation

$$\bar{\Sigma} = F[\bar{\Sigma} - \bar{\Sigma}H(H'\bar{\Sigma}H + R)^{-1}H'\bar{\Sigma}]F' + GQG' \tag{6.1}$$

(which is the optimal one-step error covariance) one obtains instead the solution of the steady-state linear equation

$$\bar{P} = F\bar{P}F' + GQG' \tag{6.2}$$

The reader will recognize that $\bar{P}$ is the state covariance of the signal model and is easier to obtain than $\bar{\Sigma}$. He should also realize or recall that $\bar{P} - \bar{\Sigma} \geq 0$. (Why?)

The usual Kalman filter gain is

$$K = F\bar{\Sigma}H(H'\bar{\Sigma}H + R)^{-1} \tag{6.3}$$

However, one uses instead

$$K^d = F\bar{P}H(H'\bar{P}H + R)^{-1} \tag{6.4}$$

which is much easier to compute. Actually, one can even use

$$K^d = F\bar{P}HR^{-1} \tag{6.5}$$

We claim that when the measurement noise is high, this is a satisfactory approximation. To see this we proceed as follows.

From (6.1) and (6.2), we obtain

$$(\bar{P} - \bar{\Sigma}) = F(\bar{P} - \bar{\Sigma})F' + F\bar{\Sigma}H(H'\bar{\Sigma}H + R)^{-1}H'\bar{\Sigma}F'$$

Now let us define a high measurement noise situation as one where R is large relative to $H'\bar{P}H$. (Thus, with $z_k = y_k + v_k$, $E[v_k v_k']$ is large relative to $E[y_k y_k']$.) Since $\bar{P} - \bar{\Sigma} \geq 0$, this means that R is large relative to $H'\bar{\Sigma}H$ and, accordingly, the above equation for $\bar{P} - \bar{\Sigma}$ yields

$$\bar{P} - \bar{\Sigma} = O(R^{-1}) \quad \text{but not} \quad O(R^{-2}) \tag{6.6}$$

Comparing (6.3) with (6.4) and (6.5), we see that this implies

$$K - K^d = O(R^{-2}) \tag{6.7}$$

Now the steady-state covariance $\bar{\Sigma}^p$ associated with use of the gain (6.4) or (6.5) in lieu of (6.3) is easily shown to satisfy

$$\bar{\Sigma}^p = (F - K^d H')\bar{\Sigma}^p(F - K^d H')' + GQG' + K^d R K^{d'}$$

and in view of (6.7), we have

$$\bar{\Sigma}^p = (F - KH')\bar{\Sigma}^p(F - KH')' + GQG' + KRK' + O(R^{-2}) \tag{6.8}$$

An alternative expression for (6.1) is, however,

$$\bar{\Sigma} = (F - KH')\bar{\Sigma}(F - KH')' + GQG' + KRK' \tag{6.9}$$

so that

$$\bar{\Sigma}^p - \bar{\Sigma} = O(R^{-2}) \tag{6.10}$$

The asymptotic stability of $F - KH'$ is crucial here. (Why?)

We see then that in high noise, the signal model covariance will be close to the optimal error covariance [Eq. (6.6)] but that the error performance associated with a suboptimal filter is an order of magnitude closer to the optimal error covariance [Eq. (6.10)]. The gain and performance of the suboptimal filter [Eqs. (6.4), (6.5), and (6.8)] can be determined in a much simpler way than that of the optimal filter [Eqs. (6.3) and (6.2)].

Obviously, as $\lambda_{\min}(R) \longrightarrow \infty$, we have $\bar{\Sigma} \longrightarrow \bar{P}$ and $\bar{\Sigma}^p \longrightarrow \bar{P}$, and with $\lambda_{\min}(R) = \infty$, there is no point in filtering. What we have shown, however, is that as $R \longrightarrow \infty$,

$$\frac{\| \bar{\Sigma}^p - \bar{\Sigma} \|}{\| \bar{P} - \bar{\Sigma} \|} \longrightarrow 0$$

which shows that, if there is a point to filtering, the additional error resulting from the suboptimal filter becomes negligible the higher the output noise is.

What of the time-varying case? Steady-state equations are obviously replaced by recursive equations: provided that bounds are imposed on various system matrices, including R_k, R_k^{-1}, P_k, and Σ_k^p, one can obtain for certain constants C_1 and C_2

$$P_k - \Sigma_k \le C_1 [\max_k \lambda_{\min}(R_k)]^{-1} \tag{6.11}$$

$$K_k - K_k^d \le C_2 [\max_k \lambda_{\min}(R_k)]^{-2} \tag{6.12}$$

and the derivation carries through much as before. Exponential stability of the signal model is normally needed (else P_k can be unbounded), as is exponential stability of the suboptimal filter.

The ideas of this section originally flow from a study of high noise filtering in Wiener's book [27] and its extension to Kalman filtering in [28].

Main Point of the Section

In high noise, simplified formulas can be used to calculate the filter gain and performance.

6.7 CHANDRASEKHAR-TYPE, DOUBLING, AND NONRECURSIVE ALGORITHMS

Of course, the Kalman filter for the case when all the signal model matrices including the noise covariance matrices are time invariant can be solved using the more general time-varying theories discussed so far. And in fact a good way to solve the algebraic equation associated with the time-invariant filter for stationary signals,* namely,

$$\bar{\Sigma} = F[\bar{\Sigma} - \bar{\Sigma}H(H'\bar{\Sigma}H + R)^{-1}H'\bar{\Sigma}]F' + GQG' \qquad (7.1)$$

given $F, H, R, G,$ and Q as time-invariant matrices, is simply to solve for $\bar{\Sigma}$ as

$$\bar{\Sigma} = \lim_{k \to \infty} \Sigma_{k+1/k} \qquad (7.2)$$

where

$$\Sigma_{k+1/k} = F[\Sigma_{k/k-1} - \Sigma_{k/k-1}H(H'\Sigma_{k/k-1}H + R)^{-1}H'\Sigma_{k/k-1}]F' + GQG' \quad (7.3)$$

for some nonnegative definite $\Sigma_{0/-1}$ (say $\Sigma_{0/-1} = 0$). Then one computes the filter gain from $K = F\bar{\Sigma}H(H'\bar{\Sigma}H + R)^{-1}$.

There are, however, other ways of proceeding when the signal mod l is time invariant and the input and measurement noise are stationary. We shall describe three different approaches.

Chandrasekhar-type Algorithms

Methods are described in [29] based on the solution of so-called Chandrasekhar-type equations rather than the usual Riccati-type equation. The advantages of this approach are that there is a reduction in computational effort (at least in the usual case where the state dimension is much greater than the output dimension), and with moderately careful programming there is an elimination of the possibility of the covariance matrix becoming nonnegative. Interestingly, it is possible to compute the filter gain recursively, without simultaneously computing the error covariance. Of course, knowing the steady-state gain, one can easily obtain the steady-state error covariance. The approach described in [29] is now briefly summarized.

Once again we will be working with the now familiar time-invariant state-space signal model

$$x_{k+1} = Fx_k + Gw_k \qquad E[w_k w_l'] = Q\delta_{kl} \qquad (7.4)$$

$$z_k = H'x_k + v_k \qquad E[v_k v_l'] = R\delta_{kl} \qquad (7.5)$$

with $E[v_k w_l'] = 0$ and the Kalman filter equations

$$K_k = F\Sigma_{k/k-1}H(H'\Sigma_{k/k-1}H + R)^{-1} \qquad (7.6)$$

$$\hat{x}_{k+1/k} = (F - K_kH')\hat{x}_{k/k-1} + K_k z_k \qquad (7.7)$$

(It is possible to cope with dependent input and measurement noise.) A lemma is now introduced.

*Independent input and measurement noises are assumed.

LEMMA 7.1. With the definition $\delta\Sigma_k = \Sigma_{k+1/k} - \Sigma_{k/k-1}$ for arbitrary $\Sigma_{0/-1} \geq 0$,

$$\delta\Sigma_{k+1} = [F - K_{k+1}H'][\delta\Sigma_k + \delta\Sigma_k H(H'\Sigma_{k/k-1}H + R)^{-1}H'\delta\Sigma_k]$$
$$\times [F - K_{k+1}H']' \tag{7.8}$$

$$= [F - K_k H'][\delta\Sigma_k - \delta\Sigma_k H(H'\Sigma_{k+1/k}H + R)^{-1}H'\delta\Sigma_k]$$
$$\times [F - K_k H']' \tag{7.9}$$

Proof. From the Riccati equation (7.3) and Kalman gain equation (7.6), we have a difference equation for the increment in error covariance as

$$\delta\Sigma_{k+1} = F\delta\Sigma_k F' - K_{k+1}(H'\Sigma_{k+1/k}H + R)K'_{k+1}$$
$$+ K_k(H'\Sigma_{k/k-1}H + R)K'_k \tag{7.10}$$

But

$$K_{k+1} = F\Sigma_{k+1/k}H(H'\Sigma_{k+1/k}H + R)^{-1}$$
$$= [K_k(H'\Sigma_{k/k-1}H + R) + F\delta\Sigma_k H](H'\Sigma_{k+1/k}H + R)^{-1} \tag{7.11a}$$
$$= [K_k(H'\Sigma_{k+1/k}H + R - H'\delta\Sigma_k H) + F\delta\Sigma_k H](H'\Sigma_{k+1/k}H + R)^{-1}$$
$$= K_k + (F - K_k H')\delta\Sigma_k H(H'\Sigma_{k+1/k}H + R)^{-1} \tag{7.11b}$$

Substitution of this expression for K_{k+1} into the above expression for $\delta\Sigma_{k+1}$ and collection of terms yields (7.9). [The derivation of (7.8) along similar lines is left to the reader.]

The formulas of this lemma underpin the various equivalent Chandrasekhar-type equation sets (of which there are a number). We shall limit our presentation here to one set only, referring the reader to [29] and the problems for other sets. All the derivations depend on certain observations. First, as shown by (7.8), rank $\delta\Sigma_{k+1} \leq$ rank $\delta\Sigma_k$, so that rank $\delta\Sigma_k \leq$ rank $\delta\Sigma_0$ for all k. Second, $\delta\Sigma_k$ may be written as $Y_k M_k Y'_k$, where M_k is a square symmetric matrix of dimension equal to rank $\delta\Sigma_0$. Third, recursions for Y_k, M_k, K_k, and $\Omega_k = H'\Sigma_{k/k-1}H + R$ tend to figure in the various equation sets. One such set is provided by:

$$\Omega_{k+1} = \Omega_k + H'Y_k M_k Y'_k H \tag{7.12a}$$

$$K_{k+1} = (K_k\Omega_k + FY_k M_k Y'_k H)\Omega_{k+1}^{-1} \tag{7.12b}$$

$$Y_{k+1} = (F - K_{k+1}H')Y_k \tag{7.12c}$$

$$M_{k+1} = M_k + M_k Y'_k H\Omega_{k+1}^{-1}H'Y_k M_k \tag{7.12d}$$

with initializations provided by $\Omega_0 = H'\Sigma_{0/-1}H + R$, $K_0 = F\Sigma_{0/-1}H\Omega_0^{-1}$, while Y_0 and M_0 are found by factoring

$$\delta\Sigma_0 = F\Sigma_{0/-1}F' + GQG' - K_0\Omega_0^{-1}K'_0 - \Sigma_{0/-1}$$

as $Y_0 M_0 Y'_0$, with M_0 square and of dimension equal to rank $\delta\Sigma_0$. In case $\Sigma_{0/-1} = 0$ and Q has full rank, one can set $M_0 = Q$, $Y_0 = G$.

The derivation of (7.12) is easy. Thus (7.12a) is immediate from the definitions of Ω_k and $\delta\Sigma_k = Y_k M_k Y'_k$, (7.12b) follows from (7.11a) and (7.12c), and (7.12d) from (7.8). An equation for $\Sigma_{k+1/k}$ is also available, as

$$\Sigma_{k+1/k} = \Sigma_{k/k-1} + Y_k M_k Y'_k \tag{7.12e}$$

Especially if $\Sigma_{k+1/k}$ is not required, there can be computational advantage in using Chandrasekhar-type algorithms. This is most easily seen by studying the way the number of scalar quantities updated by the Riccati and Chandrasekhar approaches changes as n, the state-variable dimension, changes while input and output dimension remain constant. With the Riccati approach, the number varies as n^2, while with the Chandrasekhar approach it varies with n. For values of n that are not high with respect to input and output dimensions, the Riccati equations can however be better to use.

In the remainder of this subsection, we offer a number of miscellaneous comments.

1. Information filter Chandrasekhar-type equations can be developed (see [29]).
2. Perhaps surprisingly, algorithms very like the square root algorithms dealt with earlier can be used to update the quantities Ω_k, K_k, and $L_k = Y_k M_k^{1/2}$. For details, see [24]. This idea allows the introduction of some time variation in R and Q.
3. Recall that for optimal filtering from $k = 0$, one takes $\Sigma_{0/-1} = P_0$, the initial state covariance of the signal model. The Chandrasekhar equations make no requirement that P_0 be the steady-state signal model covariance, or zero for that matter—although in both these cases M_k turns out to have low dimension. (See the text and Prob. 7.1.) This means that the algorithms provide exact filters for a class of signal models with nonstationary outputs (though the outputs are asymptotically stationary).
4. It should be emphasized that one of the main uses of the Chandrasekhar algorithm is to determine the time-invariant filter equations; thus, transient values of K_k will be thrown away and $\Sigma_{k/k-1}$ need not be computed. The easiest initialization is $\Sigma_{0/-1} = 0$ (as then the dimensions of Y_0 and M_0 can be helpfully low.)

The Doubling Algorithm

The doubling algorithm is another tool for finding the limiting solution of the Riccati equation (7.3) associated with time-invariant models and stationary noise. It allows us to pass in one iteration from $\Sigma_{k/k-1}$ to $\Sigma_{2k/2k-1}$ rather than $\Sigma_{k+1/k}$, provided that along with $\Sigma_{k/k-1}$ one updates three other matrices of the same dimension. Though it can be used for arbitrary initial conditions, we shall take $\Sigma_{0/-1} = 0$, since this allows us to get away with updating three rather than four matrices. Doubling algorithms have been part of the folklore associated with Riccati equations in linear systems problems for some time. We are unable to give any original reference containing material close to that presented here; however, more recent references include [31–33], with the latter surveying various approaches to the algorithm.

Doubling algorithm.

$$\alpha_{k+1} = \alpha_k (I + \beta_k \gamma_k)^{-1} \alpha_k \tag{7.13a}$$

$$\beta_{k+1} = \beta_k + \alpha_k (I + \beta_k \gamma_k)^{-1} \beta_k \alpha_k' \tag{7.13b}$$

$$\gamma_{k+1} = \gamma_k + \alpha_k' \gamma_k (I + \beta_k \gamma_k)^{-1} \alpha_k \tag{7.13c}$$

with

$$\alpha_1 = F' \qquad \beta_1 = HR^{-1}H' \qquad \gamma_1 = GQG' \tag{7.13d}$$

Moreover,

$$\gamma_k = \Sigma_{2^k/2^k - 1}$$

We remark that if β, γ are symmetric matrices, so are $(I + \beta\gamma)^{-1}\beta$ and $\gamma(I + \beta\gamma)^{-1}$, assuming the inverse exists. (Prove this in two lines!) This allows one to show that β_k and γ_k in (7.13) are symmetric for all k.

We turn now to a proof of the algorithm. It proceeds via several steps and may be omitted at first reading. For convenience in the proof, we shall assume F is nonsingular.

Relationship between Riccati and linear equations.

As the first step in proving (7.13), we develop a relation between Riccati and linear equations. Using (7.3), we have

$$\begin{aligned}
\Sigma_{k+1/k} &= F\Sigma_{k/k-1}(I + HR^{-1}H'\Sigma_{k/k-1})^{-1}F' + GQG' \\
&= F\Sigma_{k/k-1}[(F')^{-1} + (F')^{-1}HR^{-1}H'\Sigma_{k/k-1}]^{-1} \\
&\quad + GQG'[(F')^{-1} + (F')^{-1}HR^{-1}H'\Sigma_{k/k-1}] \\
&\quad \times [(F')^{-1} + (F')^{-1}HR^{-1}H'\Sigma_{k/k-1}]^{-1} \\
&= \{GQG'(F')^{-1} + [F + GQG'(F')^{-1}HR^{-1}H']\Sigma_{k/k-1}\} \\
&\quad \times \{(F')^{-1} + (F')^{-1}HR^{-1}H'\Sigma_{k/k-1}\}^{-1} \\
&= (C + D\Sigma_{k/k-1})(A + B\Sigma_{k/k-1})^{-1} \tag{7.14}
\end{aligned}$$

with obvious definitions of A, B, C, D. Now consider the linear equation with square X_k, Y_k:

$$\begin{bmatrix} X_{k+1} \\ Y_{k+1} \end{bmatrix} = \begin{bmatrix} A & B \\ C & D \end{bmatrix} \begin{bmatrix} X_k \\ Y_k \end{bmatrix} \tag{7.15}$$

Equation (7.14) shows that if X_k, Y_k are such that $Y_k X_k^{-1} = \Sigma_{k/k-1}$, then one must have $Y_{k+1}X_{k+1}^{-1} = \Sigma_{k+1/k}$. To accommodate $\Sigma_{0/-1} = 0$, we can take $X_0 = I, Y_0 = 0$.

Fast iteration of the linear equation. To obtain a doubling-type iteration of $\Sigma_{k/k-1}$, we shall obtain a doubling-type iteration of (7.15). It is clear that by squaring one can successively compute

$$\Phi(2) = \begin{bmatrix} A & B \\ C & D \end{bmatrix}^2, \quad \Phi(4) = \begin{bmatrix} A & B \\ C & D \end{bmatrix}^4, \quad \ldots, \quad \Phi(2^k) = \begin{bmatrix} A & B \\ C & D \end{bmatrix}^{2^k}$$

without computing intermediate powers. Then one easily obtains the matrix pairs $(X_2, Y_2), (X_4, Y_4), \ldots, (X_{2^k}, Y_{2^k})$ in sequence. We therefore need an efficient iteration for $\Phi(2^k)$. This flows from a special property of $\Phi(1)$.

Symplectic property of $\Phi(1)$. A $2n \times 2n$ matrix Z is termed symplectic if $Z'JZ = J$, where

$$J = \begin{bmatrix} 0 & -I_n \\ I_n & 0 \end{bmatrix}$$

It is easily verified from the identifications of A, B, C, D in (7.14) that

$$\Phi(1) = \begin{bmatrix} A & B \\ C & D \end{bmatrix} = \begin{bmatrix} (F')^{-1} & (F')^{-1}HR^{-1}H' \\ GQG'(F')^{-1} & F + GQG'(F')^{-1}HR^{-1}H' \end{bmatrix} \quad (7.16)$$

is symplectic. The definition of symplectic matrices shows that if Z is symplectic, so is any power of Z; therefore $\Phi(2^k)$ is symplectic for all k. Now a further property of symplectic matrices, easily verified, is the following. If a symplectic Z is written as

$$Z = \begin{bmatrix} Z_{11} & Z_{12} \\ Z_{21} & Z_{22} \end{bmatrix}$$

with Z_{11} nonsingular, then $Z_{22} = (Z'_{11})^{-1} + Z_{21}Z_{11}^{-1}Z_{12}$. This means that if $\Phi(1)$ is written in the form

$$\Phi(1) = \begin{bmatrix} \alpha_1^{-1} & \alpha_1^{-1}\beta_1 \\ \gamma_1\alpha_1^{-1} & \alpha_1' + \gamma_1\alpha_1^{-1}\beta_1 \end{bmatrix} \quad (7.17)$$

which may be done by defining α_1, β_1, γ_1 as in (7.13d), then $\Phi(2^k)$ has the form

$$\Phi(2^k) = \begin{bmatrix} \alpha_k^{-1} & \alpha_k^{-1}\beta_k \\ \gamma_k\alpha_k^{-1} & \alpha_k' + \gamma_k\alpha_k^{-1}\beta_k \end{bmatrix} \quad (7.18)$$

assuming α_k^{-1} exists.

Proof of the doubling algorithm. Using the fact that $\Phi(2^{k+1}) = \Phi(2^k)\Phi(2^k)$, together with the definitions of α_k, β_k, and γ_k in (7.18), the recursions of (7.13a) through (7.13d) are easily obtained. Since with $X_0 = I$, $Y_0 = 0$,

$$\begin{bmatrix} X_{2^k} \\ Y_{2^k} \end{bmatrix} = \Phi(2^k) \begin{bmatrix} X_0 \\ Y_0 \end{bmatrix} = \begin{bmatrix} \alpha_k^{-1} \\ \gamma_k\alpha_k^{-1} \end{bmatrix}$$

we see that $\Sigma_{2^k/2^k-1} = Y_{2^k}X_{2^k}^{-1} = \gamma_k$.

The speed of the doubling algorithm is not in doubt. Numerical stability appears not to be a problem.

Algebraic Nonrecursive Solution of the Steady-state Equation

In [30], a method is presented for solving the steady-state equation (7.1) by studying the eigenvalues and eigenvectors of the matrix $\Phi(1)$ defined in (7.16). Since any symplectic matrix Z is such that $Z^{-1} = J^{-1}Z'J$, i.e., Z^{-1} is similar to Z', both Z^{-1} and Z must have the same eigenvalues; equivalently, eigenvalues of Z occur in reciprocal pairs.

Suppose that eigenvalues of $\Phi(1)$ are distinct. Then for some square T_{ij} we have

$$\Phi(1) = \begin{bmatrix} T_{11} & T_{12} \\ T_{21} & T_{22} \end{bmatrix} \begin{bmatrix} \Lambda & 0 \\ 0 & \Lambda^{-1} \end{bmatrix} \begin{bmatrix} T_{11} & T_{12} \\ T_{21} & T_{22} \end{bmatrix}^{-1} \tag{7.19}$$

with the diagonal entries of Λ of modulus at least 1. The T_{ij} are defined by eigenvectors of $\Phi(1)$.

Suppose further that the diagonal entries of Λ have modulus strictly greater than 1. (This will in fact be the case, but we shall omit a proof.) Then the desired steady state $\bar{\Sigma}$ is given by

$$\bar{\Sigma} = T_{21}T_{11}^{-1} \tag{7.20}$$

It is not difficult to see this, subject to a qualification given below. Set

$$\begin{bmatrix} S_{11} & S_{12} \\ S_{21} & S_{22} \end{bmatrix} = \begin{bmatrix} T_{11} & T_{12} \\ T_{21} & T_{22} \end{bmatrix}^{-1}$$

Then with $X_0 = I$, $Y_0 = 0$, the solution of the linear equation (7.15) is given by

$$\begin{bmatrix} X_k \\ Y_k \end{bmatrix} = \Phi(k) \begin{bmatrix} I \\ 0 \end{bmatrix}$$

$$= \begin{bmatrix} T_{11} & T_{12} \\ T_{21} & T_{22} \end{bmatrix} \begin{bmatrix} \Lambda^k & 0 \\ 0 & \Lambda^{-k} \end{bmatrix} \begin{bmatrix} S_{11} \\ S_{21} \end{bmatrix}$$

Therefore,

$$Y_k X_k^{-1} = [T_{21}\Lambda^k S_{11} + T_{22}\Lambda^{-k}S_{21}][T_{11}\Lambda^k S_{11} + T_{12}\Lambda^{-k}S_{21}]^{-1}$$

Under the assumption on Λ and a further assumption that S_{11} is nonsingular, we have

$$\lim_{k \to \infty} \Sigma_{k/k-1} = \lim_{k \to \infty} Y_k X_k^{-1} = T_{21}T_{11}^{-1}$$

as required. It turns out that in problems of interest, S_{11} is nonsingular, so the method is valid. The theory may however run into difficulties if R is singular and one attempts to use a pseudo-inverse in place of an inverse. From the numerical point of view, it is unclear that the technique of this subsection will be preferred to those given earlier.

Via Chandrasekhar-type algorithms, recursive equations are available for the transient filter gain associated with a time-invariant signal model with constant Q, R and arbitrary P_0. Particularly for $P_0 = 0$, these equations may involve fewer quantities than the Riccati equation, and therefore be more attractive computationally. Via doubling algorithms, equations are available for recursively computing $\Sigma_{2^k/2^k-1}$ for $k = 1, 2, \ldots$. The steady state error covariance can also be determined in terms of eigenvectors of a certain $2n \times 2n$ matrix.

Problem 7.1. Show that the following equations can be used in lieu of (7.12).

$$Y_{k+1} = (F - K_k H') Y_k$$

$$M_{k+1} = M_k - M_k Y_k' H \Omega_{k+1}^{-1} H' Y_k M_k$$

$$K_{k+1} = K_k + Y_{k+1} M_k Y_k' H \Omega_{k+1}^{-1}$$

with Ω_{k+1} and $\Sigma_{k+1/k}$ given as before. Explain why, if $M_0 \geq 0$, (7.12) are to be preferred, while if $M_0 \leq 0$, these equations are to be preferred. Show that $M_0 \leq 0$ if $\Sigma_{0/-1} = \bar{P}$, where $\bar{P}$ is the signal model state covariance, i.e., the solution of $\bar{P} = F\bar{P}F' + GQG'$. Compare the dimension of M_0 in this case with that applying for arbitrary $\Sigma_{0/-1}$.

Problem 7.2. Obtain Chandrasekhar-type equations for the case when there is dependent input and output noise, thus $E[v_k w_l'] = S\delta_{kl}$, $S \neq 0$.

Problem 7.3. Derive Chandrasekhar-type equations associated with an information filter.

Problem 7.4. Why would it be unlikely that one could couple, at least usefully, Chandrasekhar-type equations and sequential processing?

Problem 7.5. Establish doubling algorithm equations to cover the case when $\Sigma_{0/-1}$ is arbitrary. (*Hint:* Use the same equations for α_k, β_k, and γ_k as in the text and one other equation.)

REFERENCES

[1] Jazwinski, A. H., *Stochastic Processes and Filtering Theory*, Academic Press, Inc., New York and London, 1970.

[2] Heffes, H., "The Effect of Erroneous Models on the Kalman Filter Response," *IEEE Trans. Automatic Control*, Vol. AC-11, No. 3, July 1966, pp. 541–543.

[3] Battin, R. H., *Astronautical Guidance*, McGraw-Hill Book Company, New York, 1964.

[4] FAGIN, S. L., "Recursive Linear Regression Theory, Optimal Filter Theory, and Error Analysis of Optimal Systems," *IEEE Intern. Conv. Record*, Vol. 12, 1964, pp. 216–240.

[5] NISHIMURA, T., "On the a Priori Information in Sequential Estimation Problems," *IEEE Trans. Automatic Control*, Vol. AC-11, No. 2, April 1966, pp. 197–204, and Vol. AC-12, No. 1, February 1967, p. 123.

[6] SAGE, A. P., and J. L. MELSA, *Estimation Theory with Applications to Communications and Control*, McGraw-Hill Book Company, New York, 1971.

[7] FITZGERALD, R. J., "Divergence of the Kalman Filter," *IEEE Trans. Automatic Control*, Vol. AC-16, No. 6, December 1971, pp. 736–747.

[8] ANDERSON, B. D. O., "Exponential Data Weighting in the Kalman-Bucy Filter," *Inform. Sciences*, Vol. 5, 1973, pp. 217–230.

[9] SORENSON, H. W., and J. E. SACKS, "Recursive Fading Memory Filtering," *Inform. Sciences*, Vol. 3, 1971, pp. 101–119.

[10] FRASER, D. C., "A New Technique for the Optimal Smoothing of Data," *M.I.T. Instrumentation Lab.*, Report T-474, January, 1967.

[11] KAMINSKI, P. G., A. E. BRYSON, and S. F. SCHMIDT, "Discrete Square Root Filtering: A Survey of Current Techniques," *IEEE Trans. Automatic Control*, Vol. AC-16, No. 6, December 1971, pp. 727–735.

[12] BIERMAN, G. J., "A Comparison of Discrete Linear Filtering Algorithms," *IEEE Trans. Aerospace and Electronic Systems*, Vol. AES-9, No. 1, January 1973, pp. 28–37.

[13] MENDEL, J., "Computational Requirements for a Discrete Kalman Filter," *IEEE Trans. Automatic Control*, Vol. AC-16, No. 6, December 1971, pp. 748–758.

[14] SINGER, R. A., and R. G. SEA, "Increasing the Computational Efficiency of Discrete Kalman Filters," *IEEE Trans. Automatic Control*, Vol. AC-16, No. 3, June 1971, pp. 254–257.

[15] RALSTON, A., *A First Course on Numerical Analysis*, McGraw-Hill Book Company, New York, 1965.

[16] BELLANTONI, J. F., and K. W. DODGE, "A Square Root Formulation of the Kalman-Schmidt Filter," *AIAA J.*, Vol. 6, June 1968, pp. 1165–1166.

[17] ANDREWS, A., "A Square Root Formulation of the Kalman Covariance Equations," *AIAA J.*, Vol. 6, June 1968, pp. 1165–1166.

[18] SCHMIDT, S. F., "Computational Techniques in Kalman Filtering," in *Theory and Applications of Kalman Filtering*, NATO Advisory Group for Aerospace Research and Development, AGARDograph 139, February 1970.

[19] GOLUB, G. H., "Numerical Methods for Solving Linear Least Squares Problems," *Numer. Math.*, Vol. 7, 1965, pp. 206–216.

[20] BUSINGER, P., and G. H. GOLUB, "Linear Least Squares Solution by Householder Transformations," *Numer. Math.*, Vol. 7, 1965, pp. 269–276.

[21] HANSON, R. J., and C. L. LAWSON, "Extensions and Applications of the House-holder Algorithms for Solving Linear Least Squares Problems," *Math. Comput.*, Vol. 23, October 1969, pp. 787–812.

[22] DYER, P., and S. McREYNOLDS, "Extension of Square-Root Filtering to Include Process Noise," *J. Optimiz. Theory Appl.*, Vol. 3, No. 6, 1969, pp. 444–459.

[23] BIERMAN, G. J., "Sequential Square Root Filtering and Smoothing of Discrete Linear Systems," *Automatica*, Vol. 10, March 1974, pp. 147–158.

[24] MORF, M., and T. KAILATH, "Square-root Algorithms for Least-squares Esti-mation," *IEEE Trans. Automatic Control*, Vol. AC-20, No. 4, August 1975, pp. 487–497.

[25] BIERMAN, G. J., "Measurement Updating Using the U-D Factorization," *Proc. 1975 IEEE Conf. on Decision and Control*, Houston, Tex., pp. 337–346.

[26] THORNTON, C. L., and G. J. BIERMAN, "Gram-Schmidt Algorithms for Covari-ance Propagation," *Proc. 1975 IEEE Conf. on Decision and Control*, Houston, Tex., pp. 489–498.

[27] WIENER, N., *Extrapolation, Interpolation and Smoothing of Stationary Time Series*, The M.I.T. Press, Cambridge, Mass., 1949.

[28] ANDERSON, B. D. O., and J. B. MOORE, "Optimal State Estimation in High Noise," *Inform. and Control*, Vol. 13, No. 4, October 1968, pp. 286–294.

[29] MORF, M., G. S. SIDHU, and T. KAILATH, "Some New Algorithms for Recur-sive Estimation in Constant, Linear, Discrete-time Systems," *IEEE Trans. Automatic Control*, Vol. AC-19, No. 4, August 1974, pp. 315–323.

[30] VAUGHAN, D. R., "A Nonrecursive Algebraic Solution for the Discrete Riccati Equation," *IEEE Trans. Automatic Control*, Vol. AC-15, No. 5, October 1970, pp. 597–599.

[31] BIERMAN, G. J., "Steady-state Covariance Computation for Discrete Linear Systems," *Proc. 1971 JACC*, paper No. 8-C3.

[32] FRIEDLANDER, B., T. KAILATH, and L. LJUNG, "Scattering Theory and Linear Least Squares Estimation," *J. Franklin Inst.*, Vol. 301, Nos. 1 and 2, January–February 1976, pp. 71–82.

[33] ANDERSON, B. D. O., "Second-order Convergent Algorithms for the Steady-state Riccati Equation," *Int. J. Control*, to appear.

[34] BURRUS, C. S., "Block Realization of Digital Filters," *IEEE Trans. Audio Electroc.*, Vol. AU-20, No. 4, pp. 230–235.

CHAPTER **7**

SMOOTHING OF
DISCRETE-TIME SIGNALS

7.1 INTRODUCTION TO SMOOTHING

Our results so far have been chiefly concerned with the Kalman filtering problem where an estimate of a signal x_k is made based on the noisy measurement set $\{z_0, z_1, \ldots, z_{k-1}\}$, denoted Z_{k-1} for short, or the set $\{z_0, z_1, \ldots, z_k\}$, denoted by Z_k. No delay need exist between the receipt of the last measurement z_{k-1} or z_k and production of the estimate $\hat{x}_{k/k-1}$ or $\hat{x}_{k/k}$. However, should a delay in the production of an estimate of x_k be permitted, then one could conceive of more measurements becoming available during the delay interval and being used in producing the estimate of x_k. Thus a delay of N time units, during which $z_{k+1}, \ldots, z_{k+N}$ appear, allows estimation of x_k by

$$\hat{x}_{k/k+N} = E[x_k | z_0, z_1, \ldots, z_{k+N}]$$

We term such an estimate a *smoothed estimate*. Any estimator producing a smoothed estimate is termed a *smoother*.

Because more measurements are used in producing $\hat{x}_{k/k+N}$ than in producing $\hat{x}_{k/k}$, one expects the estimate to be more accurate, and generally, one expects smoothers to perform better than filters, although inherent in a smoother is a delay and, as it turns out, an increase in estimator complexity. Further, the greater the delay, the greater the increase in complexity. Thus

165

it is important to examine the trade-offs between delay in processing data, improvement in performance, estimator complexity, and design difficulty. In many practical situations, a small delay is of little or no consequence, and the limiting factor is estimator complexity.

The particular classes of smoothing problems we shall consider in this chapter are those which can be solved by applying the Kalman filter results of the earlier chapters. Thus we consider least squares smoothing for discrete-time gaussian signals with additive, gaussian, and white measurement noise, or linear least squares smoothing for linear signal models where the gaussian assumption on the input and measurement noise is relaxed. The specific problem which we will look at is the computation of the conditional mean estimate

$$\hat{x}_{k-N/k} = E\{x_{k-N} | Z_k\}$$

(or, more generally, the linear least squares estimate $E^*\{x_{k-N} | Z_k\}$.

Clearly, for most smoothing applications, it is unnecessary to construct estimators which make available the estimates $\hat{x}_{k-N/k}$ for all k and for all N. Historically, three particular types of smoothing problems have been studied, each characterized by the particular subset of all possible smoothed estimates sought. *Fixed-point smoothing* is concerned with achieving smoothed estimates of a signal x_j for some fixed point j, i.e., with obtaining $\hat{x}_{j/j+N}$ for fixed j and all N. *Fixed-lag smoothing* is concerned with *on-line* smoothing of data where there is a fixed delay N between signal reception and the availability of its estimate, i.e., with obtaining $\hat{x}_{k-N/k}$ for all k and fixed N. *Fixed-interval smoothing* is concerned with the smoothing of a finite set of data, i.e., with obtaining $\hat{x}_{k/M}$ for fixed M and all k in the interval $0 \leq k \leq M$. It turns out that the various types of optimal smoothers which arise from the solution of the above problems consist of the optimal filter augmented with additional dynamics. (Actually, the term "smoother" is frequently used to denote just the system driven from the filter rather than the combination of the filter and this system, and the term "smoothing equations" is used to refer to the equations additional to the filtering equations. Both the above usages of the terms "smoother" and "smoothing equations" will be employed throughout the chapter since the particular usage intended, where this is important, can be determined from the context.) The various types of smoothers are now considered in turn.

Types of Smoothers

The *optimal fixed-point smoother* provides the optimal estimate of x_j for some critical and fixed point j based on measurement data Z_k, where $k = j + 1, j + 2, \ldots$. The fixed-point smoother output is thus the sequence $\hat{x}_{j/j+1}, \hat{x}_{j/j+2}, \ldots$ for some fixed j.

Fixed-point smoothing is useful where the *initial states* of experiments or processes are to be estimated as the experiment or process proceeds. Examples could be the estimation of the initial state of a satellite at the time of injection into orbit using the orbit tracking data, or the estimation of the initial states of a chemical process using data obtained from monitoring the process.

As already noted, smoothing will give better estimates than filtering, or, more precisely, estimates with smaller error covariance matrices. Since the achieving of these more accurate estimates is the *raison d'être* of smoothing, it is clear that the following two questions are relevant:

1. How does the improvement in use of the estimate $\hat{x}_{j/k}$ instead of $\hat{x}_{j/j}$ vary as k increases?
2. What is the maximum improvement possible, i.e., what is the improvement associated with $\lim_{k \to \infty} \hat{x}_{j/k}$ or $\hat{x}_{j/k}$ for large k?

As we shall see later in this chapter, for some estimation problems where there is a high signal-to-noise ratio, the improvement due to smoothing may be quite significant, perhaps greater than fifty percent, whereas for other estimation problems where there is a low signal-to-noise ratio, the improvement may be insignificant.

Later in this chapter, we shall also see that the improvement in estimation due to smoothing is monotone increasing as the interval $k - j$ increases, with the amount of improvement becoming effectively constant with a large enough interval. This interval is of the order of several times the dominant time constant of the filter, so that it is not necessary to introduce a delay of more than two or three times the dominant time constant of the filter to achieve essentially all the improvement due to smoothing that it is possible to achieve.

The *optimal fixed-lag smoother*, as previously noted, provides an optimal estimate of a signal or state x_{k-N}, for some fixed-lag N, based on noisy measurements of $x_0, x_1, \ldots, x_k$. Now inherent in the fixed-lag smoother is a delay between the generation of a signal x_{k-N} and its estimation as $\hat{x}_{k-N/k}$. This delay of N times the sampling interval is the same for all k, and this fact justifies the nomenclature fixed-lag smoothing. For the case when N is two or three times the dominant time constant of the optimal filter, from what has been said concerning the fixed-point smoother above, we can see that essentially as much improvement as it is possible to achieve via smoothing is achieved by the fixed-lag smoother.

What are the possible applications of fixed-lag smoothers? Most communication system applications do in fact permit a delay between signal generation and signal estimation. There is usually an inherent delay in signal transmission anyway, and so it does not appear unreasonable to permit an additional delay, possibly very small in comparison with the transmission delay, to achieve improved signal estimation. Of course, the key question which must be asked in such applications is whether or not the extra complexity in receiver design needed to achieve fixed-lag smoothing is warranted by the performance improvement due to smoothing. For example, is a fifty percent improvement in estimation worth the expense of doubling the estimator complexity? Clearly, the answer depends on the application, but the fact that there are important applications where the answer is strongly affirmative justifies investigation of this topic.

It may have been the case that a number of estimation problems in the aerospace industry, ideal for application of fixed-lag smoothing results, have been handled using the less efficient off-line fixed-interval smoothing techniques. The inefficiency of these off-line techniques should become clear in the following discussion of fixed-interval smoothing.

The *optimal fixed-interval smoother* yields the optimal smoothed estimate of a signal at each time in a fixed interval $[0, M]$, given noisy measurements of the signal at each time in the interval. Historically, fixed-interval smoothing has been a truly off-line process requiring one pass through an optimal filter (possibly on-line), storage of relevant data, and a second pass involving a time reversal through a second filter. Such fixed-interval smoothing has certainly found wide application, at least in the space program, where the off-line processing of data has been justified by the need for data smoothing of some sort. For smoothing of short data sequences, fixed-interval smoothing is an attractive proposition, but for extended sequences a quasi-optimal fixed-interval smoothing by means of fixed-lag smoothing is undoubtedly simpler to implement.

The basic idea of quasi-optimal fixed interval smoothing is as follows. Let N be chosen to be several times the dominant filter time constant and consider the case, $N \ll M$. Using a fixed-lag smoother, one evaluates $\hat{x}_{k/k+N}$ for $k = 0, 1, \ldots, M - N$; for each such k one will have $\hat{x}_{k/k+N} \doteq \hat{x}_{k/M}$. To obtain $\hat{x}_{k/M}$ for $k > M - N$, two techniques are available, as described in Sec. 7.4. One technique works by postulating that measurements are available on $[M + 1, M + N]$ with an infinite output noise covariance. Then $\hat{x}_{k/k+N} = \hat{x}_{k/M}$ for $M - N < k \leq M$, and the fixed-lag smoother can still be used to complete the fixed-interval smoothing. Description of the second technique will be deferred.

History of Discrete-time Smoothing Results

Numerous papers on the state-space approach to smoothing for linear dynamical systems have appeared since the early 1960s. Hard on the heels of solutions to various filtering problems via these techniques came corresponding solutions to smoothing problems. Reference [1] is a survey of many of the smoothing results now available, with reference to over 100 papers.

Smoothing as a significant topic in textbooks is more recent [2, 3, 4]. However, in all the works on the subject of smoothing there are but a relatively few devoted to applications, and indeed the application of smoothing algorithms to control and communication system problems is still a wide-open field.

Among the early references are those of Rauch [5] and Rauch *et al.* [6] who developed sequential algorithms for discrete-time, optimal fixed-interval, fixed-point, and fixed-lag smoothing. Weaver [7] and Lee [8] looked at the fixed-point and fixed-interval problems, respectively, using an alternative approach. Mayne [9] showed that numerical advantages in computation accrue in calculating the fixed-interval estimates as the sum of a forward-time and reverse-time optimal filtered estimate, rather than as a correction to filtered estimates as in earlier results. Fraser and Potter [10] further developed this particular approach and Bryson and Henrikson [11] looked at the time-correlated measurement noise case.

More recently, Kelly and Anderson [12] showed that earlier fixed-lag smoothing algorithms were in fact computationally unstable. Stable algorithms have since been demonstrated by a number of authors including Chirarattananon and Anderson [13] and Moore [14]. The approach taken in [14] is used to develop the fixed-lag smoothing results of this chapter. The three types of smoothing problems are viewed in this chapter as Kalman filtering problems associated with an *augmented* signal model, i.e., a signal model of which the original signal model forms a component part. The development parallels to some extent earlier work by Zachrisson [15] and Willman [16] on fixed-point smoothing, by Premier and Vacroux [17] on filtering associated with systems having delay elements, and by Farooq and Mahalanabis [18] on fixed-lag smoothing.

High signal-to-noise ratio results, and the significance of the filter dominant time constant in defining smoothing lags are described in [19, 20], in the context of continuous-time results. The ideas however are equally applicable to discrete time. (Close parallels between discrete-time and continuous-time linear-quadratic problems are normally standard; for the case of the fixed-lag smoother, however, this is not the case, since the continuous-time optimal smoother is infinite dimensional, while the discrete-time smoother is finite dimensional.)

Most recent work on smoothers has been concerned with finding alternative formulas and algorithms. As examples, we might quote [21–23]. Examples of applications can be found in [24–28].

In [27], for example, the problem is considered of constructing the track taken by a submarine during an exercise in the post-exercise phase, the reconstruction being to permit analysis and evaluation of the tactics used by the participants in the exercise. Various sensors provide positional information at discrete instants of time. In rough terms, a smoothing version of the filtering problem of Sec. 3.4 is tackled.

The Kalman filter itself turns out to be a good fixed-lag smoother for a wide range of applications as pointed out by Hedelin [29]. The estimate $\hat{x}_{k/k-1}$ in some instances is a suboptimal estimate of x_{k-N} for some N. This observation is explored more fully in [30].

7.2 FIXED-POINT SMOOTHING

In this section, we shall extend the Kalman filtering results of earlier chapters to provide fixed-point smoothing results. In fact we shall show that, from one point of view, *the fixed-point smoothing problem is a Kalman filtering problem in disguise and therefore may be solved by direct application of the Kalman filtering results.* First we define the fixed-point smoothing problem of interest, then develop the structure of the optimal fixed-point smoother and study its properties.

Discrete-time fixed-point smoothing problem. For the usual state-space signal model, determine the estimate

$$\hat{x}_{j/k} = E[x_j \,|\, Z_k] \tag{2.1}$$

and the associated error covariance

$$\Sigma_{j/k} = E\{[x_j - \hat{x}_{j/k}][x_j - \hat{x}_{j/k}]' \,|\, Z_k\} \tag{2.2}$$

for some fixed j and all $k > j$.

Derivation of Fixed-point Smoothing Results

Let us consider an augmenting state vector x_k^a for the signal model satisfying for $k \geq j$ the recursive equation

$$x_{k+1}^a = x_k^a \tag{2.3}$$

initialized at time instant j by $x_j^a = x_j$ as depicted in Fig. 7.2-1. From (2.3) we have immediately that

$$x_{k+1}^a = x_j \tag{2.4}$$

for all $k \geq j$ and thus from the definitions of conditional estimates and conditional covariances the following identifications can be made:

$$\hat{x}_{k+1/k}^a = \hat{x}_{j/k} \tag{2.5}$$

$$\Sigma_{k+1/k}^{aa} = \Sigma_{j/k} \tag{2.6}$$

Here, $\Sigma_{k+1/k}^{aa}$ denotes the covariance of the error $(\hat{x}_{k+1/k}^a - x_{k+1}^a)$.

The strategy we adopt is to simply apply Kalman filter results to the augmented model of Fig. 7.2-1 to obtain the filtered estimate $\hat{x}_{k+1/k}^a$ and its

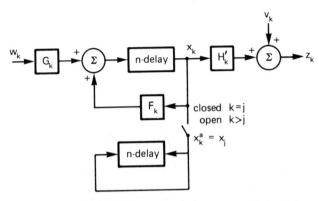

Fig. 7.2-1 Augmented signal model. The blocks labelled *n*-delay comprise an *n*-vector of parallel, not cascade, unit delays, with *n* the state-vector dimension.

error covariance $\Sigma^{aa}_{k+1/k}$ or, equivalently, the desired fixed-point estimate $\hat{x}_{j/k}$ and its error covariance $\Sigma_{j/k}$. (This approach was first pointed out in [15]. Alternative approaches are described in other texts [2–4]).

The augmented signal model is as follows:

$$\begin{bmatrix} x_{k+1} \\ x^a_{k+1} \end{bmatrix} = \begin{bmatrix} F_k & 0 \\ 0 & I \end{bmatrix}\begin{bmatrix} x_k \\ x^a_k \end{bmatrix} + \begin{bmatrix} G_k \\ 0 \end{bmatrix}w_k \tag{2.7}$$

$$z_k = [H'_k \quad 0]\begin{bmatrix} x_k \\ x^a_k \end{bmatrix} + v_k \tag{2.8}$$

with the state vector at $k = j$ satisfying $[x'_j \quad x^{a\prime}_j] = [x'_j \quad x'_j]$.

Formally then, the Kalman filter for the augmented signal model (2.7) and (2.8) is described for $k \geq j$ by the equations

$$\begin{bmatrix} \hat{x}_{k+1/k} \\ \hat{x}^a_{k+1/k} \end{bmatrix} = \left\{ \begin{bmatrix} F_k & 0 \\ 0 & I \end{bmatrix} - \begin{bmatrix} K_k \\ K^a_k \end{bmatrix}[H'_k \quad 0] \right\}\begin{bmatrix} \hat{x}_{k/k-1} \\ \hat{x}^a_{k/k-1} \end{bmatrix} + \begin{bmatrix} K_k \\ K^a_k \end{bmatrix}z_k$$

with the state at $k = j$ specified in terms of $\hat{x}_{j/j-1}$ since $x^a_j = x_j$ and thus $\hat{x}^a_{j/j-1} = \hat{x}_{j/j-1}$. The gain matrix is given for $k \geq j$ after a minor simplification as

$$\begin{bmatrix} K_k \\ K^a_k \end{bmatrix} = \begin{bmatrix} F_k & 0 \\ 0 & I \end{bmatrix}\begin{bmatrix} \Sigma_{k/k-1} & [\Sigma^a_{k/k-1}]' \\ \Sigma^a_{k/k-1} & \Sigma^{aa}_{k/k-1} \end{bmatrix}\begin{bmatrix} H_k \\ 0 \end{bmatrix}[H'_k\Sigma_{k/k-1}H_k + R_k]^{-1}$$

where the error covariance matrix of the augmented state vector is given from

$$\begin{bmatrix} \Sigma_{k+1/k} & [\Sigma^a_{k+1/k}]' \\ \Sigma^a_{k+1/k} & \Sigma^{aa}_{k+1/k} \end{bmatrix} = \begin{bmatrix} F_k & 0 \\ 0 & I \end{bmatrix}\begin{bmatrix} \Sigma_{k/k-1} & [\Sigma^a_{k/k-1}]' \\ \Sigma^a_{k/k-1} & \Sigma^{aa}_{k/k-1} \end{bmatrix}$$

$$\times \left\{ \begin{bmatrix} F'_k & 0 \\ 0 & I \end{bmatrix} - \begin{bmatrix} H_k \\ 0 \end{bmatrix}[K'_k \quad K^{a\prime}_k] \right\} + \begin{bmatrix} G_k \\ 0 \end{bmatrix}Q_k[G'_k \quad 0]$$

Notice that the matrix $\Sigma_{k/k-1}$ appearing here is precisely that associated with the usual Kalman filter equation; this is so since the first subvector of the state vector of the augmented filter is $\hat{x}_{k/k-1}$. The covariance at $k = j$ is given in terms of the filter covariance $\Sigma_{j/j-1}$ as

$$\begin{bmatrix} \Sigma_{j/j-1} & [\Sigma^a_{j/j-1}]' \\ \Sigma^a_{j/j-1} & \Sigma^{aa}_{j/j-1} \end{bmatrix} = \begin{bmatrix} \Sigma_{j/j-1} & \Sigma_{j/j-1} \\ \Sigma_{j/j-1} & \Sigma_{j/j-1} \end{bmatrix}$$

(In writing down this boundary condition for the above difference equation, one must use the fact that $x^a_j = x_j$.) These equations appear cumbersome, but, as one might expect, one can separate out the Kalman filter equations for the original signal model and the fixed-point smoothing equations. These latter equations are now extracted directly from the above augmented filter equations using (2.5) and (2.6).

The Fixed-point Smoothing Equations

The fixed-point smoother is as depicted in Fig. 7.2-2 and is described for $k \geq j$ by the equations

$$\hat{x}_{j/k} = \hat{x}_{j/k-1} + K^a_k \tilde{z}_k \tag{2.9}$$

with initial state $\hat{x}_{j/j-1}$. The gain matrix is given by

$$K^a_k = \Sigma^a_{k/k-1} H_k [H'_k \Sigma_{k/k-1} H_k + R_k]^{-1} \tag{2.10}$$

where

$$\Sigma^a_{k+1/k} = \Sigma^a_{k/k-1}[F_k - K_k H'_k]' \tag{2.11}$$

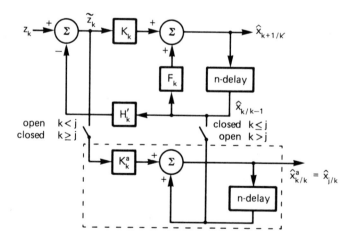

Fig. 7.2-2   Kalman filter for augmented signal model or equivalently Fixed-Point Smoother.

with initial value $\Sigma^a_{j/j-1} = \Sigma_{j/j-1}$. The covariance of the error term $(x_j - \hat{x}_{j/k})$ is given from

$$\Sigma_{j/k} = \Sigma_{j/k-1} - \Sigma^a_{k/k-1} H_k [K^a_k]' \tag{2.12}$$

or equivalently

$$\Sigma_{j/k} = \Sigma_{j/j-1} - \sum_{i=j}^{k} \{\Sigma^a_{i/i-1} H_i [K^a_i]'\}$$

$$= \Sigma_{j/j-1} - \sum_{i=j}^{k} \{\Sigma^a_{i/i-1} H_i [H'_i \Sigma_{i/i-1} H_i + R_i]^{-1} H'_i [\Sigma^a_{i/i-1}]'\} \tag{2.13}$$

The improvement from smoothing is $[\Sigma_{j/k} - \Sigma_{j/j-1}]$, as is given from (2.13). Sometimes it is helpful to express the improvement from smoothing in a rough way as a percentage:

$$\% \text{ improvement from smoothing} = \frac{tr\,[\Sigma_{j/k} - \Sigma_{j/j-1}]}{tr\,[\Sigma_{j/j-1}]} \times 100\%$$

Properties of the Fixed-point Smoother

Some properties of the fixed-point smoother follow immediately from the above fixed-point smoothing equations.

1. The fixed-point smoother is driven from the innovations process $\tilde{z}_k = (z_k - H'_k \hat{x}_{k/k-1})$ of the Kalman filter for the nonaugmented signal model. The smoother is a linear discrete-time system of dimension equal to that of the filter.

2. As in the case of the Kalman filter, the smoother parameters and error covariance are independent of the measurements and therefore can be computed beforehand. Also, as in the case of the filter, the relevant conditional probability density (of x_j given Z_k) is gaussian and is therefore defined by the conditional mean $\hat{x}_{j/k}$ and the conditional covariance $\Sigma_{j/k}$. A further point to notice is that in the event that $[H'_k \Sigma_{k/k-1} H_k + R_k]$ is singular, the inverse operation on this matrix, as in the filter calculations, may be replaced by the pseudo-inverse operation.

3. For the usual cases considered where F_k and $\Sigma_{k/k-1}$ are nonsingular for all k, the fixed-point smoother gain K^a_k may be expressed nonrecursively in terms of the Kalman filter gain K_k as follows:

$$K^a_k = \Sigma^a_{k/k-1} [F_k \Sigma_{k/k-1}]^{-1} K_k \tag{2.14}$$

where $\Sigma^a_{k/k-1}$ is expressed nonrecursively as

$$\Sigma^a_{k/k-1} = \Sigma_{j/j-1} \prod_{l=j}^{k-1} [F_l - K_l H'_l]' \tag{2.15}$$

4. For the time-invariant case when $H_k, K_k, \Sigma_{k/k-1}$, and F_k are indepen-

dent of k and are denoted by H, K, $\bar{\Sigma}$, and F, respectively, the above expression for $\Sigma^a_{k+1/k}$ and K^a_k and $\Sigma_{j/k}$ simplify as

$$K^a_k = \Sigma^a_{k/k-1}[F\bar{\Sigma}]^{-1}K \tag{2.16}$$

where

$$\Sigma^a_{k/k-1} = \bar{\Sigma}[F' - HK']^{k-j} \tag{2.17}$$

and

$$\Sigma_{j/k} = \bar{\Sigma} - \sum_{l=j}^{k}\{\Sigma^a_{l/l-1}H[H'\bar{\Sigma}H + R]^{-1}H'[\Sigma^a_{l/l-1}]'\}$$

$$= \bar{\Sigma} - \bar{\Sigma}\left\{\sum_{l=j}^{k}[\tilde{F}']^{l-j}H[H'\bar{\Sigma}H + R]^{-1}H'\tilde{F}^{l-j}\right\}\bar{\Sigma} \tag{2.18}$$

where the notation $\tilde{F} = [F - KH']$ is used for convenience. Notice that even for the case of a time-invariant signal model and a time-invariant Kalman filter, the fixed-point smoother is time varying since K^a_k is time varying. Notice also that further manipulations of (2.18) yield a linear matrix equation for the improvement due to smoothing $[\bar{\Sigma} - \Sigma_{j/k}]$ as

$$[\bar{\Sigma} - \Sigma_{j/k}] - \bar{\Sigma}\tilde{F}'\bar{\Sigma}^{-1}[\bar{\Sigma} - \Sigma_{j/k}]\bar{\Sigma}^{-1}\tilde{F}\bar{\Sigma}$$

$$= \bar{\Sigma}H[H'\bar{\Sigma}H + R]^{-1}H'\bar{\Sigma} - \bar{\Sigma}[\tilde{F}']^{k-j+1}H[H'\bar{\Sigma}H + R]^{-1}H'\tilde{F}^{k-j+1}\bar{\Sigma}$$

$$\tag{2.19}$$

This equation is derived by expanding its left-hand side using (2.18). Notice that in the limit as k approaches infinity, we have that $\lim_{k\to\infty}\tilde{F}^{k-j+1} = 0$ and thus (2.19) reduces to the linear matrix equation

$$[\bar{\Sigma} - \Sigma_{j/\infty}] - \bar{\Sigma}\tilde{F}'\bar{\Sigma}^{-1}[\bar{\Sigma} - \Sigma_{j/\infty}]\bar{\Sigma}^{-1}\tilde{F}\bar{\Sigma} = \bar{\Sigma}H[H'\bar{\Sigma}H + R]^{-1}H'\bar{\Sigma}$$

$$\tag{2.20}$$

The quantity $[\bar{\Sigma} - \Sigma_{j/\infty}]$ is the maximum improvement possible due to smoothing. This equation may be solved using standard techniques to yield a solution for $\bar{\Sigma} - \Sigma_{j/\infty}$ and thereby $\Sigma_{j/\infty}$. The equation also helps provide a rough argument to illustrate the fact that as *the signal-to-noise ratio decreases, the improvement due to smoothing disappears*. For suppose that with F, G, H, and Q fixed, R is increased. As we know, as $R \to \infty$, $\bar{\Sigma} \to \bar{P}$, the covariance of the state of the signal model, $K \to 0$, and $\tilde{F} \to F$. The left side of (2.20) therefore approaches

$$(\bar{P} - \Sigma_{j/\infty}) - \bar{P}F'\bar{P}^{-1}(\bar{P} - \Sigma_{j/\infty})\bar{P}^{-1}F\bar{P}$$

while the right side approaches zero. Therefore, $\bar{P} - \Sigma_{j/\infty}$ approaches zero, or, as $R \to \infty$, $\Sigma_{j/\infty} \to \bar{\Sigma}$. Rigorous analysis for continuous-time signal smoothing appears in [19], and Prob. 2.1 illustrates the claim.

5. The improvement due to smoothing $[\bar{\Sigma} - \Sigma_{j/k}]$ increases monotonic-

ally with increasing k, as indicated by (2.13). The eigenvalues of the Kalman filter matrix $\tilde{F} = [F - KH']$, and in particular the dominant eigenvalue of $[F - KH']$, govern the rate of increase. As a rule of thumb, we can say that *essentially all the smoothing that it is possible to achieve can be achieved in two or three dominant time constants of the Kalman filter.*

6. The stability question for the fixed-point smoother is trivial. Equation (2.9) is relevant, and shows that any initial error in the smoother state persists; i.e., the smoother is Lyapunov stable, but not asymptotically stable. Equation (2.9) also shows that $\hat{x}_{j/k}$ for large k will be computed as the sum of a large number of quantities; it is conceivable that round-off errors could accumulate. The solution, at least in the time-invariant case, is to set K_k^a, an exponentially decaying quantity, to be zero from some value of k onwards.

Fixed-point Signal Smoothers

Frequently, smoothing problems arise when an estimate of the entire state vector is not required, as in the problem of signal smoothing when the output signal $y_k = H'_k x_k$ is of lower dimension than that of the state vector. For such problems one might obtain the smoothed estimate of y_j using the smoothed state estimates as $\hat{y}_{j/k} = H'_j \hat{x}_{j/k}$, or more directly by premultiplying (2.9) by H'_j and using the identifications $\hat{y}_{j/k} = H'_j \hat{x}_{j/k}$ and $\hat{y}_{j/k-1} = H'_j \hat{x}_{j/k-1}$. Instead of an n-dimensional smoother driven from the Kalman filters, where n is the dimension of the state vector x_k, we have a p-dimensional signal smoother driven from the Kalman filter, where p is the dimension of the signal vector y_k.

The approach just described may also be applied to yield smoothed estimates of any other linear combination of states, say, $\bar{y}_k = \bar{H}'_k x_k$ for some specified $\bar{H}_k$.

Main Points of the Section

A study of fixed-point smoothing points up the fact that improvement due to smoothing is monotonic increasing as more measurement data becomes available. The time constant of this increase is dependent on the dominant time constant of the Kalman filter. As a rule of thumb, the smoother achieves essentially all the improvement possible from smoothing after two or three times the dominant time constant of the Kalman filter. This maximum improvement from smoothing is dependent on the signal-to-noise ratio and the signal model dynamics and can vary from zero improvement to approaching one hundred percent improvement for some signal models at high signal-to-noise ratios.

Problem 2.1. Suppose there is given a scalar, time-invariant signal model where $F_k = 0.95$, $H_k = 1$, $Q_k = 1$, and $R_k = 10$. Show that the maximum improvement due to fixed-point smoothing of the initial state $(\bar{\Sigma} - \Sigma_{0/\infty})$ is greater than thirty percent of $\bar{\Sigma}$. Show that the improvement is considerably less for the case $F_k = 0.95$, $H_k = 1$, $Q_k = 1$, and $R_k = 100$, and is negligible for the case $F_k = 0.1$, $H_k = 1$, $Q_k = 1$, and $R_k = 1$.

Problem 2.2. For the signal model as in Problem 2.1, where $F_k = 0.95$, $H_k = 1$, $Q_k = 1$, and $R_k = 10$, determine the improvement due to fixed-point smoothing of the initial state $(\bar{\Sigma} - \Sigma_{0/k})$ as a percentage of $\bar{\Sigma}$ for $k = 1, 2, 5$, and 10. Estimate the value of k for which the improvement due to smoothing is ninety percent of that which is possible. How does this value of k relate to the closed-loop Kalman filter eigenvalue?

Problem 2.3. Let $\bar{\Phi}(l, j)$ denote $\bar{F}_{l-1}\bar{F}_{l-2} \dots \bar{F}_j$, with $\bar{F}_l = F_l - K_l H_l'$. Show that

$$\hat{x}_{j/k} = \hat{x}_{j/j-1} + \sum_{l=j}^{k} \Sigma_{j/j-1}\bar{\Phi}'(l, j)H_l(H_l'\Sigma_{l/l-1}H_l + R_l)\tilde{z}_l$$

$$= \hat{x}_{j/j} + \sum_{l=j+1}^{k} \Sigma_{j/j-1}\bar{\Phi}'(l, j)H_l(H_l'\Sigma_{l/l-1}H_l + R_l)\tilde{z}_l.$$

These formulas will be used in Sec. 7.4.

7.3 FIXED-LAG SMOOTHING

The fixed-point smoothing results of the previous section certainly are able to provide insights into the nature of smoothing, but the fixed-point smoother has a rather limited range of applications. On the other hand, the fixed-lag smoother is a very useful device since it allows "on-line" production of smoothed estimates. It is possible to derive fixed-lag smoothing results directly from the fixed-point smoothing results of the previous section, or alternatively to derive the results using Kalman filter theory on an augmented signal model somewhat along the lines of the derivation of fixed-point smoothing. Since the second approach is perhaps the easier from which to obtain recursive equations, we shall study this approach in detail. We shall also point out very briefly the relationships between the fixed-point and fixed-lag smoothing equations. We begin our study with a precise statement of the fixed-lag smoothing problem.

Discrete-time fixed-lag smoothing problem. For the usual state-space signal model, determine for all k and some fixed-lag N recursive equations for the estimate

$$\hat{x}_{k-N/k} = E[x_{k-N}|Z_k] \tag{3.1}$$

and the associated error covariance

$$\Sigma_{k-N/k} = E\{[x_{k-N} - \hat{x}_{k-N/k}][x_{k-N} - \hat{x}_{k-N/k}]' \mid Z_k\} \qquad (3.2)$$

Derivation of Fixed-lag Smoothing Results

Consider the model of Fig. 7.3-1. It has the state equations

$$\begin{bmatrix} x_{k+1} \\ x_{k+1}^{(1)} \\ x_{k+1}^{(2)} \\ \cdot \\ \cdot \\ \cdot \\ x_{k+1}^{(N+1)} \end{bmatrix} = \begin{bmatrix} F_k & 0 & \cdots & 0 & 0 \\ I & 0 & & 0 & 0 \\ 0 & I & & 0 & 0 \\ \cdot & \cdot & \cdot & & \cdot \\ \cdot & \cdot & & \cdot & \cdot \\ \cdot & \cdot & & \cdot & \cdot \\ 0 & 0 & \cdots & I & 0 \end{bmatrix} \begin{bmatrix} x_k \\ x_k^{(1)} \\ x_k^{(2)} \\ \cdot \\ \cdot \\ \cdot \\ x_k^{(N+1)} \end{bmatrix} + \begin{bmatrix} G_k \\ 0 \\ 0 \\ \cdot \\ \cdot \\ \cdot \\ 0 \end{bmatrix} w_k \qquad (3.3)$$

$$z_k = [H_k' \quad 0 \quad \cdots \quad 0 \quad 0] \begin{bmatrix} x_k \\ x_k^{(1)} \\ x_k^{(2)} \\ \cdot \\ \cdot \\ \cdot \\ x_k^{(N+1)} \end{bmatrix} + v_k \qquad (3.4)$$

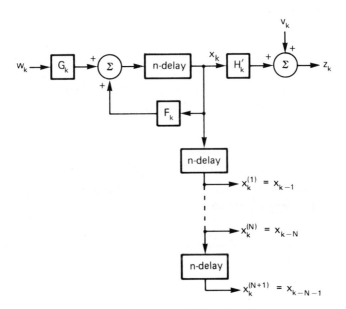

Fig. 7.3-1 Augmented signal model.

The model evidently is an augmentation of the basic signal model we have been dealing with. As is clear from the figure (or the equations),

$$x^{(1)}_{k+1} = x_k, \qquad x^{(2)}_{k+1} = x_{k-1}, \qquad \ldots, \qquad x^{(N+1)}_{k+1} = x_{k-N}$$

Therefore, conditioning on Z_k, we have

$$E[x^{(i)}_{k+1} \,|\, Z_k] = \hat{x}_{k+1-i/k} \qquad \text{for } i = 1, \ldots, N+1$$

In other words, *the one-step prediction estimate of the entire state of the augmented signal model contains within it smoothed estimates of the state of the basic signal model for lags up to N*. Kalman filter results applied to this augmented signal model lead to equations involving the following state estimates, augmented covariance matrix, and augmented Kalman gain matrix. The notation we adopt for these quantities is as follows:

$$\begin{bmatrix} \hat{x}_{k+1/k} \\ \hat{x}^{(1)}_{k+1/k} \\ \cdot \\ \cdot \\ \cdot \\ \hat{x}^{(N+1)}_{k+1/k} \end{bmatrix}, \quad \begin{bmatrix} \Sigma_{k+1/k} & [\Sigma^{(1)}_{k+1/k}]' & \cdots & [\Sigma^{(N+1)}_{k+1/k}]' \\ \Sigma^{(1)}_{k+1/k} & \Sigma^{(1,1)}_{k+1/k} & \cdots & [\Sigma^{(1,N+1)}_{k+1/k}]' \\ \cdot & \cdot & & \cdot \\ \cdot & \cdot & & \cdot \\ \cdot & \cdot & & \cdot \\ \Sigma^{(N+1)}_{k+1/k} & \Sigma^{(N+1,1)}_{k+1/k} & \cdots & \Sigma^{(N+1,N+1)}_{k+1/k} \end{bmatrix}, \quad \begin{bmatrix} K_k \\ K^{(1)}_k \\ \cdot \\ \cdot \\ \cdot \\ K^{(N+1)}_k \end{bmatrix}$$

For convenience, we make the identifications

$$\Sigma^{(i,0)}_{k+1/k} = \Sigma^{(i)}_{k+1/k} \quad \text{and} \quad \Sigma^{(0,0)}_{k+1/k} = \Sigma^{(0)}_{k+1/k} = \Sigma_{k+1/k}$$

As noted above, we have for $i = 0, 1, \ldots, N$

$$\hat{x}^{(i+1)}_{k+1/k} = \hat{x}_{k-i/k} \tag{3.5}$$

so that

$$\Sigma^{(i+1,i+1)}_{k+1/k} = \Sigma_{k-i/k} \tag{3.6}$$

A little reflection shows that the correct initializations of the augmented state estimate and covariance are given for $i, j = 0, 1, \ldots, N+1$ by

$$\hat{x}_{0/-1} = \bar{x}_0 \qquad \hat{x}^{(i)}_{0/-1} = 0 \quad \text{for } i \neq 0 \tag{3.7}$$

and

$$\Sigma_{0/-1} = P_0 \qquad \Sigma^{(i,j)}_{0/-1} = 0 \quad \text{for } i, j \text{ not both zero} \tag{3.8}$$

The smoothing equations will now be derived, with one or two intermediate steps relegated to a problem.

The Fixed-lag Smoothing Equations

These equations are extracted directly from the augmented Kalman filter equations as

$$\hat{x}^{(i+1)}_{k+1/k} = \hat{x}^{(i)}_{k/k-1} + K^{(i+1)}_k \tilde{z}_k$$

for $0 \leq i \leq N$ or, equivalently, by virtue of (3.5),

$$\hat{x}_{k-i/k} = \hat{x}_{k-i/k-1} + K_k^{(i+1)} \tilde{z}_k \tag{3.9}$$

where $\hat{x}_{0/-1} = \bar{x}_0$ and $\hat{x}_{-i/-1} = 0$. The gain matrices are given by

$$K_k^{(i+1)} = \Sigma_{k/k-1}^{(i)} H_k [H_k' \Sigma_{k/k-1} H_k + R_k]^{-1} \tag{3.10}$$

where

$$\Sigma_{k+1/k}^{(i+1)} = \Sigma_{k/k-1}^{(i)} [F_k - K_k H_k']' \tag{3.11}$$

initialized by $\Sigma_{k/k-1}^{(0)} = \Sigma_{k/k-1}$. It turns out that the only other covariance submatrices of interest are the diagonal ones

$$\Sigma_{k+1/k}^{(i+1,i+1)} = \Sigma_{k/k-1}^{(i,i)} - \Sigma_{k/k-1}^{(i)} H_k [K_k^{(i+1)}]'$$

or equivalently by virtue of (3.6),

$$\Sigma_{k-i/k} = \Sigma_{k-i/k-1} - \Sigma_{k/k-1}^{(i)} H_k [K_k^{(i+1)}]' \tag{3.12}$$

The fixed-lag smoothing covariance $\Sigma_{k-N/k}$ may be expressed nonrecursively using (3.12) as follows

$$\Sigma_{k-N/k} = \Sigma_{k-N/k-N-1} - \sum_{l=k-N}^{k} [\Sigma_{l/l-1}^{(q)} H_l [K_l^{(q+1)}]'$$

$$= \Sigma_{k-N/k-N-1} - \sum_{l=k-N}^{k} \{\Sigma_{l/l-1}^{(q)} H_l [H_l' \Sigma_{l/l-1} H_l + R_l]^{-1} H_l' [\Sigma_{l/l-1}^{(q)}]'\} \tag{3.13}$$

where $q \equiv l - k + N$.

We refer to Eqs. (3.9) through (3.13) as the fixed-lag smoothing equations. The fixed-lag smoother as a dynamical system is illustrated in Fig. 7.3-2; it is obtained by stacking together implementations of (3.9) for different i. A single set of state-space equations corresponding to the figure are given in the next section.

Properties of the Fixed-lag Smoother

The equations for the fixed-lag smoother bear some resemblance to the fixed-point smoothing equations of the previous sections, and the properties we now list for the fixed-lag smoother run parallel to the properties listed in the previous section for the fixed-point smoother.

1. The fixed-lag smoother described by (3.9) may be viewed as being driven from the innovations process $\tilde{z}_k = (z_k - H_k' \hat{x}_{k/k-1})$ and the states $\hat{x}_{k/k-1}$ of the Kalman filter for the original signal model. The smoother defined in this way is a linear discrete-time system of dimension N times the dimension of the filter. Its state-space equations are

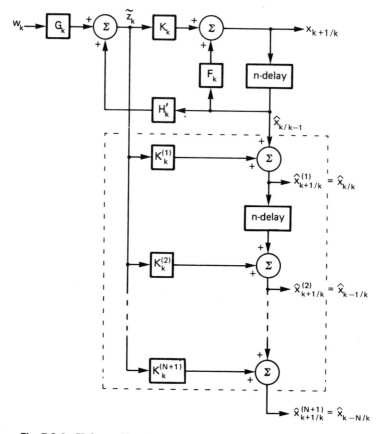

Fig. 7.3-2 Kalman filter for augmented signal model or equivalently Fixed-Lag Smoother.

$$
\begin{bmatrix} \hat{x}_{k/k} \\ \hat{x}_{k-1/k} \\ \hat{x}_{k-2/k} \\ \cdot \\ \cdot \\ \cdot \\ \hat{x}_{k+1-N/k} \end{bmatrix} = \begin{bmatrix} 0 & 0 & 0 & \cdots & 0 & 0 \\ I & 0 & 0 & \cdots & 0 & 0 \\ 0 & I & 0 & \cdots & 0 & 0 \\ \cdot & \cdot & \cdot & & \cdot & \cdot \\ \cdot & \cdot & \cdot & & \cdot & \cdot \\ \cdot & \cdot & \cdot & & \cdot & \cdot \\ 0 & 0 & 0 & \cdots & I & 0 \end{bmatrix} \begin{bmatrix} \hat{x}_{k-1/k-1} \\ \hat{x}_{k-2/k-1} \\ \hat{x}_{k-3/k-1} \\ \cdot \\ \cdot \\ \cdot \\ \hat{x}_{k-N/k-1} \end{bmatrix}
$$

$$
+ \begin{bmatrix} K_k^{(1)} \\ K_k^{(2)} \\ K_k^{(3)} \\ \cdot \\ \cdot \\ \cdot \\ K_k^{(N)} \end{bmatrix} \tilde{z}_k + \begin{bmatrix} I \\ 0 \\ 0 \\ \cdot \\ \cdot \\ \cdot \\ 0 \end{bmatrix} \hat{x}_{k/k-1} \qquad (3.14a)
$$

$$\hat{x}_{k-N/k} = \begin{bmatrix} 0 & 0 & \cdots & I \end{bmatrix} \begin{bmatrix} \hat{x}_{k-1/k-1} \\ \hat{x}_{k-2/k-1} \\ \cdot \\ \cdot \\ \cdot \\ \hat{x}_{k-N/k-1} \end{bmatrix} + K_k^{(N+1)} \tilde{z}_k \qquad (3.14b)$$

with zero (or arbitrary) initial states. Of course, it should be noted that the smoother outputs for $k < N$ are uninteresting quantities. Also the initial state of the smoother (at $k = 0$) affects only the smoother outputs for $k < N$, but no later ones. This fact should become clear from studies requested in one of the problems.

2. It is evident from (3.14) that with no extra effort, smoothed estimates for all lags $i = 1, \ldots, N - 1$ can be obtained. This property is not shared by other forms of smoother discussed later.

3. As in the case of the Kalman filter, the smoother parameters and error covariance are independent of the measurements and therefore can be computed beforehand. Also, as in the case of the filter, the relevant conditional probability density (of x_{k-N} given Z_k) is gaussian and is therefore defined by the conditional mean $\hat{x}_{k-N/k}$ and the conditional variance $\Sigma_{k-N/k}$. If $[H_k'\Sigma_{k/k-1}H_k + R_k]$ is singular, the inverse operation on this matrix, as in the filter equations, may be replaced by the pseudo-inverse operation.

4. For the time-invariant case when H_k, K_k, $\Sigma_{k/k-1}$, and F_k are independent of k and are denoted by H, K, $\bar{\Sigma}$, and F, respectively, the quantities $\Sigma_{k+1/k}^{(i)}$, $K_k^{(i)}$, and $\Sigma_{k-N/k}$ are also independent of k and are denoted by $\Sigma^{(i)}$, $K^{(i)}$, and Σ_N, respectively. *This means that the fixed-lag smoother is time invariant.* For $0 \le i \le N$ we have

$$K^{(i+1)} = \Sigma^{(i)} H[H'\bar{\Sigma}H + R]^{-1}$$
$$\Sigma^{(i)} = \bar{\Sigma}[F' - HK']^i \qquad (3.15)$$

and

$$[\bar{\Sigma} - \Sigma_N] = \sum_{i=0}^{N} \{\Sigma^{(i)} H[H'\bar{\Sigma}H + R]^{-1} H'[\Sigma^{(i)}]'\}$$
$$= \bar{\Sigma}\Big(\sum_{i=0}^{N} \{[\tilde{F}']^i H[H'\bar{\Sigma}H + R]^{-1} H' \tilde{F}^i\}\Big)\bar{\Sigma} \qquad (3.16)$$

where we have used the notation $\tilde{F} = [F - KH']$. Further manipulation leads to a linear matrix equation for $[\bar{\Sigma} - \Sigma_N]$ as

$$[\bar{\Sigma} - \Sigma_N] - \bar{\Sigma}\tilde{F}'\bar{\Sigma}^{-1}[\bar{\Sigma} - \Sigma_N]\bar{\Sigma}^{-1}\tilde{F}\bar{\Sigma}$$
$$= \bar{\Sigma}H[H'\bar{\Sigma}H + R]^{-1}H'\bar{\Sigma}$$
$$- \bar{\Sigma}[\tilde{F}']^{N+1}H[H'\bar{\Sigma}H + R]^{-1}H'\tilde{F}^{N+1}\bar{\Sigma} \qquad (3.17)$$

If one identifies N with $k - j$ in the fixed-point smoothing equations,

(3.17) becomes identical with (2.19). (Why?) Letting N approach infinity, we obtain

$$[\bar{\Sigma} - \Sigma_\infty] - \bar{\Sigma}\tilde{F}'\bar{\Sigma}^{-1}[\bar{\Sigma} - \Sigma_\infty]\bar{\Sigma}^{-1}\tilde{F}\bar{\Sigma} = \bar{\Sigma}H[H'\bar{\Sigma}H + R]^{-1}H'\bar{\Sigma} \quad (3.18)$$

which is naturally equivalent to (2.20). Therefore as for fixed-point smoothing, the greater the signal-to-noise ratio, the greater the possible improvement from fixed-lag smoothing.

5. From (3.16), it is evident that the improvement due to smoothing, viz., $[\bar{\Sigma} - \Sigma_N]$, increases monotonically as N increases. The eigenvalues of the Kalman filter matrix $\tilde{F} = [F - KH']$, and in particular the dominant eigenvalues of $[F - KH']$, govern the rate of increase. As a rule of thumb, we can say that essentially all the smoothing that it is possible to achieve can be achieved with N selected to be two or three times the dominant time constant of the Kalman filter. Though this remark has been made with reference to time-invariant filters, with some modification it also holds for time-varying filters.

6. The fixed-lag smoother inherits the stability properties of the original filter since the only feedback paths in the fixed-lag smoother are those associated with the subsystem comprising the original filter. [Stability is also evident from (3.14); all eigenvalues of the system matrix are zero.] The smoother is therefore free of the computational instability problems of the smoothers in the early literature.

7. The storage requirements and computational effort to implement the fixed-lag smoothing algorithm (3.9)–(3.11) are readily assessed. For the time-varying case, storage is required for at least

$$[F_{k-i} - K_{k-i}H'_{k-i}] \qquad \Sigma_{k-i/k-i-1} \quad \text{and} \quad \hat{x}_{k-i/k}$$

for $i = 1, 2, \ldots, N$ and for $H_k[H'_k\Sigma_{k/k-1}H_k + R_k]^{-1}\tilde{z}_k$. For the time-invariant case, storage is required for $\hat{x}_{k-i/k}$ for $i = 1, 2, \ldots, N$ and for $[F - KH']$, $\bar{\Sigma}$, and $H[H'\bar{\Sigma}H + R]^{-1}\tilde{z}_k$. Computational effort is reasonably assessed by the number of multiplications involved. Here the key cost for the time-varying case is the calculation of $\Sigma^{(i)}_{k/k-1}$ for $i = 1, 2, \ldots, N$ requiring (Nn^3) multiplications. For the time-invariant case, the key cost is the on-line calculation of $K^{(i+1)}_k\tilde{z}_k$ for $i = 1, 2, \ldots, N$ requiring but (mnN) multiplications.

8. A special class of problem for which filtered state estimates yield fixed-lag smoothed estimates is studied in Prob. 3.3. Generalizations of this simplification are explored in [30].

An Efficient Time-varying Fixed-lag Smoothing Algorithm

One rearrangement of (3.9) and (3.10) is

$$\hat{x}_{k-i/k} = \hat{x}_{k-i/k-1} + \Sigma^{(i)}_{k/k-1}e^{(1)}_k \quad (3.19)$$

where

$$e_k^{(1)} = H_k[H_k'\Sigma_{k/k-1}H_k + R_k]^{-1}\tilde{z}_k \qquad (3.20)$$

This arrangement together with the explicit formula for $\Sigma_{k/k-1}^{(i)}$ obtained by successive application of (3.11), namely,

$$\Sigma_{k/k-1}^{(i)} = \Sigma_{k-i/k-i-1}\prod_{l=k-i}^{k-1}[F_l - K_l H_l']' \qquad (3.21)$$

suggests the further restructuring of the smoothing equations for $1 \le i \le N$ as

$$\hat{x}_{k-i/k} = \hat{x}_{k-i/k-1} + \Sigma_{k-i/k-i-1}e_k^{(i+1)} \qquad (3.22)$$

where

$$e_k^{(i+1)} = [F_{k-i} - K_{k-i}H_{k-i}']e_k^{(i)} \qquad (3.23)$$

Successive application of (3.22) and (3.23) for $i = 0, 1, \ldots, N$ yields the efficient algorithm

$$\hat{x}_{k-N/k} = \hat{x}_{k-N/k-N} + \Sigma_{k-N/k-N-1}\left[\sum_{i=1}^{N} e_{k+i-n}^{(i+1)}\right] \qquad (3.24)$$

This algorithm reduces the key calculation cost from (Nn^3) multiplications for calculating $\Sigma_{k/k-1}^{(i)}$ to (Nn^2) multiplications for calculating $e_k^{(i+1)}$ for $i = 1, 2, \ldots, N$. Also an (Nn^2p) cost is eliminated in the algorithm. This improvement in calculation efficiency is at the expense of requiring storage for $e_{k+1-n}^{(2)}, e_{k+2-n}^{(3)}, \ldots, e_k^{(n+1)}$.

Further reduction in calculation complexity may be achieved by processing the measurement vector one component at a time; a development is requested in a problem.

Reduced Dimension Time-invariant Fixed-lag Smoothers

The fixed-lag smoothers derived earlier are not necessarily of minimal dimension when viewed as finite-dimensional linear systems. Standard techniques could be readily employed, at least in the time-invariant case, to pass from any unreachable and/or unobservable realization to a minimal realization of the smoothing equations. Alternatively, we shall now consider two different realizations of the fixed-lag (state) smoothing equations which may be of reduced dimension. Any reduction in the dimension of the smoother depends on the dimension p of the signal vector and the dimension m of the signal model input. The technique used is not one of casting out unreachable or unobservable parts of the smoother, as will be seen.

Successive applications of the smoothing equations (3.9) for $i = 0$, $1, \ldots, N$ yield

$$\hat{x}_{k-N/k} = \hat{x}_{k-N/k-N-1} + \sum_{l=k-N}^{k} K^{(q+1)}\tilde{z}_l \qquad (3.25)$$

where $q \equiv l - k + N$. The delayed filtered states $\hat{x}_{k-N/k-N-1}$ can be obtained from the Kalman filter equations

$$\hat{x}_{k-N+1/k-N} = F\hat{x}_{k-N/k-N-1} + K\tilde{z}_{k-N} \qquad (3.26)$$

We see that the smoother, driven from the relevant Kalman filter, requires N delays of dimension p to achieve $\tilde{z}_l$ for $l = k - N, k - N + 1, \ldots, k$ in (3.25), and a *further* Kalman filter (3.26). Thus we have a fixed-lag smoother of dimension $(n + Np)$ rather than (Nn) as we had earlier in the section. For the case of large N and n, and small p, this clearly leads to a considerable reduction in smoother dimension.

It is also possible to achieve a fixed-lag smoother of dimension $(n + Nm)$. The derivations will be left to the reader to work out in one of the problems, but the method of approach is simply described as follows. Instead of the augmented signal model used in earlier derivations, where the particular augmented states $x_k^N = x_{k-N}$ are achieved by feeding the original signal model *states* directly into N cascaded delay elements, each of dimension n, an augmented signal model is used where the augmented states $x_k^N = x_{k-N}$ are achieved by feeding the original signal model *inputs* directly into N cascaded delay elements, each of dimension m; then the input sequence delayed by N time units is fed into a system which is identical to the original signal model, except that the discrete-time arguments of the system matrices are appropriately delayed. The augmentations to the original signal model are therefore of dimension $(n + Nm)$, and as a consequence, when derivations paralleling those described earlier in the section are applied to the augmented signal model, the resulting fixed-lag smoother driven by the Kalman filter for the original signal model is of dimension $(n + Nm)$.

Fixed-lag Signal Smoothers

As in the case of fixed-point smoothing, when the dimension of the signal $y_k = H'_k x_k$ is less than the dimension of the state x_k, it is possible to construct a signal smoother with less storage requirements. The details will be omitted.

Fixed-lag Smoothers Driven by Filtered Estimates

Recall that the fixed-lag smoothers above are driven by the innovations process $\tilde{z}_k$ and the states $\hat{x}_{k/k-1}$ of the Kalman filter. There is no difficulty reorganizing the various equations so that the smoother is driven by the filtered estimates alone. To see this, note that from the Kalman filter equations we have

$$K_k \tilde{z}_k = \hat{x}_{k+1/k} - F_k \hat{x}_{k/k-1}$$

and as a consequence

$$\tilde{z}_k = (K_k' K_k)^{-1} K_k' (\hat{x}_{k+1/k} - F_k \hat{x}_{k/k-1})$$

The latter equation of course requires that K_k has rank p where p is the dimension of z_k. Alternative fixed-lag smoothing equations are, thus, for $0 \le i \le N$,

$$\hat{x}_{k-i/k} = \hat{x}_{k-i/k-1} + K_k^{(i+1)}(K_k' K_k)^{-1} K_k'(\hat{x}_{k+1/k} - F_k \hat{x}_{k/k-1})$$

or for the case when $[F_k \Sigma_{k/k-1}]^{-1}$ exists for all k, we have that

$$K_k^{(i+1)} \tilde{z}_k = \Sigma_{k/k-1}^{(i)} [F_k \Sigma_{k/k-1}]^{-1} K_k \tilde{z}_k$$

and thus that

$$\hat{x}_{k-i/k} = \hat{x}_{k-i/k-1} + \Sigma_{k/k-1}^{(i)} [F_k \Sigma_{k/k-1}]^{-1}(\hat{x}_{k+1/k} - F_k \hat{x}_{k/k-1})$$

Clearly, the disadvantage of this smoothing equation is that more computations are involved in its implementation than for the earlier algorithm (3.9).

Relationships Between Fixed-lag and Fixed-point Smoothing Equations

So far in this chapter, we have derived the fixed-point smoothing equations and the fixed-lag smoothing equations by application of Kalman filter theory to appropriately augmented signal models. It is an interesting observation that each of these smoothing equations can be derived directly from the other, and so we devote this subsection to indicating how this can be done.

First we shall specialize the fixed-lag smoothing equations to yield the fixed-point smoothing equations. Setting

$$i = k - j, \ K_k^a \equiv K_k^{(k-j)}, \ \Sigma_{k+1/k}^a \equiv \Sigma_{k+1/k}^{(k-j)}, \text{ and } \Sigma_{k+1/k}^{aa} \equiv \Sigma_{k+1/k}^{(k-j,k-j)}$$

we may substitute directly into the fixed-lag smoothing equations (3.9)–(3.13) to achieve immediately the fixed-point smoothing equations (2.9)–(2.13).

This idea is easily reversible to allow derivation of the fixed-lag smoothing equations from the fixed-point equations. One replaces j in the fixed-point equations by $k - N$, and then rearranges them in recursive form. The details are omitted.

Suboptimal Fixed-lag Smoothers

The work of P. Hedelin [29, 30] suggests that the filtered estimate $\hat{x}_{k/k}$ can sometimes provide a better estimate of x_{k-N} than does $\hat{x}_{k-N/k-N}$; this surprising result suggests that the filter can sometimes be used as a suboptimal smoother, and sometimes an optimal smoother designed for a lag N can be used as a suboptimal smoother for a lag greater than N.

Main Points of the Section

Stable, finite-dimensional, fixed-lag smoothers can be constructed driven from a Kalman filter and derived by means of Kalman filter theory applied to the original signal model augmented with delay elements. Manipulations of the various equations result in more efficient algorithms. For example, for the time-invariant smoother, the dimension of the smoother may be either Nn, $n + Np$, or $n + Nm$, or even lower as illustrated in a problem, depending on the method of derivation. (Here, n is the signal model dimension, p is the signal dimension, and m is the signal model input dimension.) The theory of the fixed-lag smoother is quite tidy, but the engineering application requiring tradeoffs between improvement due to smoothing, smoother complexity, and amount of fixed-lag requires the solution of some matrix equations.

Problem 3.1. Show that the outputs of a fixed-lag smoother driven from a Kalman filter with the smoother states instantaneously reset to zero (or some arbitrary values) at $k = j$ yield the correct fixed-lag estimates for $k \geq j + N$.

Problem 3.2. Write down the Kalman filter equations for the augmented signal model (3.3)–(3.4) and extract the fixed-lag smoothing equations (3.9)–(3.12). Derive the reduced order fixed-lag smoother of dimension $(n + Nm)$ described in the text, where m is the signal model input vector dimension. Sketch block diagrams of the augmented signal model used and the resulting smoother.

Problem 3.3. Suppose there is given a single-input signal model in completely reachable canonical form with

$$
\begin{bmatrix} x_{k+1}^{(1)} \\ x_{k+1}^{(2)} \\ \cdot \\ \cdot \\ \cdot \\ x_{k+1}^{(n)} \end{bmatrix} = \begin{bmatrix} 0 & & & \\ 0 & & & \\ \cdot & & I & \\ \cdot & & & \\ \cdot & & & \\ -a_1 & -a_2 & \cdots & -a_n \end{bmatrix} \begin{bmatrix} x_k^{(1)} \\ x_k^{(2)} \\ \cdot \\ \cdot \\ \cdot \\ x_k^{(n)} \end{bmatrix} + \begin{bmatrix} 0 \\ 0 \\ \cdot \\ \cdot \\ \cdot \\ 1 \end{bmatrix} w_k
$$

Show that it is possible to build a (state) fixed-lag smoother of dimension N, where it is assumed that the fixed-lag is N and the smoother is driven from the Kalman filter. These ideas are generalized in [30].

Problem 3.4. Write down transfer functions for two of the fixed-lag smoothers referred to in the section (of course, the time-invariant case must be assumed).

Problem 3.5. Work out the storage requirements and the number of multiplications required for the fixed-lag smoother (3.24) for the case of sequential processing with $R = \text{diag}\{r_1, r_2, \ldots, r_p\}$.

Experimental data is often noisy and available only over a fixed time interval. In this section, we consider the smoothing of such data. We shall first define the fixed-interval smoothing problem and then develop algorithms for its optimal and quasi-optimal solution.

Discrete-time fixed-interval smoothing problem. For the usual state-space signal model, determine for fixed M and all l in the interval $0 \leq l \leq M$ the estimate

$$\hat{x}_{l/M} = E[x_l \,|\, Z_M] \tag{4.1}$$

and the associated error covariance

$$\Sigma_{l/M} = E\{[x_l - \hat{x}_{l/M}][x_l - \hat{x}_{l/M}]' \,|\, Z_M\} \tag{4.2}$$

As a first step to providing a practical solution to this problem, we shall define a solution that is unlikely to be of great utility. It can, however, be modified to obtain a useful solution.

The Optimal Fixed-interval Smoothing Equations

Let us first consider the fixed-lag smoother equations (3.14) specialized to the case when the fixed-lag N is chosen such that $N = M + 1$. These are:

$$
\begin{bmatrix} \hat{x}_{k/k} \\ \hat{x}_{k-1/k} \\ \cdot \\ \cdot \\ \cdot \\ \hat{x}_{k-M/k} \end{bmatrix}
=
\begin{bmatrix} 0 & 0 & \cdots & 0 & 0 \\ I & 0 & \cdots & 0 & 0 \\ \cdot & & & & \cdot \\ \cdot & & & & \cdot \\ \cdot & & & & \cdot \\ 0 & 0 & \cdots & I & 0 \end{bmatrix}
\begin{bmatrix} \hat{x}_{k-1/k-1} \\ \hat{x}_{k-2/k-1} \\ \cdot \\ \cdot \\ \cdot \\ \hat{x}_{k-M-1/k-1} \end{bmatrix}
+
\begin{bmatrix} K_k^{(1)} \\ K_k^{(2)} \\ \cdot \\ \cdot \\ \cdot \\ K_k^{(M+1)} \end{bmatrix} \tilde{z}_k
+
\begin{bmatrix} I \\ 0 \\ \cdot \\ \cdot \\ \cdot \\ 0 \end{bmatrix} \hat{x}_{k/k-1}
\tag{4.3}
$$

When $k = M$, the smoothed estimates $\hat{x}_{l/M}$ are all available as subvectors of the vector on the left side of (4.3). We conclude that *the fixed-interval smoothed estimates $\hat{x}_{l/M}$ are available from a fixed-lag smoother by taking $k = M$ and choosing the fixed-lag as $N = M + 1$.*

Other forms of fixed-lag smoother do not provide smoothed estimates for all values of lag; it is also possible to use such forms to achieve fixed-interval estimates. We postulate that measurements are available over the interval $[M + 1, 2M]$, but that the associated output noise covariance is infinity (making the measurements worthless). Then one builds a fixed-lag smoother to operate over $[0, 2M]$. (Note that it cannot be time invariant in view of the change of noise covariance.) At time M, $\hat{x}_{0/M}$ is available, at time

$M + 1$, $\hat{x}_{1/M+1}$ is available, at time $M + 2$, $\hat{x}_{2/M+2}$ is available, and so on. But because of the worthlessness of measurements past time M, one has $\hat{x}_{1/M+1} = \hat{x}_{1/M}$, $\hat{x}_{2/M+2} = \hat{x}_{2/M}$, and so on.

In the state-estimation equation, the effect of the infinite noise covariance matrix on $[M + 1, 2M]$ is simply to disconnect the $\bar{z}_k$ input to the smoother for $k > M$.

As stated earlier, the procedure outlined in this subsection is unlikely to be of great utility; the main reason is that M may be large, so that the dimension of the optimal smoother becomes very large indeed. Some reduction in dimension is possible using the reduced dimension smoother ideas developed in the last section, but then one may have to wait till time $2M$ to recover all the smoothed estimates. Even with such reduction in dimension, the reduced dimension smoother has dimension proportional to M.

Let us now consider a way round these difficulties.

The Quasi-optimal Fixed-interval Smoother

Suppose that instead of identifying N with M, we take N as several times the dominant time constant of the Kalman filter. Then, as we know, $\hat{x}_{k/k+N} \doteqdot \hat{x}_{k/M}$ for $k + N \leq M$; this means that a fixed lag smoother (of any realization) with lag N will produce estimates $\hat{x}_{0/M}, \hat{x}_{1/M}, \ldots, \hat{x}_{M-N/M}$, $N < M$; in fact, it is frequently the case that $N \ll M$.

It remains to be seen how to obtain $\hat{x}_{k/M}$ for $k > M - N$. Two approaches are possible. If the fixed-lag smoother is of a form containing fixed-lag smoothed estimates for all lags less than N, $\hat{x}_{k/M}$ for $k > M - N$ will be available at time M, simultaneously with $\hat{x}_{M-N/M}$. Alternatively, and irrespective of the form of the fixed-lag smoother, one may postulate output noise of infinite variance on $[M + 1, M + N]$ and run the smoother through till time $M + N$. The output will be $\hat{x}_{M-N+1/M}$ at time $M + 1$, $\hat{x}_{M-N+2/M}$ at time $M + 2$, etc.

Other Approaches to Fixed-interval Smoothing

As remarked in the introductory section, there are other approaches to fixed-interval smoothing. These approaches involve running a Kalman filter forward in time over the interval $[0, M]$, storing either the state estimates or the measurements. Then these stored quantities are run backwards to obtain the fixed-lag smoothed estimates, in reverse sequence as $\hat{x}_{M-1/M}, \hat{x}_{M-2/M}, \ldots$. The storage requirements and delay in processing compare unfavourably with the quasi-optimal smoother, unless N and M are comparable, and in this instance the fixed-lag smoother approach would seem effective and no more demanding of storage requirements and processing delays.

We can derive the equations as follows. Using Eqs. (2.9) through (2.11), it is easy to establish, as requested in Prob. 2.3, the following equations:

$$\hat{x}_{j/k} = \hat{x}_{j/j-1} + \sum_{l=j}^{k} \Sigma_{j/j-1} \Phi'(l,j) H_l (H_l' \Sigma_{l/l-1} H_l + R_l)^{-1} \tilde{z}_l \qquad (4.4a)$$

$$= \hat{x}_{j/j} + \sum_{l=j+1}^{k} \Sigma_{j/j-1} \Phi'(l,j) H_l (H_l' \Sigma_{l/l-1} H_l + R_l)^{-1} \tilde{z}_l \qquad (4.4b)$$

Here, $\Phi(l,j)$ denotes $\bar{F}_{l-1} \bar{F}_{l-2} \ldots \bar{F}_j$ and $\bar{F}_l = F_l - K_l H_l'$. With j replaced by $(j-1)$ in (4.4b), we have

$$\hat{x}_{j-1/k} = \hat{x}_{j-1/j-1} + \sum_{l=j}^{k} \Sigma_{j-1/j-2} \Phi'(l, j-1) H_l (H_l' \Sigma_{l/l-1} H_l + R_l)^{-1} \tilde{z}_l$$

$$= \hat{x}_{j-1/j-1} + \Sigma_{j-1/j-2} \bar{F}_{j-1}' \Sigma_{j/j-1}^{-1} \sum_{l=j}^{k} \Sigma_{j/j-1} \Phi'(l,j)$$

$$\times H_l (H_l' \Sigma_{l/l-1} H_l + R_l)^{-1} \tilde{z}_l$$

$$= \hat{x}_{j-1/j-1} + \Sigma_{j-1/j-2} \bar{F}_{j-1}' \Sigma_{j/j-1}^{-1} (\hat{x}_{j/k} - \hat{x}_{j/j-1})$$

with the last equation following using (4.4a). Notice incidentally that, as is easily shown,

$$\Sigma_{j-1/j-2} \bar{F}_{j-1}' = \Sigma_{j-1/j-1} F_{j-1}'$$

so that, replacing k by M, we have

$$\hat{x}_{j-1/M} = \hat{x}_{j-1/j-1} + \Sigma_{j-1/j-1} F_{j-1}' \Sigma_{j/j-1}^{-1} (\hat{x}_{j/M} - \hat{x}_{j/j-1}) \qquad (4.5)$$

This equation yields smoothed estimates by taking $j = M, M-1, M-2, \ldots$. The initialization is evidently with a filtered estimate.

We leave to the reader the derivation of a recursion for the error covariance:

$$\Sigma_{j-1/M} = \Sigma_{j-1/j-1} + A_{j-1} [\Sigma_{j/M} - \Sigma_{j/j-1}] A_{j-1}' \qquad (4.6)$$

where

$$A_{j-1} = \Sigma_{j-1/j-1} F_{j-1}' \Sigma_{j/j-1}^{-1} \qquad (4.7)$$

Formulas for smoothed estimates which involve computation of $\hat{x}_{M/M}$ for initializing purposes and storage of the measurements rather than filtered estimates for running a recursive algorithm can be found in [6].

An alternative derivation of the fixed-lag equations can be obtained in the following way. Because $p(x_0, x_1, \ldots, x_M | Z_M)$ is a gaussian density, the sequence $\hat{x}_{0/M}, \hat{x}_{1/M}, \ldots, \hat{x}_{M/M}$ maximizes the density for fixed Z_M. One can show this maximization is equivalent to the minimization (assuming nonsingular P_0, Q_i, and R_i) of

$$J = \tfrac{1}{2}(x_0 - \bar{x}_0)' P_0^{-1} (x_0 - \bar{x}_0) + \sum_{i=0}^{M-1} [\tfrac{1}{2} w_i Q_i^{-1} w_i$$

$$+ \tfrac{1}{2}(z_i - H_i' x_i)' R_i^{-1} (z_i - H_i' x_i)] + \tfrac{1}{2}(z_M - H_M' x_M)' R_M^{-1} (z_M - H_M' x_M)$$

subject to $x_{k+1} = F_k x_k + G_k w_k$. Various approaches to this minimization problem exist (see, e.g., [3]).

Finally, we comment that there exists a group of smoothing formulas [9, 10, 21] based on the principle of combining $\hat{x}_{k/k}$ and $E[x_k | z_{k+1}, z_{k+2}, \ldots, z_M]$, the latter quantity being computed by a "backwards-time" filter.

Main Points of the Section

Fixed-interval smoothing can be achieved either optimally or quasi-optimally by direct application of the fixed-lag smoothing equations, which in turn are Kalman filtering equations in disguise. The computations involved in quasi-optimal fixed-interval smoothing may be considerably less than in optimal smoothing for large data sets.

Problem 4.1. Use (4.5) and (4.7) to show that with $\tilde{x}_{k/j} = x_k - \hat{x}_{k/j}$,

$$\tilde{x}_{j-1/M} + A_{j-1}\hat{x}_{j/M} = \tilde{x}_{j-1/j-1} + A_{j-1}F_{j-1}\hat{x}_{j-1/j-1}$$

Show that

$$E[\tilde{x}_{j-1/M}\hat{x}'_{j/M}] = 0 \quad \text{and} \quad E[\tilde{x}_{j-1/j-1}\hat{x}'_{j-1/j-1}] = 0$$

and thus obtain (4.6).

Problem 4.2. Obtain a formula expressing $\Sigma_{j-1/j-1}$ in terms of $\Sigma_{j/j}$; explain how it might be of assistance in fixed-interval smoothing calculations, assuming the formula to be numerically stable.

Problem 4.3. Obtain fixed-interval smoothing equations corresponding to the case when $Q = 0$. Show that the smoother can be thought of as a backward predictor, i.e.,

$$\hat{x}_{j/M} = F_j^{-1}\hat{x}_{j+1/M} \qquad P_{j/M} = F_j^{-1}P_{j+1/M}(F_j')^{-1}$$

provided F_j^{-1} exists.

REFERENCES

[1] MEDITCH, J. S., "A Survey of Data Smoothing for Linear and Nonlinear Dynamic Systems," *Automatica*, Vol. 9, No. 2, March 1973, pp. 151–162.

[2] MEDITCH, J. S., *Stochastic Optimal Linear Estimation and Control*, McGraw-Hill Book Company, New York, 1969.

[3] BRYSON, JR., A. E., and Y. C. HO, *Applied Optimal Control*, Blaisdell Publishing Company, Waltham, Mass., 1969.

[4] SAGE, A. P., and J. L. MELSA, *Estimation Theory with Applications to Communications and Control*, McGraw-Hill Book Company, New York, 1971.

[5] RAUCH, H. E., "Solutions to the Linear Smoothing Problem," *IEEE Trans. Automatic Control*, Vol. AC-8, No. 4, October 1963, pp. 371–372.

[6] RAUCH, H. E., F. TUNG, and C. T. STRIEBEL, "Maximum Likelihood Estimates of Linear Dynamic Systems," *AIAA J.*, Vol. 3, 1965, pp. 1445–1450.

[7] WEAVER, C. S., "Estimating the Output of a Linear Discrete System with Gaussian Inputs," *IEEE Trans. Automatic Control*, Vol. AC-8, October 1963, pp. 372–374.

[8] LEE, R. C. K., *Optimal Estimation, Identification and Control*, The M.I.T. Press, Cambridge, Mass., 1964.

[9] MAYNE, D. Q., "A Solution to the Smoothing Problem for Linear Dynamic Systems," *Automatica*, Vol. 4, 1966, pp. 73–92.

[10] FRASER, D. C., and J. E. POTTER, "The Optimum Linear Smoother as a Combination of Two Optimum Linear Filters," *IEEE Trans. Automatic Control*, Vol. AC-14, No. 4, August 1969, pp. 387–390.

[11] BRYSON, JR., A. E., and L. J. HENRIKSON, "Estimation Using Sampled-data Containing Sequentially Correlated Noise," *Technical Report No. 533*, Division of Engineering and Applied Physics, Harvard University, Cambridge, Mass., June 1967.

[12] KELLY, C. N., and B. D. O. ANDERSON, "On the Stability of Fixed-lag Smoothing Algorithms," *J. Franklin Inst.*, Vol. 291, No. 4, April 1971, pp. 271–281.

[13] MEHRA, R. K., "On the Identification of Variances and Adaptive Kalman Filtering," *IEEE Trans. Automatic Control*, Vol. AC-15, No. 2, April 1970, pp. 175–184.

[14] MOORE, J. B., "Discrete-time Fixed-lag Smoothing Algorithms," *Automatica*, Vol. 9, No. 2, March 1973, pp. 163–174.

[15] ZACHRISSON, L. E., "On Optimal Smoothing of Continuous-time Kalman Processes," *Inform. Sciences*, Vol. 1, 1969, pp. 143–172.

[16] WILLMAN, W. W., "On the Linear Smoothing Problem," *IEEE Trans. Automatic Control*, Vol. AC-14, No. 1, February 1969, pp. 116–117.

[17] PREMIER, R., and A. G. VACROUX, "On Smoothing in Linear Discrete Systems with Time Delays," *Int. J. Control*, Vol. 13, No. 2, 1971, pp. 299–303.

[18] FAROOQ, M., and A. K. MAHALANABIS, "A Note on the Maximum Likelihood State Estimation of Linear Discrete Systems with Multiple Time Delays," *IEEE Trans. Automatic Control*, Vol. AC-16, No. 1, February 1971, pp. 105–106.

[19] ANDERSON, B. D. O., and S. CHIRARATTANANON, "Smoothing as an Improvement on Filtering: A Universal Bound," *Electronics Letters*, Vol. 7, No. 18, September 1971, p. 524.

[20] ANDERSON, B. D. O., "Properties of Optimal Linear Smoothing," *IEEE Trans. Automatic Control*, Vol. AC-14, No. 1, February 1969, pp. 114–115.

[21] FRIEDLANDER, B., T. KAILATH, and L. LJUNG, "Scattering Theory and Linear Least Squares Estimation II: Discrete Time Problems," *J. Franklin Inst.*, Vol. 301, Nos. 1 and 2, January–February 1976, pp. 71–82.

[22] LAINIOTIS, D. G., and K. S. GOVINDARAJ, "A Unifying Approach to Linear Estimation via the Partitioned Algorithm, II: Discrete Models," *Proc. IEEE 1975 Conf. on Decision and Control*, pp. 658–659.

[23] MORF, M., and T. KAILATH, "Square-root Algorithms for Least-squares Estimation," *IEEE Trans. Automatic Control*, Vol. AC-20, No. 4, August 1975, pp. 487–497.

[24] RAUCH, H. E., "Optimum Estimation of Satellite Trajectories including Random Fluctuations in Drag," *AIAA J.*, Vol. 3, No. 4, April 1965, pp. 717–722.

[25] NORTON, J. P., "Optimal Smoothing in the Identification of Linear Time-varying Systems," *Proc. IEE*, Vol. 122, No. 6, June 1975, pp. 663–668.

[26] NASH, R. A., J. F. KASPER, B. S. CRAWFORD, and S. A. LEVINE, "Application of Optimal Smoothing to the Testing and Evaluation of Inertial Navigation Systems and Components," *IEEE Trans. Automatic Control*, Vol. AC-16, No. 6, December 1971, pp. 806–816.

[27] CRANE, R. N., "An Application of Nonlinear Smoothing to Submarine Exercise Track Reconstruction," *Proc. 3rd Symp. on Nonlinear Estimation Theory and its Applications*, 1972, pp. 36–44.

[28] CANTONI, A., and P. BUTLER, "The Linear Minimum Mean Square Error Estimator Applied to Channel Equalization," *IEEE Trans. Communications*, Vol. COM 25, No. 4, April 1977, pp. 441–446.

[29] HEDELIN, P., "Can the Zero-lag Filter be a good smoother?" *IEEE Trans. Information Theory.* Vol. IT-23, No. 4, July 1977, pp. 490–499.

[30] HEDELIN, P., and I. JÖNSSON, "Applying a Smoothing Criterion to the Kalman Filter," submitted for publication.

CHAPTER 8

APPLICATIONS IN NONLINEAR FILTERING

8.1 NONLINEAR FILTERING

So far in this text we have seen that the optimal linear filtering theorems and algorithms are clean and powerful. The fact that the filter equation and the performance calculations together with the filter gain calculations are decoupled is particularly advantageous, since the performance calculations and filter gain calculations can be performed off line; and as far as the on-line filter calculations are concerned, the equations involved are no more complicated than the signal model equations. The filtered estimates and the performance measures are simply the means and covariances of the a posteriori probability density functions, which are gaussian. The vector filtered estimates together with the matrix performance covariances are clearly sufficient statistics* of these a posteriori state probability densities.

By comparison, optimal *nonlinear* filtering is far less precise, and we must work hard to achieve even a little. The most we attempt in this book is to see what happens when we adapt some of the linear algorithms to non-linear environments.

---

*Sufficient statistics are collections of quantities which uniquely determine a probability density in its entirety.

So as not to depart very far from the linear gaussian signal model, in the first instance we will work with the model

$$x_{k+1} = f_k(x_k) + g_k(x_k)w_k \tag{1.1}$$

$$z_k = h_k(x_k) + v_k \tag{1.2}$$

where the quantities $F_k x_k$, $H_k x_k$, and G_k of earlier linear models are replaced by $f_k(x_k)$, $h_k(x_k)$, and $g_k(x_k)$, with $f_k(\cdot)$, $h_k(\cdot)$ nonlinear (in general) and $g_k(\cdot)$ nonconstant (in general). The subscript on $f_k(\cdot)$, etc., is included to denote a possible time dependence. Otherwise the above model is identical to the linear gaussian models of earlier chapters. In particular, $\{v_k\}$ and $\{w_k\}$ are zero mean, white gaussian processes, and x_0 is a gaussian random variable. We shall assume $\{v_k\}$, $\{x_k\}$, and x_0 are mutually independent, that $E[v_k v_k'] = R_k$, $E[w_k w_k'] = Q_k$, and x_0 is $N(\bar{x}_0, P_0)$. Throughout the chapter we denote

$$F_k = \left.\frac{\partial f_k(x)}{\partial x}\right|_{x=\hat{x}_{k/k}}, \qquad H_k' = \left.\frac{\partial h_k(x)}{\partial x}\right|_{x=\hat{x}_{k/k-1}}, \qquad G_k = g_k(\hat{x}_{k/k}) \tag{1.3}$$

[This means that the ij component of F_k is the partial derivative with respect to x_j of the ith component of $f_k(\cdot)$, and similarly for H_k', each derivative being evaluated at the point indicated.]

In the next section, approximations are introduced to derive a clearly suboptimal filter for the signal model above, known as an *extended Kalman filter*. The filter equations are applied to achieve quasi-optimal demodulation of FM (frequency modulation) signals in low noise. A special class of extended Kalman filters is defined in Sec. 8.3 involving cone-bounded nonlinearities, and upper bounds on performance are derived. In Sec. 8.4, a more sophisticated "gaussian sum" nonlinear estimation theory is derived, where, as the name suggests, the a posteriori densities are approximated by a sum of gaussian densities. The nonlinear filter algorithms involve a bank of extended Kalman filters, where each extended Kalman filter keeps track of one term in the gaussian sum. The gaussian sum filter equations are applied to achieve quasi-optimal demodulation of FM signals in high noise. Other nonlinear filtering techniques outside the scope of this text use different means for keeping track of the a posteriori probability distributions than the gaussian sum approach of Sec. 8.4. For example, there is the point-mass approach of Bucy and Senne [1], the spline function approach of de Figueiredo and Jan [2], and the Fourier series expansion approach used successfully in [3], to mention just a few of the many references in these fields.

Problem 8.1. (Formal Approach to Nonlinear Filtering). Suppose that

$$x_{k+1} = f(x_k) + g(x_k)w_k \qquad z_k = h(x_k) + v_k$$

with $\{w_k\}$, $\{v_k\}$ independent gaussian, zero mean sequences. Show that $p(x_{k+1} | x_k, Z_k)$ is known and, together with $p(x_k | Z_k)$, determines $p(x_{k+1} | Z_k)$ by integration. (This is the time-update step.) Show that $p(z_{k+1} | x_{k+1}, Z_k)$ is known and, together with

$p(x_{k+1}|Z_k)$, determines $p(x_{k+1}|Z_{k+1})$ by integration. (This is the measurement-update step.) A technical problem which can arise is that if $g(x_k)$ is singular, $p(x_{k+1}|x_k, Z_k)$ is not well defined; in this case, one needs to work with characteristic functions rather than density functions.

8.2 THE EXTENDED KALMAN FILTER

We retain the notation introduced in the last section. The nonlinear functions $f_k(x_k)$, $g_k(x_k)$, and $h_k(x_k)$, if sufficiently smooth, can be expanded in Taylor series about the conditional means $\hat{x}_{k/k}$ and $\hat{x}_{k/k-1}$ as

$$f_k(x_k) = f_k(\hat{x}_{k/k}) + F_k(x_k - \hat{x}_{k/k}) + \cdots$$
$$g_k(x_k) = g_k(\hat{x}_{k/k}) + \cdots = G_k + \cdots$$
$$h_k(x_k) = h_k(\hat{x}_{k/k-1}) + H'_k(x_k - \hat{x}_{k/k-1}) + \cdots$$

Neglecting higher order terms and assuming knowledge of $\hat{x}_{k/k}$ and $\hat{x}_{k/k-1}$ enables us to approximate the signal model (1.1) and (1.2) as

$$x_{k+1} = F_k x_k + G_k w_k + u_k \tag{2.1}$$
$$z_k = H'_k x_k + v_k + y_k \tag{2.2}$$

where u_k and y_k are calculated on line from the equations

$$u_k = f_k(\hat{x}_{k/k}) - F_k \hat{x}_{k/k} \qquad y_k = h_k(\hat{x}_{k/k-1}) - H'_k \hat{x}_{k/k-1} \tag{2.3}$$

The Kalman filter for this approximate signal model is a trivial variation of that derived in earlier chapters. Its equations are as follows:

EXTENDED KALMAN FILTER EQUATIONS:

$$\hat{x}_{k/k} = \hat{x}_{k/k-1} + L_k[z_k - h_k(\hat{x}_{k/k-1})] \tag{2.4}$$
$$\hat{x}_{k+1/k} = f_k(\hat{x}_{k/k}) \tag{2.5}$$
$$L_k = \Sigma_{k/k-1} H_k \Omega_k^{-1} \qquad \Omega_k = H'_k \Sigma_{k/k-1} H_k + R_k \tag{2.6}$$
$$\Sigma_{k/k} = \Sigma_{k/k-1} - \Sigma_{k/k-1} H_k [H'_k \Sigma_{k/k-1} H_k + R_k]^{-1} H'_k \Sigma_{k/k-1} \tag{2.7}$$
$$\Sigma_{k+1/k} = F_k \Sigma_{k/k} F'_k + G_k Q_k G'_k \tag{2.8}$$

Initialization is provided by $\Sigma_{0/-1} = P_0$, $\hat{x}_{0/-1} = \bar{x}_0$.

The significance of $\hat{x}_{k+1/k}$ *and* $\Sigma_{k+1/k}$. The above extended Kalman filter is nothing other than a standard and exact Kalman filter for the signal model (2.1)–(2.3). When applied to the original signal model (1.1) and (1.2), it is no longer linear or optimal and the notations $\hat{x}_{k+1/k}$ and $\Sigma_{k/k-1}$ are now loose and denote approximate conditional means and covariances, respectively.

Coupling of conditional mean, filter gain, and filter performance equations. The equations for calculating the filter gain L_k are coupled to the filter equations since H_k and F_k are functions of $\hat{x}_{k/k-1}$. The same is true for the approximate performance measure $\Sigma_{k/k-1}$. We conclude that, *in general, the calculation of L_k and $\Sigma_{k/k-1}$ cannot be carried out off line.* Of course in any particular application it may be well worthwhile to explore approximations which would allow decoupling of the filter and filter gain equations. In the next section, a class of filters is considered of the form of (2.4) and (2.5), where the filter gain L_k is chosen as a result of some off-line calculations. For such filters, there is certainly no coupling to a covariance equation.

Quality of approximation. The approximation involved in passing from (1.1) and (1.2) to (2.1) and (2.2) will be better the smaller are $\|x_k - \hat{x}_{k/k}\|^2$ and $\|x_k - \hat{x}_{k/k-1}\|^2$. Therefore, we would expect that in high signal-to-noise ratio situations, there would be fewer difficulties in using an extended Kalman filter. When a filter is actually working, so that quantities trace $(\Sigma_{k/k})$ and trace $(\Sigma_{k/k-1})$ become available, one can use these as guides to $\|x_k - \hat{x}_{k/k}\|^2$ and $\|x_k - \hat{x}_{k/k-1}\|^2$, and this in turn allows review of the amount of approximation involved. Another possibility for determining whether in a given situation an extended Kalman filter is or is not working well is to check how white the pseudo-innovations are, for the whiter these are the more nearly optimal is the filter. Again off-line Monte Carlo simulations can be useful, even if tedious and perilous, or the application of performance bounds such as described in the next section may be useful in certain cases when there exist cone-bounded conditions on the nonlinearities.

Selection of a suitable co-ordinate basis. We have already seen that for a certain nonlinear filtering problem—the two-dimensional tracking problem discussed in Chap. 3, Sec. 4,—one coordinate basis can be more convenient than others. This is generally the case in nonlinear filtering, and in [4], an even more significant observation is made. For some coordinate basis selections, the extended Kalman filter may diverge and be effectively useless, whereas for other selections it may perform well. This phenomenon is studied further in [5], where it is seen that $V_k = \hat{x}'_{k/k-1}\Sigma^{-1}_{k/k-1}\hat{x}_{k/k-1}$ is a Lyapunov function ensuring stability of the autonomous filter for certain coordinate basis selections, but not for others.

Variations of the extended Kalman filter. There are a number of variations on the above extended Kalman filter algorithm, depending on the derivation technique employed and the assumptions involved in the derivation. For example, filters can be derived by including more terms in the Taylor series expansions of $f_k(x_k)$ and $h_k(x_k)$; the filters that result when two terms are involved are called second order extended Kalman filters. Again, there are algorithms (see problems) in which the reference trajectory is improved by iteration techniques, the resulting filters being termed iterated extended Kalman filters. Any one of these algorithms may be superior to the standard extended Kalman filter in a particular filtering application, but there are no real guidelines here, and each case has to be studied separately using Monte Carlo simulations. Other texts [6, 7] should be consulted for derivations and examples. For the case when cone-bounded nonlinearities are involved in an extended Kalman filter, it may well be, as shown in [5], that the extended Kalman filter performs better if the nonlinearities in the filter are modified by tightening the cone bounds. This modification can be conveniently effected by introducing dither signals prior to the nonlinearities, and compensating for the resulting bias using a filtered version of the error caused by the cone bound adjustment.

Gaussian sum approach. There are nonlinear algorithms which involve collections of extended Kalman filters, and thereby become both more powerful and more complex than the algorithm of this section. In these algorithms, discussed in a later section, the a posteriori density function $p(x_k|Z_k)$ is approximated by a sum of gaussian density functions, and assigned to each gaussian density function is an extended Kalman filter. In situations where the estimation error is small, the a posteriori density can be approximated adequately by one gaussian density, and in this case the gaussian sum filter reduces to the extended Kalman filter of this section.

The following theorem gives some further insight into the quality of the approximations involved in the extended Kalman filter algorithm and is of key importance in demonstrating the power of the gaussian sum algorithms of later sections. In particular, the theorem shows that under certain conditions, the notion that the errors $\|x_k - \hat{x}_{k/k}\|^2$ and $\|x_k - \hat{x}_{k/k-1}\|^2$ have to be small to ensure that the extended Kalman filter is near optimal can be relaxed to requiring only that $\Sigma_{k/k}$ and $\Sigma_{k/k-1}$ (or their traces) be small. With $\gamma[x - \bar{x}, \Sigma]$ denoting the gaussian density,

$$\gamma[x - \bar{x}, \Sigma] = \frac{\exp\{-\frac{1}{2}(x - \bar{x})'\Sigma^{-1}(x - \bar{x})\}}{\{(2\pi)^{n/2}|\Sigma|^{1/2}\}}$$

the following result can be established:

THEOREM 2.1. For the signal model (1.1) and (1.2) and filter of (2.4) through (2.8), if

$$p(x_k|Z_{k-1}) = \gamma[x_k - \hat{x}_{k/k-1}, \Sigma_{k/k-1}] \tag{2.9}$$

then for fixed $h_k(\cdot)$, $\hat{x}_{k/k-1}$, and R_k

$$p(x_k|Z_k) \longrightarrow \gamma[x_k - \hat{x}_{k/k}, \Sigma_{k/k}]$$

uniformly in x_k and z_k as $\Sigma_{k/k-1} \longrightarrow 0$. Again if

$$p(x_k|Z_k) = \gamma[x_k - \hat{x}_{k/k}, \Sigma_{k/k}] \tag{2.10}$$

then for fixed $f_k(\cdot)$, $g_k(\cdot)$, $\hat{x}_{k/k}$, and Z_k,

$$p(x_{k+1}|Z_k) \longrightarrow \gamma[x_{k+1} - \hat{x}_{k+1/k}, \Sigma_{k+1/k}]$$

as $\Sigma_{k/k} \longrightarrow 0$. In these equations, it is assumed that the relevant probability densities exist, otherwise characteristic functions must be used.

*Proof.** By Bayes' rule, we have

$$p(x_k, z_k|Z_{k-1}) = p(x_k|Z_k)p(z_k|Z_{k-1}) = p(z_k|x_k, Z_{k-1})p(x_k|Z_{k-1})$$

or

$$p(x_k|Z_k) = \frac{p(z_k|x_k)p(x_k|Z_{k-1})}{\int p(z_k|x_k)p(x_k|Z_{k-1})dx_k}$$

Denote the denominator (which is evidently independent of x_k) by δ^{-1}. Then from (1.2) and (2.9),

$$p(x_k|Z_k) = \delta\gamma[x_k - \hat{x}_{k/k-1}, \Sigma_{k/k-1}]\gamma[z_k - h_k(x_k), R_k]$$
$$= \delta\{\gamma[x_k - \hat{x}_{k/k-1}, \Sigma_{k/k-1}]\gamma[z_k - H'_k(x_k - \hat{x}_{k/k-1})$$
$$- h_k(\hat{x}_{k/k-1}), R_k] + \epsilon_k\}$$

where

$$\epsilon_k = \gamma[x_k - \hat{x}_{k/k-1}, \Sigma_{k/k-1}]\{\gamma[z_k - h_k(x_k), R_k]$$
$$- \gamma[z_k - H'_k(x_k - \hat{x}_{k/k-1}) - h_k(\hat{x}_{k/k-1}), R_k]\} \quad (2.11)$$

Now tedious algebraic manipulations involving completion of the square arguments (see the problems) give the relationship

$$\gamma[x_k - \hat{x}_{k/k-1}, \Sigma_{k/k-1}]\gamma[z_k - H'_k(x_k - \hat{x}_{k/k-1}) - h_k(\hat{x}_{k/k-1}), R_k]$$
$$= \gamma[x_k - \hat{x}_{k/k}, \Sigma_{k/k}]\gamma[z_k - h_k(\hat{x}_{k/k-1}), \Omega_k] \quad (2.12)$$

where $\hat{x}_{k/k}$, $\Sigma_{k/k}$, and Ω_k are as defined in (2.4), (2.6), and (2.7). Without the replacement of $\gamma(z_k - h_k(x_k), R_k]$ by an approximating quantity involving x_k linearly, an identity like (2.12) cannot be obtained. Notice that the approximating quantity involves approximation of $z_k - h_k(x_k)$ by the same quantity as in the approximate signal model of (2.2), for which (2.4) and the following equations constitute the Kalman filter. Notice also that the approximation error is wrapped up in the quantity ϵ_k.

Using (2.12), there results

$$p(x_k|Z_k) = \frac{\gamma[z_k - h_k(\hat{x}_{k/k-1}), \Omega_k]\gamma[x_k - \hat{x}_{k/k}, \Sigma_{k/k}] + \epsilon_k}{\int \{\gamma[z_k - h_k(\hat{x}_{k/k-1}), \Omega_k]\gamma[x_k - \hat{x}_{k/k}, \Sigma_{k/k}] + \epsilon_k\} dx_k}$$
$$= \frac{\gamma[z_k - h_k(\hat{x}_{k/k-1}), \Omega_k]\gamma[x_k - \hat{x}_{k/k}, \Sigma_{k/k}] + \epsilon_k}{\gamma[z_k - h_k(\hat{x}_{k/k-1}), \Omega_k] + \int \epsilon_k dx_k}$$

*The proof may be omitted on a first reading.

In the problems, derivation is requested of the facts that with $h_k(\cdot)$, $\hat{x}_{k/k-1}$ and Z_{k-1} all fixed, then $\Sigma_{k/k-1} \to 0$ implies $\epsilon_k \to 0$ uniformly in x_k and z_k, and $\int \epsilon_k dx_k \to 0$ uniformly in z_k. The first part of the theorem then follows.

For the proof of the second part of the theorem, we must work with characteristic functions in case $p(x_{k+1}|x_k)$ is not well defined. Thus in lieu of

$$p(x_{k+1}|Z_k) = \int p(x_{k+1}|x_k)p(x_k|Z_k)\, dx_k$$

we write

$$\phi_{x_{k+1}|Z_k}(s) = \int \phi_{x_{k+1}|x_k}(s)p(x_k|Z_k)\, dx_k$$

Here,

$$\phi_{x_{k+1}|x_k}(s) = \exp\left[js'f_k(x_k) - \tfrac{1}{2}s'g(x_k)Q_kg'(x_k)s\right]$$

The following quantity approximates $\phi_{x_{k+1}|x_k}(s)$:

$$\psi_{x_{k+1}|x_k}(s) = \exp\left[js'[f_k(\hat{x}_{k/k}) + F_k(x_k - \hat{x}_{k/k})] - \tfrac{1}{2}s'G_kQ_kG_k's\right]$$

One can show that

$$\psi_{x_{k+1}|Z_k}(s) = \int \psi_{x_{k+1}|x_k}(s)p(x_k|Z_k)\, dx_k = \exp\left[js'\hat{x}_{k+1/k} - \tfrac{1}{2}s'\Sigma_{k+1/k}s\right]$$

$$(2.13)$$

and that

$$\int [\phi_{x_{k+1}|x_k}(s) - \psi_{x_{k+1}|x_k}(s)]p(x_k|Z_k)\, dx_k \qquad (2.14)$$

can be made arbitrarily small, for fixed real s, by choosing $\Sigma_{k/k}$ arbitrarily small. Proof of the first claim is requested in Prob. 2.5, and of the second claim in Prob. 2.6. Therefore, for any fixed s,

$$\phi_{x_{k+1}|Z_k}(s) \longrightarrow \psi_{x_{k+1}|Z_k}(s)$$

and this means that

$$p(x_{k+1}|Z_k) \longrightarrow \gamma(x_{k+1} - \hat{x}_{k+1/k}, \Sigma_{k+1/k})$$

The question of whether the convergence may be uniform is not clear, although if $g_k(x_k)Q_kg(x_k)'$ is bounded above and below for all x_k by a nonsingular matrix, one can prove uniformity of convergence. The theorem is now established.

As it stands, the theorem is of very limited applicability if (2.9), (2.10) are to hold approximately for $k = 0, 1, 2, \ldots$. First, the a priori density $p(x_0)$ must be gaussian. Second, in view of the formula $\Sigma_{k+1/k} = F_k\Sigma_{k/k}F_k' + G_kQ_kG_k'$, it is necessary that $G_kQ_kG_k'$ be very small if $\Sigma_{k+1/k}$ is to be very small; i.e. the input noise variance must be very small. Even then, one is only guaranteed that (2.9), (2.10) will hold approximately. As it turns out, the gaussian sum filter of a later section provides a way round these difficulties.

EXAMPLE: *Demodulation of Angle-Modulated Signals* An example of the application of extended Kalman filter theory is now discussed to illustrate its usefulness and limitations. We consider the communications task of the demodulation of frequency- or phase-modulated signals in additive gaussian white noise, with the modulating signal assumed gaussian [8–11].

Let us consider the case of frequency modulation (FM) where the message $\lambda(t)$ has a first order Butterworth spectrum, being modeled as the output of a first order, time-invariant linear system with one real pole driven by continuous-time "white" noise. This message is then passed through an integrator to yield $\theta(t) = \int_0^t \lambda(\tau)d\tau$, which then is employed to phase modulate a carrier signal with carrier frequency ω_c rad/sec. The model state equations can then be written as

$$\begin{bmatrix} \dot{\lambda}(t) \\ \dot{\theta}(t) \end{bmatrix} = \begin{bmatrix} -\dfrac{1}{\beta} & 0 \\ 1 & 0 \end{bmatrix} \begin{bmatrix} \lambda(t) \\ \theta(t) \end{bmatrix} + \begin{bmatrix} 1 \\ 0 \end{bmatrix} w(t)$$

$$z(t) = \sqrt{2} \sin[\omega_c t + \theta(t)] + v(t)$$

for some noise disturbances $v(t)$ and $w(t)$ and some $\beta > 0$. Of course, higher order state models can be constructed for messages with more sophisticated spectra.

Continuous-time signal models such as above have not been discussed in the text to this point, nor is there a need to fully understand them in this example. After $z(t)$ is bandpass filtered in an intermediate frequency [IF] filter at the receiver and sampled, a discrete-time signal model for this sampled signal can be employed for purposes of demodulator design.* The equations for the state of such a model are linear, but the measurement equations are nonlinear.

Two methods of sampling are now described, and the measurement-update equations of the extended Kalman filter derived for each case. A discussion of the time-update equations is not included, as these are given immediately from the equations for the state vector, which are linear.

(i) *Uniform scalar sampling.* A periodic sampling of the received signal after bandlimiting on the IF filter is carried out so that the sampled signal can be represented by the following nonlinear scalar measurement equation:

$$z_k = \sqrt{2} \sin(\omega_0 k + l'x_k) + v_k \qquad (2.15)$$

Here, x_k is the state at time k of a linear discrete-time system driven by white gaussian noise, ω_0 is the intermediate frequency, and $l'x_k$ is the phase. For the FM case, $x_k' = [\lambda_k \quad \theta_k]$ and $l'x_k = \theta_k$, and thus $l' = [0 \quad 1]$. The message we denote as $c'x_k$, and for the FM case we see that $c' = [1 \quad 0]$, yielding $c'x_k = \lambda_k$.

We introduce the assumption† that $\{v_k\}$ is white gaussian noise with mean zero and covariance $E\{v_k v_l\} = r\delta_{kl}$, and is independent of $\{x_k\}$. Evidently,

---

*The development of discrete-time models by sampling of continuous-time systems is described in Appendix C.

†To give a more rigorous derivation of a discrete-time signal model from a sampled bandlimited process turns out to be more trouble than it is worth.

$$h_k(x_k) = \sqrt{2} \sin(\omega_0 k + l'x_k), \; H_k = \sqrt{2} \, l \cos(\omega_0 k + l'\hat{x}_{k/k-1})$$

We also define

$$\Omega_k = H_k' \Sigma_{k/k-1} H_k + r = 2l' \Sigma_{k/k-1} l \cos^2(\omega_0 k + l'\hat{x}_{k/k-1}) + r$$

Then the extended Kalman filter measurement-update equations are

$$\hat{x}_{k/k} = \hat{x}_{k/k-1} + \Sigma_{k/k-1} \sqrt{2} l (\cos \omega_0 k$$
$$+ l'\hat{x}_{k/k-1}) \Omega_k^{-1} [z_k - \sqrt{2} \sin(\omega_0 k + l'\hat{x}_{k/k-1})] \tag{2.16}$$

$$\Sigma_{k/k} = \Sigma_{k/k-1} - 2\Sigma_{k/k-1} l l' \Sigma_{k/k-1} \Omega_k^{-1} \cos^2(\omega_0 k + l'\hat{x}_{k/k-1}) \tag{2.17}$$

(The time-update equations are as in a standard Kalman filter and need not be restated here.) In this particular example it is clear that there is coupling between the processor equation (2.16) and the covariance equation (2.17). Decoupling of the above nonlinear filtering equations is possible if it is assumed that components at twice the carrier frequency will not propagate through the low-pass filtering of a combination of the time-update equations and (2.16). The term

$$2 \cos(\omega_0 k + l'\hat{x}_{k/k-1}) \sin(\omega_0 k + l'\hat{x}_{k/k-1}) = \sin 2(\omega_0 k + l'\hat{x}_{k/k-1})$$

can be dropped from the above equations, and the term

$$2 \cos^2(\omega_0 k + l'\hat{x}_{k/k-1}) = 1 + \cos 2(\omega_0 k + l'\hat{x}_{k/k-1})$$

can be replaced by 1. The equations are now

$$\hat{x}_{k/k} = \hat{x}_{k/k-1} + \Sigma_{k/k-1}(l'\Sigma_{k/k-1}l + r)^{-1} \sqrt{2} \, l z_k \cos(\omega_0 k + l'\hat{x}_{k/k-1}) \tag{2.18}$$

$$\Sigma_{k/k} = \Sigma_{k/k-1} - \Sigma_{k/k-1} l l' \Sigma_{k/k-1}(l'\Sigma_{k/k-1}l + r)^{-1} \tag{2.19}$$

and it is clear that the covariance equation (2.19) is *decoupled* from the processor equation (2.18). For those familiar with phase-locked loops, it is not hard to see that the processor (2.18) can be realized as a digital phase-locked loop.

Simulation studies in [8] show that at least for transient responses, the performance of the uncoupled estimators is not as good as that of the coupled estimators. From the limited data available it also appears that the same is true in high noise environments under steady-state conditions. However, performance of the uncoupled estimators in low noise environments and under steady state is as good as the performance of the coupled estimators.

(ii) In-phase and quadrature-phase sampling. The in-phase and quadrature-phase sampling technique, described in [9] and explored in the problems, translates the scalar analog FM signal to a two-dimensional, discrete-time, baseband process with measurement equations.

$$z_k = \sqrt{2} \begin{bmatrix} \sin l'x_k \\ \cos l'x_k \end{bmatrix} + v_k \tag{2.20}$$

It is assumed* that the noise term $\{v_k\}$ is now a gaussian white vector noise

---

*As with scalar sampling, the derivation of a more sophisticated signal model is not warranted.

process with mean zero and covariance

$$E[v_k v_l'] = \begin{bmatrix} r & 0 \\ 0 & r \end{bmatrix} \delta_{kl}$$

and is independent of $\{x_k\}$.

Evidently, with

$$h_k(x) = \sqrt{2} \begin{bmatrix} \sin l' x_k \\ \cos l' x_k \end{bmatrix}$$

we have

$$H_k' = \frac{\partial h_k(x)}{\partial x} \bigg|_{x = \hat{x}_{k/k-1}} = \sqrt{2} \begin{bmatrix} \cos l' \hat{x}_{k/k-1} \\ -\sin l' \hat{x}_{k/k-1} \end{bmatrix} l'$$

Some routine calculations then yield that

$$H_k(H_k' \Sigma_{k/k-1} H_k + R_k)^{-1}$$

$$= \sqrt{2} l [\cos l' \hat{x}_{k/k-1} \quad -\sin l' \hat{x}_{k/k-1}](r + 2l' \Sigma_{k/k-1} l)^{-1}$$

$$= H_k(r + 2l' \Sigma_{k/k-1} l)^{-1}$$

The measurement-update equations are therefore

$$\hat{x}_{k/k} = \hat{x}_{k/k-1} + (r + 2l' \Sigma_{k/k-1} l)^{-1} \Sigma_{k/k-1} H_k[z_k - h_k(\hat{x}_{k/k-1})] \quad (2.21)$$

$$\Sigma_{k/k} = \Sigma_{k/k-1} - 2(r + 2l' \Sigma_{k/k-1} l)^{-1} \Sigma_{k/k-1} l l' \Sigma_{k/k-1} \quad (2.22)$$

and by making the same approximations as for the scalar sampling case, we obtain

$$\hat{x}_{k/k} = \hat{x}_{k/k-1} + (r + 2l' \Sigma_{k/k-1} l)^{-1} \Sigma_{k/k-1} H_k z_k \quad (2.23)$$

Note that the error covariance equation is decoupled from the processor equation which, in turn, may be realized as a digital multivariable phase-locked loop.

Simulation results. Consider the case of FM demodulation where the signal model states and transmitted signal are

$$x(t) = \begin{bmatrix} \lambda(t) \\ \theta(t) \end{bmatrix} \quad s(t) = \sqrt{2} \sin [\omega_c t + \theta(t)]$$

Here, $\lambda(t)$ is the message and $\theta(t) = \int_0^t \lambda(\tau) \, d\tau$.

For simulation purposes we assume that $\lambda(\cdot)$ is gaussian with mean zero, unit variance, and that it has a first order Butterworth spectrum with a bandwidth $(1/\beta)$ radians/second. Now FM theory tells us that the power spectral density of a wide-band frequency-modulated waveform $s(t)$ is determined by, and has approximately the same form as, the probability density function of the message $\lambda(t)$; and, in particular, for the case here when the message is gaussian with unit variance, we also have that the baseband spectrum has a root-mean-square bandwidth of 1 rad/sec.* The bandwidth expansion ratio is the

*An adequate heuristic derivation is given in many undergraduate texts (see, e.g., [18]).

ratio of the bandwidth of the baseband of $s(t)$, here 1 rad/sec, and the bandwidth of $\lambda(t)$, here $(1/\beta)$ rad/sec, and is thus β.

The state model in continuous time which has the above properties is

$$
\begin{bmatrix} \dot{\lambda}(t) \\ \dot{\theta}(t) \end{bmatrix} = \begin{bmatrix} -\dfrac{1}{\beta} & 0 \\ 1 & 0 \end{bmatrix} \begin{bmatrix} \lambda(t) \\ \theta(t) \end{bmatrix} + \begin{bmatrix} 1 \\ 0 \end{bmatrix} w(t)
$$

where $w(t)$ is a zero mean noise process with $E[w(t)w(\tau)] = (2/\beta)\delta(t - \tau)$ and $\delta(t - \tau)$ is the Dirac delta function. All we say here about this model is that when it is sampled at $t_k = kT$, we have

$$
x_{k+1} \triangleq \begin{bmatrix} \lambda_{k+1} \\ \theta_{k+1} \end{bmatrix} = \begin{bmatrix} \exp\left(-\dfrac{T}{\beta}\right) & 0 \\ -\beta\left[\exp\left(-\dfrac{T}{\beta}\right) - 1\right] & 1 \end{bmatrix} \begin{bmatrix} \lambda_k \\ \theta_k \end{bmatrix} + \begin{bmatrix} 1 \\ -\beta \end{bmatrix} w_k \quad (2.24)
$$

$$
E[w_j w_k] = \left[1 - \exp\left(-2\dfrac{T}{\beta}\right)\right]\delta_{jk} \qquad E[w_k] = 0 \quad (2.25)
$$

We take this discrete-time system with output (2.15) or (2.20) where $E[v_j w_k] = 0*$ to be the signal model.

Further details on the derivation of discrete-time models for FM demodulation are given in [8], including a rationale for selecting a sampling rate of greater than eight times (say sixteen times) the baseband bandwidth, here 1 rad/sec.

The performance measure frequently taken for FM systems is the steady-state inverse of the message error covariance ξ_λ^{-1}, where $\xi_\lambda = \lim\limits_{k \to \infty} E[\lambda_k - \hat{\lambda}_{k/k}]^2$.

In Fig. 8.2-1, this measure is plotted against the ratio of the carrier signal energy to the noise energy (carrier-to-noise ratio CNR), which in our case is $(2\beta/rT)$ with $T = 2\pi/16$.

Monte Carlo simulations are employed to calculate the curves for two different bandwidth expansion ratios $\beta = 25$ and $\beta = 100$. The filter for quadrature- and in-phase sampling is a shade more complex than that for the standard sampling technique, but there is a significant performance improvement. Both performance curves exhibit a knee (threshold) as channel noise increases. Above threshold (i.e., in low noise) the performance turns out to be near optimum, but below threshold the low error assumptions used in the derivations of the extended Kalman filter are no longer valid and the performance is far from optimal. A comparison with more nearly optimal filters for the quadrature- and in-phase sampling case is given in a later section.

Main Points of the Section

Kalman filter theory can be applied to yield a useful nonlinear estimator (an extended Kalman filter) for certain nonlinear estimation problems.

---

*This assumption appears reasonable even though a strict derivation of a sampled model from the above continuous-time model has $E[v_j w_k] \neq 0$ for all j, k.

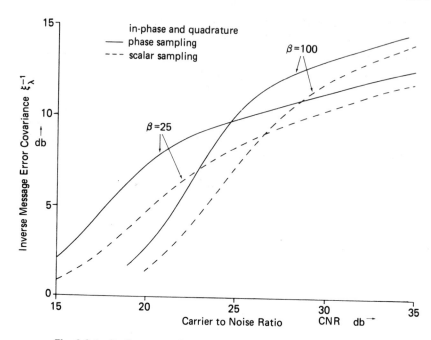

Fig. 8.2-1 Performance of extended Kalman filter FM demodulators.

Application of the extended Kalman filter to a specific problem, as illustrated in the example of FM demodulation, may require ingenuity and appropriate simplifications to achieve a reasonable tradeoff between performance and algorithm complexity. Simulations may be required to determine under what conditions the filter is working well. Normally it is necessary that the magnitude of the state error be comparable with the dimension of the region over which the nonlinear system behaves linearly. Under certain conditions, it is sufficient for the magnitude of the mean square state error to be small.

Problem 2.1. Formulate an iterated extended Kalman filter algorithm as follows. Use the idea that once an estimate $\hat{x}_{k/k}$ is calculated via the usual extended Kalman filter equations, then $h_k(x_k)$ can be linearized about the estimate $\hat{x}_{k/k}$ rather than $\hat{x}_{k/k-1}$. Generally this should be an improved estimate and thus give a more accurate linearization and thereby a more accurate signal model than (2.3). With this "improved" model, an "improved" calculation for h_k and $\hat{x}_{k/k}$ can be achieved. Is there any guarantee that the iterations will converge and achieve an improved estimate?

Problem 2.2. Derive extended Kalman fixed-lag smoothing algorithms.

Problem 2.3. Establish the identity (2.12) in two ways: (a) Express

$$\| x_k - \hat{x}_{k/k-1} \|^2_{\Sigma^{-1}_{k/k-1}} + \| z_k - H'_k(x_k - \hat{x}_{k/k-1}) - h_k(\hat{x}_{k/k-1}) \|^2_{R^{-1}_k}$$

in the form $\| x_k - a \|^2_B + b$ by a completion of the square type argument. (b) Alternatively, postulate that (2.1) and (2.2) are exact, and apply the usual Kalman filter theory. Likewise establish the identity (2.14).

Problem 2.4. With ϵ_k defined as in (2.11), show that for fixed $\hat{x}_{k/k-1}$, $h_k(\cdot)$, and R_k, $\epsilon_k \longrightarrow 0$ uniformly in x_k and z_k as $\Sigma_{k/k-1} \longrightarrow 0$, and $\int \epsilon_k \, dx_k \longrightarrow 0$ uniformly in z_k as $\Sigma_{k/k-1} \longrightarrow 0$. [*Hint:* Show first that $e^{-y^2} - e^{-(y-\delta)^2} \longrightarrow 0$ uniformly for $y \in (-\infty, \infty)$ as $\delta \longrightarrow 0$. Now take arbitrary $\eta > 0$, and show that

$$\gamma[z_k - h_k(x_k), R_k] - \gamma[z_k - H'_k(x_k - \hat{x}_{k/k-1}) - h_k(\hat{x}_{k/k-1}), R_k] < \eta$$

provided that $\| x_k - \hat{x}_{k/k-1} \| < \text{some } \phi(\eta)$; the continuity of $h_k(\cdot)$ must be used. Show that if $\| x_k - \hat{x}_{k/k-1} \| > \phi$, one can choose $\Sigma_{k/k-1}$ sufficiently small that

$$\gamma(x_k - \hat{x}_{k/k-1}, \Sigma_{k/k-1}) < \frac{\eta}{2}[(2\pi)^{m/2}|R_k|^{1/2}]^{-1}$$

Conclude that $\epsilon_k < \eta$ for all x_k, z_k. To obtain the integral result, express the integral as a sum of two integrals, one over $\| x_k - \hat{x}_{k/k-1} \| < \phi(\eta)$, the other over $\| x_k - \hat{x}_{k/k-1} \| \geq \phi(\eta)$. Let $\Sigma_{k/k-1} \longrightarrow 0$ to show that the second integral converges to zero uniformly with respect to z_k.]

Problem 2.5. Check the evaluation of (2.13) by direct calculation.

Problem 2.6. Show that for fixed real s, the integral in (2.14) can be made arbitrarily small. [*Hint:* For fixed s, and $\epsilon > 0$, there exists δ such that $|\phi_{x_{k+1}|x_k}(s) - \psi_{x_{k+1}|x_k}(s)| < \epsilon$ when $\| x_k - \hat{x}_{k/k} \| < \delta$. Break up the integral into two integrals, one over $\| x_k - \hat{x}_{k/k} \| < \delta$, the other over $\| x_k - \hat{x}_{k/k} \| \geq \delta$. Show that the second integral can be made arbitrarily small by choosing $\Sigma_{k/k}$ suitably small.]

Problem 2.7. Let $\xi(t) = \cos[2\pi f_0 t + \theta(t)]$ be a bandpass process with bandwidth W such that $W \ll f_0$. Sample at $t_k = \alpha k/W$, where α is a scalar satisfying $0 < \alpha \leq 1$, k is the sampling index $(0, 1, \ldots)$, and also at $t'_k = t_k + \Delta$ for Δ a small fraction of the sampling interval $(t_{k+1} - t_k)$ to obtain the vector sampled process $[\xi(t_k) \; \xi(t'_k)]'$ for $k = 0, 1, \ldots$. Establish that with the above assumptions $\xi(\cdot)$ can be reconstructed from the sequences of $\xi(t_k)$ and $\xi(t'_k)$. For the case when $t_k = km/f_0$, where m is an integer $\leq f_0/W$ and $t'_k = t_k + 1/4f_0$, show that $\xi(t_k) = \cos[\theta(t_k)]$ and $\xi(t'_k) = -\sin[\theta(t_k)]$, the in-phase and quadrature-phase components.

8.3 A BOUND OPTIMAL FILTER

In this section, we examine a specialization of the signal model of (1.1) and (1.2) by assuming that

$$x_{k+1} = f(x_k) + G_k w_k \tag{3.1}$$

$$z_k = h(x_k) + v_k \tag{3.2}$$

As earlier, $\{w_k\}$ and $\{v_k\}$ are independent, zero mean, white gaussian sequences with covariances $E[w_k w_l'] = Q_k \delta_{kl}$ and $E[v_k v_l'] = R_k \delta_{kl}$. Here G_k is independent of x_k, and the nonlinearities $f(\cdot)$ and $h(\cdot)$ are assumed to satisfy the cone bounds

$$\| f_k(x + \delta) - f_k(x) - \bar{F}_k \delta \| \leq \| \Delta \bar{F}_k \delta \| \qquad (3.3a)$$

$$\| h_k(x + \delta) - h_k(x) - \bar{H}_k' \delta \| \leq \| \Delta \bar{H}_k' \delta \| \qquad (3.3b)$$

for all x and δ, and for some matrices $\bar{F}_k$, $\bar{H}_k$, $\Delta \bar{F}_k$, and $\Delta \bar{H}_k$ independent of x and δ. Here $\| y \|$ denotes the Euclidean norm $(y'y)^{1/2}$. Evidently if $f_k(\cdot)$ is differentiable, its slope lies between $\bar{F}_k - \Delta \bar{F}_k$ and $\bar{F}_k + \Delta \bar{F}_k$, and likewise for $h_k(\cdot)$.

We shall study filters of the following class:

$$\hat{x}_{k/k} = \hat{x}_{k/k-1} + L_k[z_k - h_k(\hat{x}_{k/k-1})] \qquad (3.4a)$$

$$\hat{x}_{k+1/k} = f_k(\hat{x}_{k/k}) \qquad (3.4b)$$

Here, as in the extended Kalman filter equations, $\hat{x}_{k/k}$ denotes a state estimate at time k given measurements $z_0, z_1, \ldots, z_k$ that is *not* necessarily a conditional mean estimate. The structure of the filter (3.4) is the same as that of the extended Kalman filter, and it is evidently a heuristically reasonable structure. In contrast, however, to the extended Kalman filter, the filter gain L_k does not depend on the state estimate; for the moment we shall leave the sequence $\{L_k\}$ unspecified, but subsequently we shall pin it down.

One reason for studying the above somewhat restricted class of non-linear filters is to gain some insight into what happens when a linear filter has unintentional cone-bounded nonlinearities. More importantly, we are able to obtain performance bounds for this class of nonlinear filter and, more-over, we are able to derive a *bound-optimal filter* as the filter for which the error variance upper bounds are minimized. From the theoretical point of view, then, the filters of this section are more significant than extended Kalman filters for which no general performance or stability results other than approximate ones are known. Of course, it should be recognized that when a signal model satisfies the restrictions (3.3) and the extended Kalman filter gain is decoupled from the state estimate, then the extended Kalman filter itself belongs to the class of filters studied in this section.

As we develop results for the class of filters described above, it is well to keep in mind that the looseness of the sector bound (3.3) determines the looseness of the performance bounds to be derived. For signal models with loose bounds (3.3), as when $h_k(x)$ is $\sin(\omega_0 k + l'x)$ in the FM signal model of the previous section, the bounds will not be of much use.

The theory developed below for performance bounds and bound optimal filters is a discrete-time version of continuous-time results in [12].

First we introduce the following definitions and recursive relationships. In the following equations $\{\alpha_k\}$ and $\{\beta_k\}$ are sequences of positive scalars which will be specialized later. Also $\{L_k\}$ is to be considered a fixed sequence; later, it too will be specialized.

$$\tilde{x}_{k/k} = x_k - \hat{x}_{k/k} \qquad \Sigma_{k/k} = E\{\tilde{x}_{k/k}\tilde{x}'_{k/k}\} \tag{3.5}$$

and similarly for $\hat{x}_{k+1/k}$ and $\Sigma_{k+1/k}$.

$$\bar{\Sigma}_{k+1/k} = (1 + \alpha_k)\bar{F}_k\bar{\Sigma}_{k/k}\bar{F}'_k + \left(1 + \frac{1}{\alpha_k}\right)\operatorname{tr}(\Delta\bar{F}_k\bar{\Sigma}_{k/k}\Delta\bar{F}'_k)I + G_kQ_kG'_k \tag{3.6a}$$

$$\bar{\Sigma}_{k/k} = (1 + \beta_k)(I - L_k\bar{H}'_k)\bar{\Sigma}_{k/k-1}(I - L_k\bar{H}'_k)'$$
$$+ L_kR_kL'_k + \left(1 + \frac{1}{\beta_k}\right)\operatorname{tr}(\Delta\bar{H}'_k\bar{\Sigma}_{k/k-1}\Delta\bar{H}_k)L_kL'_k \tag{3.6b}$$

initialized by $\bar{\Sigma}_{0/-1} = \Sigma_{0/-1} = P_0$. With these definitions, we have the first main result.

THEOREM 3.1 (Performance Bounds). With the signal model (3.1)–(3.3), the filter (3.4), and with the definitions (3.5) and (3.6), the filter error covariances are bounded as

$$\Sigma_{k/k} \leq \bar{\Sigma}_{k/k} \qquad \Sigma_{k+1/k} \leq \bar{\Sigma}_{k+1/k} \tag{3.7}$$

for all $\alpha_k > 0$ and $\beta_k > 0$. For a fixed sequence $\{L_k\}$, the bounds are minimized with α_k and β_k chosen as

$$\alpha_k^* = \left[n\frac{\operatorname{tr}(\Delta\bar{F}_k\bar{\Sigma}_{k/k}\Delta\bar{F}'_k)}{\operatorname{tr}(\bar{F}_k\bar{\Sigma}_{k/k}\bar{F}'_k)}\right]^{1/2}$$

$$\beta_k^* = \left[\frac{\operatorname{tr}(L_kL'_k)\operatorname{tr}(\Delta\bar{H}'_k\bar{\Sigma}_{k/k-1}\Delta\bar{H}_k)}{\operatorname{tr}[(I - L_k\bar{H}'_k)\bar{\Sigma}_{k/k-1}(I - L_k\bar{H}'_k)']}\right]^{1/2} \tag{3.8}$$

*Proof.** With the definitions

$$p_k = f_k(x_k) - f_k(\hat{x}_{k/k}) - \bar{F}_k\tilde{x}_{k/k} \qquad \tilde{f}_k = f_k(x_k) - f_k(\hat{x}_{k/k})$$
$$q_k = h_k(x_k) - h_k(\hat{x}_{k/k-1}) - \bar{H}'_k\tilde{x}_{k/k-1} \qquad \tilde{h}_k = h_k(x_k) - h_k(\hat{x}_{k/k-1})$$

we have

$$\Sigma_{k+1/k} = E\{\tilde{f}_k\tilde{f}'_k\} + G_kQ_kG'_k$$
$$\Sigma_{k/k} = E\{[\tilde{x}_{k/k-1} - L_k\tilde{h}_k][\tilde{x}_{k/k-1} - L_k\tilde{h}_k]'\} + L_kR_kL'_k \tag{3.9}$$

Now

$$\tilde{f}_k\tilde{f}'_k = \bar{F}_k\tilde{x}_{k/k}\tilde{x}'_{k/k}\bar{F}'_k + p_kp'_k + \bar{F}_k\tilde{x}_{k/k}p'_k + p_k\tilde{x}'_{k/k}\bar{F}'_k \tag{3.10a}$$

*The proof may be omitted at first reading.

$$[\tilde{x}_{k/k-1} - L_k\tilde{h}_k][\tilde{x}_{k/k-1} - L_k\tilde{h}_k]' = [I - L_k\bar{H}'_k]\tilde{x}_{k/k-1}\tilde{x}'_{k/k-1}[I - L_k\bar{H}'_k]'$$
$$+ L_kq_kq'_kL'_k + [I - L_k\bar{H}'_k]\tilde{x}_{k/k-1}q'_kL'_k + L_kq_k\tilde{x}'_{k/k-1}[I - L_k\bar{H}'_k]'$$

$$(3.10b)$$

Taking expectations of both sides of (3.10) and substitution into (3.9) does not immediately lead to a recursive relationship in $\Sigma_{k/k}$ and $\Sigma_{k+1/k}$. However, with some manipulation, we can obtain the desired recursive relationships. Expectations of the first summands on the right-hand sides in (3.10) are easily dealt with. We deal next with the second term; observe that the cone-bound inequalities (3.3) may be rewritten as

$$\|p_k\| \leq \|\Delta\bar{F}_k\tilde{x}_{k/k}\| \qquad \|q_k\| \leq \|\Delta\bar{H}'_k\tilde{x}_{k/k-1}\| \qquad (3.11)$$

For arbitrary y, we then have by the Cauchy-Schwarz inequality

$$|y'p_k|^2 \leq \|y\|^2\|p_k\|^2 \quad \text{and} \quad |y'L_kq_k|^2 \leq \|L'_ky\|^2\|q_k\|^2$$

Therefore, by application of (3.11)

$$E[p_kp'_k] \leq E[\tilde{x}'_{k/k}\Delta\bar{F}'_k\Delta\bar{F}_k\tilde{x}_{k/k}]I = \text{tr}\,(\Delta\bar{F}_k\Sigma_{k/k}\Delta\bar{F}'_k)I \qquad (3.12a)$$

$$E[L_kq_kq'_kL_k] \leq E[\tilde{x}'_{k/k-1}\Delta\bar{H}_k\Delta\bar{H}'_k\tilde{x}_{k/k-1}]L_kL'_k = \text{tr}\,(\Delta\bar{H}'_k\Sigma_{k/k-1}\Delta\bar{H}_k)L_kL'_k$$

$$(3.12b)$$

We deal with the last two terms on the right side of (3.10a) as follows. Since

$$y'\bar{F}_k\tilde{x}_{k/k}p'_ky \leq \alpha_k|y'\bar{F}_k\tilde{x}_{k/k}|\left|\frac{p'_ky}{\alpha_k}\right|$$

$$\leq \frac{1}{2}\alpha_k\left[y'\bar{F}_k\tilde{x}_{k/k}\tilde{x}'_{k/k}\bar{F}'_ky + \frac{1}{\alpha_k^2}y'p_kp'_ky\right]$$

for any $\alpha_k > 0$, we have the bound

$$E[\bar{F}_k\tilde{x}_{k/k}p'_k + p_k\tilde{x}'_{k/k}\bar{F}'_k] \leq \alpha_k[\bar{F}_k\Sigma_{k/k}\bar{F}'_k + \frac{1}{\alpha_k^2}\text{tr}\,(\Delta\bar{F}_k\Sigma_{k/k}\Delta\bar{F}'_k)I]$$

Likewise, in relation to (3.10b), we have the following bound for any $\beta_k < 0$:

$$E[(I - L_k\bar{H}'_k)\tilde{x}_{k/k-1}q'_kL'_k + L_kq_k\tilde{x}'_{k/k-1}(I - L_k\bar{H}'_k)']$$

$$\leq \beta_k\left[(I - L_k\bar{H}'_k)\Sigma_{k/k-1}(I - L_k\bar{H}'_k)' + \frac{1}{\beta_k^2}\text{tr}\,(\Delta\bar{H}'_k\Sigma_{k/k-1}\Delta\bar{H}_k)L_kL'_k\right]$$

Substitution of these inequalities into (3.9) and (3.10) yields the desired recursive inequalities:

$$\Sigma_{k+1/k} \leq (1 + \alpha_k)\bar{F}_k\Sigma_{k/k}\bar{F}'_k + \left(1 + \frac{1}{\alpha_k}\right)\text{tr}\,(\Delta\bar{F}_k\Sigma_{k/k}\Delta\bar{F}'_k)I + G_kQ_kG'_k$$

$$\Sigma_{k/k} \leq (1 + \beta_k)(I - L_k\bar{H}'_k)\Sigma_{k/k-1}(I - L_k\bar{H}'_k)'$$

$$+ L_k\left[R_k + \left(1 + \frac{1}{\beta_k}\right)\text{tr}\,(\Delta\bar{H}'_k\Sigma_{k/k-1}\Delta\bar{H}_k)I\right]L'_k \qquad (3.13)$$

Subtracting (3.13) from (3.6) results in difference equations in $[\bar{\Sigma}_{k/k} - \Sigma_{k/k}]$ and $[\bar{\Sigma}_{k/k-1} - \Sigma_{k/k-1}]$ which carry nonnegative definite initial values into such values for all k, thereby establishing the bounds (3.7). Optimal values for α and β are obtained by searching the $\alpha\beta$ space to minimize tr $(\bar{\Sigma}_{k/k})$ and tr $(\bar{\Sigma}_{k+1/k})$. Setting the relevant partial derivatives to zero yields the optimal values of (3.8).

Remarks

The recursive equations for the bounds $\bar{\Sigma}_{k/k}$ and $\bar{\Sigma}_{k+1/k}$ can be viewed as standard filtering equations for a linear signal model related to the original nonlinear model (3.1) and (3.2), but with additional noise at the inputs and outputs. We see that the effect of the nonlinearities is taken into account by the addition of noise to a linear signal model. Notice, however, that the calculation of the covariances of the additional noise terms requires the solution of the recursive equations for $\bar{\Sigma}_{k/k}$ and $\bar{\Sigma}_{k+1/k}$. As the nonlinearities become more linear in the sense that the cone-bound parameters $\Delta \bar{H}_k$ and $\Delta \bar{F}_k$ approach zero, the standard filtering equations for a linear signal model defined by matrices $\{\bar{F}_k, \bar{H}_k, G_k, Q_k, R_k\}$ are obtained.

A Bound Optimal Filter

So far, we have given no indication of how the gain L_k might be selected. Here, we shall show that it is possible to select L_k in order to minimize the bound on the error covariance (of course, this is different from choosing L_k to minimize the error covariance itself). The resulting filter is termed a bound-optimal filter.

One of the advantages of the error covariance bound equation is that the bound is computable in advance; the same is true of the bound-optimal sequence $\{L_k^*\}$, which has the property that

$$\bar{\Sigma}_{k/k}^* = \bar{\Sigma}_{k/k}(L_k^*) \leq \bar{\Sigma}_{k/k}(L_k) \quad \text{and} \quad \bar{\Sigma}_{k+1/k}^* = \bar{\Sigma}_{k+1/k}(L_k^*) \leq \bar{\Sigma}_{k+1/k}(L_k)$$

for all other gain sequences $\{L_k\}$.

The precise result is as follows.

THEOREM 3.2 (Bound Optimal Gain). With notation as earlier, and with the $\{\alpha_k\}$, $\{\beta_k\}$ sequences arbitrary but fixed, the bound optimal gain sequence is given by

$$L_k^* = (1 + \beta_k)\bar{\Sigma}_{k/k-1}^* \bar{H}_k(\bar{W}_k^*)^{-1} \tag{3.14}$$

where

$$\bar{W}_k^* = \left[R_k + (1 + \beta_k)\bar{H}_k'\bar{\Sigma}_{k/k-1}^*\bar{H}_k + \left(1 + \frac{1}{\beta_k}\right) \text{tr } (\Delta \bar{H}_k'\bar{\Sigma}_{k/k-1}^*\Delta \bar{H}_k)I \right] \tag{3.15a}$$

$$\bar{\Sigma}_{k+1/k}^* = (1 + \alpha_k)\bar{F}_k\bar{\Sigma}_{k/k}^*\bar{F}_k' + \left(1 + \frac{1}{\alpha_k}\right) \text{tr } (\Delta \bar{F}_k\bar{\Sigma}_{k/k}^*\Delta \bar{F}_k')I + G_kQ_kG_k' \tag{3.15b}$$

$$\bar{\Sigma}_{k/k}^* = (1 + \beta_k)\bar{\Sigma}_{k/k-1}^* - (1 + \beta_k)^2 \bar{\Sigma}_{k/k-1}^* \bar{H}_k(\bar{W}_k^*)^{-1} \bar{H}_k' \bar{\Sigma}_{k/k-1}^*$$

$$= (1 + \beta_k)(I - L_k^* \bar{H}_k') \bar{\Sigma}_{k/k-1}^* \tag{3.15c}$$

and

$$\bar{\Sigma}_{0/-1}^* = \Sigma_{0/-1} = P_0.$$

Proof. Lengthy manipulations of (3.6) and the definitions (3.14) and (3.15) immediately yield

$$\bar{\Sigma}_{k+1/k} - \bar{\Sigma}_{k+1/k}^* = (1 + \alpha_k)\bar{F}_k(\bar{\Sigma}_{k/k} - \bar{\Sigma}_{k/k}^*)\bar{F}_k'$$

$$+ \left(1 + \frac{1}{\alpha_k}\right) \text{tr} \, [\Delta \bar{F}_k(\bar{\Sigma}_{k/k} - \bar{\Sigma}_{k/k}^*)\Delta \bar{F}_k']I$$

$$\bar{\Sigma}_{k/k} - \bar{\Sigma}_{k/k}^* = \left(1 + \frac{1}{\beta_k}\right) \text{tr} \, [\Delta \bar{H}_k'(\bar{\Sigma}_{k/k-1} - \bar{\Sigma}_{k/k-1}^*)\Delta \bar{H}_k]L_k L_k'$$

$$+ (1 + \beta_k)(I - L_k \bar{H}_k')(\bar{\Sigma}_{k/k-1} - \bar{\Sigma}_{k/k-1}^*)(I - L_k \bar{H}_k')'$$

$$+ (L_k - L_k^*)\bar{W}_k^*(L_k - L_k^*)'$$

With $(\bar{\Sigma}_{0/0} - \bar{\Sigma}_{0/0}^*)$ nonnegative definite, clearly $(\bar{\Sigma}_{k/k} - \bar{\Sigma}_{k/k}^*)$ and $(\bar{\Sigma}_{k+1/k} - \bar{\Sigma}_{k+1/k}^*)$ are nonnegative for all k and L_k, as required.

Observe that as the cone bounds collapse, the bound optimal filter becomes the Kalman filter. Also observe that the bound optimal filter is achieved for a specified $\{\alpha_k\}$ and $\{\beta_k\}$ sequence. Different sequences lead to different $\{L_k^*\}$, and it is not straightforward to simultaneously optimize α_k, β_k, and L_k.

Long-term Performance

Questions arise as to what happens as $k \to \infty$: Can one obtain a time-invariant filter? Will the error bound tend to a finite limit? Will the filter be stable? It is awkward to state quantitative answers to these questions. Nevertheless, certain comments can be made. Consider (3.6), and suppose that $\bar{F}_k, G_k, L_k$, etc., are all constant. Then the mappings for constructing $\bar{\Sigma}_{k/k}$ from $\bar{\Sigma}_{k/k-1}$ and $\bar{\Sigma}_{k+1/k}$ from $\bar{\Sigma}_{k/k}$ are linear; this means that one could arrange the entries of $\bar{\Sigma}_{k/k-1}$ and $\bar{\Sigma}_{k/k}$ in a vector σ_k, say, and find constant matrices B and C for which $\sigma_{k+1} = B\sigma_k + C$. A sufficient, and virtually necessary, condition for σ to approach a finite limit is that $|\lambda_i(B)| < 1$.

For the bound optimal filter, the situation is a little less clear, in that the update equation for passing from σ_k to σ_{k+1} is nonlinear. We can still, however, make several comments.

1. If for some L in (3.6), one has $\bar{\Sigma}_{k/k}$ and $\bar{\Sigma}_{k+1/k}$ approaching a finite limit, then with L replaced by L_k^*, $\bar{\Sigma}_{k/k}$ and $\bar{\Sigma}_{k+1/k}$ will be bounded.
2. As the nonlinearities become more and more linear, $\bar{\Sigma}_{k/k}$ and $\bar{\Sigma}_{k+1/k}$ approach the usual linear filtering quantities; thus one could reason-

ably conjecture that for small enough $\Delta \bar{F}$ and $\Delta \bar{H}$, one would have $\bar{\Sigma}^*_{k/k}$ and $\bar{\Sigma}^*_{k+1/k}$ approaching a finite limit as $k \longrightarrow \infty$ if the corresponding optimal quantities in the linear situation approached a finite limit.

As far as stability is concerned, we can again make qualitative remarks. An exponentially stable linear system retains its exponential stability if a small amount of nonlinearity is introduced [13]. We should therefore expect that the bound optimal filter would be asymptotically stable for small enough nonlinearities if the corresponding optimal filter in the linear situation was exponentially stable.

Main Points of the Section

The Kalman filter can be imbedded in a class of nonlinear filters which are designed for application when the signal models are nonlinear with the restriction that the linearities are cone bounded. The cone-bounded non-linearity assumption allows a derivation of performance bounds for the non-linear filters, and if desired, a bound optimal filter can be derived. In the limit as the cone bounds collapse, the filters become linear, the performance bounds approach the actual system performance, and the bound optimal filter approaches the Kalman filter.

The theory of this section is perhaps most significant for the perspective it allows on the linear filtering theory. We are assured that in applying linear filtering theory to a process which has mild sector nonlinearities, we cannot go far wrong if we model the nonlinearities as additional process and measurement noise.

Problem 3.1. For the case of a scalar state signal model, is it possible to derive tighter performance bounds than those of this section?

Problem 3.2. Carry out the manipulations required to demonstrate that L^*_k as defined in (3.14) is the bound optimal filter gain.

8.4 GAUSSIAN SUM ESTIMATORS

In more sophisticated nonlinear estimation schemes than those of the previous two sections, an attempt is usually made to calculate, at least approximately, the relevant a posteriori probability density functions, or sufficient statistics of these functions. For the case of interest in this chapter where the state vector x_k is a first order Markov process, knowledge of the signal model equations, the a posteriori density function $p(x_k | Z_k)$ at time t_k, and the new measurement at time $k + 1$ is actually sufficient for an update of the a posteriori density function to $p(x_{k+1} | Z_{k+1})$. (See Prob. 1.1.) With knowledge of $p(x_k | Z_k)$ for each k, either a MAP (maximum a posteriori), conditional mean estimate, or other type of estimate can be calculated.

Though this approach is conceptually appealing, the difficulties when no appoximation is used can be very great. In general, the storage of $p(x_k | Z_k)$ requires a large number of bits, since for each value of x_k, the corresponding $p(x_k | Z_k)$ must be stored. If $p(x_k | Z_k)$ is definable via a finite number of parameters, this problem is alleviated, but such a description is not possible in general. Further, even assuming storage difficulties can be overcome, there is another difficulty, that of computation, because in each iteration, an integration is required. Again, for special forms of density, the integration problem can be circumvented.

Approximation of some description is, in effect, needed to overcome the storage and computation problem. One type of approximation depends on making the assumption that the low order moments of $p(x_k | Z_k)$ are, at least approximately, a set of sufficient statistics, and then near optimal estimators for a limited class of problems can be derived. The assumptions may be sound when the estimation error is small or the nonlinearities mild. The extended Kalman filter of the previous section can be derived in this way.

The extended Kalman filter involves working by and large with approximations to densities defined using first and second order moments. One way to refine this idea would be to work with higher order moments, but such a refinement tends to concentrate attention on approximation of the density in the vicinity of its mean. In this section, we examine a different type of refinement, in a sense involving collections of first and second order moments that do not concentrate attention on only one part of the density. More precisely, we work with the signal model of (1.1) and (1.2) restated as

$$x_{k+1} = f_k(x_k) + g_k(x_k)w_k \qquad (4.1)$$

$$z_k = h_k(x_k) + v_k \qquad (4.2)$$

with $\{w_k\}$ and $\{v_k\}$ having the usual properties, including independence. We develop Bayesian estimation algorithms using *gaussian sum approximations* for the densities $p(x_k | Z_k)$, $k = 0, 1, 2, \ldots$. (See [14–16].) In the gaussian sum approach, the key idea is to approximate the density $p(x_k | Z_k)$ as a sum of gaussian densities where the covariance of each separate gaussian density is sufficiently small for the time evolution of its mean and covariance to be calculated accurately using the extended Kalman filter algorithm. The resulting estimators consist of a bank of extended Kalman filters, with each filter tracking the evolution of its assigned gaussian density. The signal estimate is a weighted sum of the filter outputs, where the weightings are calculated from the residuals (or nominal innovations) of the extended Kalman filters. In a low noise environment, the resulting estimator can be very nearly optimal.

In a high noise environment, it is necessary to frequently reinitialize the algorithm in such a way that the conditional error covariance associated with each separate gaussian density, and thus each filter in the filter bank, is always sufficiently small for the filter to be operating near optimally. In other words, instead of working directly with an estimation problem in which there is inevitably a large error variance, as in a high noise environment, we work with an appropriately assembled collection of somewhat contrived problems in which the error variance is small for each problem of the collection, as in a low noise environment, and for which optimal or near optimal solutions exist. The solutions of the subproblems are so orchestrated as to yield an optimal or near optimal solution to the original problem.

Gaussian Sum Approximations

Let $\gamma(x - m_i, B_i)$ denote the normal (gaussian) density

$$\gamma[x - m_i, B_i] = (2\pi)^{-n/2}|B_i|^{-1/2}\exp\{-\tfrac{1}{2}(x - m_i)'B_i^{-1}(x - m_i)\}$$

The mean is the n-vector m_i and the covariance is the nonsingular matrix B_i. The following lemma, quoted from [17], sums up the approximation property.

LEMMA 4.1. Any probability density $p(x)$ can be approximated as closely as desired in the space* $L_1(R^n)$ by a density of the form

$$p_A(x) = \sum_{i=1}^{m}\alpha_i\gamma[x - m_i, B_i] \qquad (4.3)$$

for some integer m, positive scalars α_i with $\sum_{i=1}^{m}\alpha_i = 1$, n-vectors m_i, and positive definite matrices B_i.

A proof is requested for this intuitively reasonable result in the problems. Notice that $p_A(\cdot)$ itself satisfies the two crucial properties of a probability density: it is nonnegative for all x and integrates over R^n to 1.

There are numerous approaches to the numerical task of approximating an arbitrary probability density by gaussian sums [14] using nonlinear optimization techniques, but in any given application the chances are that simplifications can be achieved by taking into account the class of densities investigated, as illustrated in the FM demodulation example considered later in this section. A general approximation procedure is also outlined in the problems and shows that an arbitrarily small bound on the covariances of the summands may be imposed.

To help understand the importance of the gaussian sum description of a density, let us assume that $p(x_k|Z_k)$ is expressed as the gaussian sum

$$p(x_k|Z_k) = \sum_{i=1}^{m}\alpha_{ik}\gamma[x_k - m_{ik}, B_{ik}] \qquad (4.4)$$

*The approximation is such that $\int_{R^n}|p(x) - p_A(x)|\,dx$ can be made arbitrarily small.

Then $\hat{x}_{k/k} = E[x_k | Z_k]$ and $\Sigma_{k/k} = E[x_k - \hat{x}_{k/k})(x_k - \hat{x}_{k/k})']$ are readily calculated, and this is a crucial advantage of using gaussian sums. The calculation is easy, yielding

$$\hat{x}_{k/k} = \sum_{i=1}^{m} \alpha_{ik} m_{ik} \tag{4.5a}$$

$$\Sigma_{k/k} = \sum_{i=1}^{m} \alpha_{ik}\{B_{ik} + (\hat{x}_{k/k} - m_{ik})(\hat{x}_{k/k} - m_{ik})'\} \tag{4.5b}$$

Measurement-update Equation

We now examine the question of passing from $p(x_k | Z_{k-1})$ to $p(x_k | Z_k)$ when a new measurement z_k becomes available. More precisely, we shall assume that $p(x_k | Z_{k-1})$ is a weighted sum of gaussian densities, and we shall show how $p(x_k | Z_k)$ can be similarly expressed. Suppose then that

$$p(x_k | Z_{k-1}) = \sum_{i=1}^{m} \alpha_{i,k-1} \gamma[x_k - \bar{m}_{ik}, \bar{B}_{ik}] \tag{4.6}$$

If $p(x_k | Z_{k-1})$ were simply $\gamma[x_k - \bar{m}_{ik}, \bar{B}_{ik}]$, then our knowledge of the extended Kalman filter suggests it would be reasonable to approximate $p(x_k | Z_k)$ by $\gamma[x_k - m_{ik}, B_{ik}]$, where

$$m_{ik} = \bar{m}_{ik} + K_{ik}[z_k - h_k(\bar{m}_{ik})] \tag{4.7a}$$

$$B_{ik} = \bar{B}_{ik} - \bar{B}_{ik} H_{ik} \Omega_{ik}^{-1} H'_{ik} \bar{B}_{ik} \qquad H'_{ik} = \frac{\partial h_k(x)}{\partial x}\bigg|_{x = \bar{m}_{ik}} \tag{4.7b}$$

$$K_{ik} = \bar{B}_{ik} H_{ik} \Omega_{ik}^{-1} \tag{4.7c}$$

$$\Omega_{ik} = H'_{ik} \bar{B}_{ik} H_{ik} + R_k \tag{4.7d}$$

However, $p(x_k | Z_{k-1})$ is not simply one gaussian density, but a linear combination of such. It turns out that $p(x_k | Z_k)$ is then approximately a linear combination of the densities $\gamma[x_k - m_{ik}, B_{ik}]$ with weights determined by

$$\alpha_{ik} = \frac{\alpha_{i,k-1} \gamma[z_k - h_k(\bar{m}_{ik}), \Omega_{ik}]}{\sum_{j=1}^{m} \alpha_{j,k-1} \gamma[z_k - h_k(\bar{m}_{jk}), \Omega_{jk}]} \tag{4.7e}$$

More precisely, we have the following result.

THEOREM 4.1. With $z_k = h_k(x_k) + v_k$ as in (4.2), and with $p(x_k | Z_{k-1})$ given by (4.6), the updated density $p(x_k | Z_k)$ approaches the gaussian sum

$$\sum_{i=1}^{m} \alpha_{ik} \gamma[x_k - m_{ik}, B_{ik}]$$

calculated via (4.7) uniformly in x_k and z_k as $\bar{B}_{ik} \rightarrow 0$ for $i = 1, 2, \ldots, m$.

Proof. Following the pattern of proof for Theorem 2.1, we have

$$p(x_k \mid Z_k) = \delta p(z_k \mid x_k) p(x_k \mid Z_{k-1})$$

$$= \delta \sum_{i=1}^{m} \alpha_{i, k-1} \{ \gamma[x_k - \bar{m}_{ik}, \bar{B}_{ik}] \gamma[z_k - h_k(x_k), R_k] \}$$

where δ^{-1} is a normalizing constant independent of x_k. Results within the proof of Theorem 2.1, tell us that the term $\{\cdot\}$ above approaches

$$\gamma[z_k - h_k(\bar{m}_{ik}), \Omega_{ik}] \gamma[x_k - m_{ik}, B_{ik}]$$

uniformly in x_k and z_k as $\bar{B}_{ik} \to 0$ for $i = 1, 2, \ldots, m$. Moreover, as $\bar{B}_{ik} \to 0$, one also has

$$\int \gamma[x_k - \bar{m}_{ik}, \bar{B}_{ik}] \gamma[z_k - h_k(x_k), R_k] \, dx_k \longrightarrow$$

$$\int \gamma[z_k - h_k(\bar{m}_{ik}), \Omega_{ik}] \gamma[x_k - m_{ik}, B_{ik}] \, dx_k = \gamma[z_k - h_k(\bar{m}_{ik}), \Omega_{ik}]$$

uniformly in z_k. Since

$$\delta = \sum_{i=1}^{m} \alpha_{i, k-1} \int \gamma[x_k - \bar{m}_{ik}, \bar{B}_{ik}] \gamma[z_k - h_k(x_k), R_k] \, dx_k$$

it follows that

$$\delta \longrightarrow \sum_{i=1}^{m} \alpha_{i, k-1} \gamma[z_k - h_k(\bar{m}_{ik}), \Omega_{ik}]$$

The claim of the theorem is then immediate.

Time-update Equations

With knowledge of the density $p(x_k \mid Z_k)$ expressed as the sum of gaussian terms and knowledge of the dynamic equations of the state vector x_k, it is possible to calculate approximately the one-step-ahead prediction density $p(x_{k+1} \mid Z_k)$, also expressed as a sum of gaussian terms. More precisely, we consider $p(x_k \mid Z_k)$ as the summation

$$p(x_k \mid Z_k) = \sum_{i=1}^{m} \alpha_{ik} \gamma[x_k - m_{ik}, B_{ik}] \qquad (4.8)$$

and apply our extended Kalman filter theory to yield an approximate expression for the one-step predicted estimate of each gaussian distribution $\gamma[x_k - m_{ik}, B_{ik}]$ to the gaussian distribution $\gamma[x_{k+1} - m_{i, k+1}, \bar{B}_{i, k+1}]$, where

$$\bar{m}_{i, k+1} = f_k(m_{ik}) \qquad F_{ik} = \frac{\partial f_k(x)}{\partial x} \bigg|_{m_{ik}} \qquad (4.9a)$$

$$\bar{B}_{i, k+1} = F_{ik} B_{ik} F'_{ik} + G_k(m_{ik}) Q_k G'_k(m_{ik}) \qquad (4.9b)$$

We are led to the following theorem.

THEOREM 4.2. With $x_{k+1} = f_k(x_k) + g_k(x_k) w_k$ as in (4.1) and $p(x_k \mid Z_k)$ expressed as the gaussian sum (4.8), the one-step-ahead a posteriori density $p(x_{k+1}|Z_k)$ approaches the gaussian sum

$$\sum_{i=1}^{m} \alpha_{ik}\gamma[x_{k+1} - \bar{m}_{i,k+1}, \bar{B}_{i,k+1}]$$

uniformly in x_k as $B_{ik} \rightarrow 0$ for $i = 1, 2, \ldots, m$.

The proof is a straightforward extension of that for Theorem 2.1, and is omitted.

Filter Implementation

In both the measurement- and time-update equations described above, the updated mean and covariance of each summand follow from the usual extended Kalman filter equations. Thus, on the proviso that the m separate extended Kalman filter covariances are small, one implements the overall filter by running m separate extended filters and adjusting the weights $\alpha_{l,k}$ at each measurement update. (The weights are not changed at each time update.)

The conditional minimum mean-square-error estimate $\hat{x}_{k/k}$ is simply the weighted sum of the states of the bank of extended Kalman filters as in (4.5).

Whenever the individual filter covariance matrices get too large, it is necessary to reinitialize in the sense that one reexpresses $p(x_k|Z_k)$ or $p(x_{k+1}|Z_k)$, as the case may be, as a sum of gaussian densities, all with small covariance. (This may have to be done after each time update, if the input noise variance is large.)

In summary, the above results suggest the following recursive estimation algorithm for updating an a posteriori density $p(x_k|Z_k)$, initialized by $k = 0$.

$\{1\} =$ Approximate $p(x_k|Z_{k-1})$ with a sum of m gaussian densities as in (4.6) satisfying some constraint $\bar{B}_{ik} < \epsilon I$ for some small $\epsilon > 0$.

$\{2\} =$ Apply the measurement-update equations as in Theorem 4.1.

$\{3\} =$ Apply the time-update equations as in Theorem 4.2.

$\{4\} =$ Set $k + 1 = k$.

$\{5\} =$ Check that $\bar{B}_{ik} < \epsilon I$. If the constraint is satisfied, go to $\{2\}$; if not, go to $\{1\}$.

In view of the earlier theorems, we have:

THEOREM 4.3. For the algorithm composed of steps $\{1\}$ through $\{5\}$, as $m \rightarrow \infty$ then ϵ can be chosen such that $\epsilon \rightarrow 0$, resulting in the property that $B_{ik} \rightarrow 0$ and $\bar{B}_{ik} \rightarrow 0$. Moreover, the gaussian sum approximation to $p(x_k|Z_k)$ approaches $p(x_k|Z_k)$.

Remarks

1. Simulation results on some examples suggest that the above algorithm still works satisfactorily even when m is surprisingly small (say $m = 6$).

For the case when $m = 1$, the algorithm of course reduces to the extended Kalman filter, which can be satisfactory on occasions.

2. In many cases, an optimal approximation of $p(x_k | Z_{k-1})$ by a gaussian sum with covariance constraints would require too much on-line calculation to be practical. However, it may be possible to construct simple and suboptimal methods which work reasonably well, as the following example shows.

EXAMPLE. *Demodulation of Angle-Modulated Signals* Referring to our example of FM demodulation of an earlier section, we consider a single-pole message model corresponding to the first order Butterworth message spectrum with state equations.

$$\begin{bmatrix} \dot{\lambda} \\ \dot{\theta} \end{bmatrix} = \begin{bmatrix} -\beta^{-1} & 0 \\ 1 & 0 \end{bmatrix} \begin{bmatrix} \lambda \\ \theta \end{bmatrix} + \begin{bmatrix} 1 \\ 0 \end{bmatrix} w(t) \qquad (4.10)$$

Thus the state of the FM model is a two-dimensional vector composed of the message $\lambda(t)$ and the phase $\theta(t)$. The transmitted signal is $s(t) = \sqrt{2} \sin [\omega_c t + \theta(t)]$, where ω_c is the carrier frequency, and is corrupted by additive gaussian noise assumed to be white in the bandwidth of interest. The received signal is frequency shifted to an intermediate frequency ω_0, where it is bandlimited and sampled via in-phase and quadrature-phase sampling as in Sec. 8.2. (See also [9] for further details.) There results a baseband process which can be modeled in discrete time with a sampling period of T, as

$$x_{k+1} = Fx_k + w_k$$

and

$$z_k = \sqrt{2} \begin{bmatrix} \sin \theta_k \\ \cos \theta_k \end{bmatrix} + v_k$$

In these equations we have $x'_k = [\lambda_k \quad \theta_k] = [\lambda(kT) \quad \theta(kT)]$, while F_k is the transition matrix $\Phi(T)$ of (4.10) and defined from

$$\dot{\Phi}(t) = \begin{bmatrix} -\beta^{-1} & 0 \\ 1 & 0 \end{bmatrix} \Phi(t) \qquad \Phi(0) = I$$

and $E[w_k w'_l] = Q\delta_{kl}$, where

$$Q = \int_0^T \Phi(t) \begin{bmatrix} 2\beta^{-1} & 0 \\ 0 & 0 \end{bmatrix} \Phi'(t) \, dt$$

Also $\{w_k\}$ and $\{v_k\}$ are assumed to be independent, zero mean, white gaussian noise processes with $E[v_k v'_l] = rI\delta_{kl}$. [If the received signal before sampling is assumed to be white with variance $\eta\delta(t - \tau)$, then to a first order of approximation $r = \eta/T$.] We set $\lambda(t_0) = \lambda_0$, a random variable with probability density $N[0, 1]$, and $\theta(t_0) = \theta_0$ is uniformly distributed in $[-\pi, \pi]$.

For this FM demodulation example, the step $\{1\}$ of the gaussian sum algorithm above, where $p(x_k | Z_{k-1})$ is approximated by a gaussian sum in which the covariance $\bar{B}_{ik}$ of each gaussian term $(i = 1, 2, \ldots)$ must satisfy $\bar{B}_{ik} < \epsilon I$ for some small $\epsilon > 0$, may be relaxed. In fact, only the phase variance

$\bar{B}_{ik}^{22}$ (the 2-2 component of $\bar{B}_{ik}$) need satisfy $\bar{B}_{ik}^{22} < \epsilon$ for some small $\epsilon < 0$, as explored in one of the problems. Of course this relaxation represents a considerable reduction in implementation effort since any selection of a grid of means $\bar{m}_{ik}$ in the x space (here the $\theta\lambda$ space) will be less dense as λ varies than would otherwise be the case.

To achieve a practical realization of the gaussian sum algorithm for the FM demodulation problem, three simplifying decisions are made as a matter of engineering judgment. They are now listed:

1. The densities in the gaussian sum approximations are initialized or reinitialized as having the same phase variance. This leads to considerable simplification in the measurement update equations. In effect, the grid of $\bar{m}_{ik}$ in the $\theta\lambda$ space is equispaced as θ varies, with the spacing dependent in a simple way on ϵ, as defined above.

2. The a posteriori density is approximated by a gaussian sum with means $\bar{m}_{ik}$ along a one-dimensional grid in the $\theta\lambda$ (phase-frequency) space usually not parallel to the θ axis. Such an approximation allows a considerable reduction in the number of terms in a gaussian sum approximation and, in effect, reduces the dimension of the problem. The underlying assumption here is that the a posteriori densities conditioned on phase are approximately gaussian. This assumption appears a reasonable one from the simulation results and is supported to some extent by the relaxation of the covariance constraints discussed above and in Prob. 4.3.

3. In approximating $p(x_k | Z_k)$ by a gaussian sum with covariance constraints, we consider the phase θ to be modulo 2π. Thus the $\theta\lambda$ space is cylindrical. The advantage of considering the probability density $p(\theta | \lambda, Z_k)$ as a folded density on a circle rather than the unfolded density on a real line should be clear. Such a density could be represented reasonably accurately by a gaussian sum on a finite grid of, say, six points evenly spaced around the circle. In support of this simplification we note that the conditional innovations $z_k - h_k(\bar{m}_{ik})$ in (4.7) associated with the extended Kalman filter depend on the phase error modulo 2π rather than on the actual phase error.

The above engineering decisions, when taken together, lead to the grid points $\bar{m}_{ik}$ being evenly spaced around an ellipse on the $\theta\lambda$ cylinder, not necessarily orthogonal to the λ axis but unlikely to be parallel or near parallel to it. In any reinitialization, the orientation of the ellipse on the $\theta\lambda$ cylinder must be determined and the phase angle of a reference grid point must also be determined.

Employing the above simplifications and also certain variations not discussed here, [16] develops a more detailed algorithm than that of the previous subsection, designed specifically for the FM demodulation problem. However, in essence it is a bank of m extended Kalman filters where the state estimate is taken to be the conditional mean, calculated as a weighted sum (4.5) of the outputs of the filters.

In the digital computer simulations, $\lim_{t \to \infty} E\{\lambda^2(t)\} = 1$ (as for the example of Sec. 8.2), the root-mean-square bandwidth of the FM baseband spectrum = 1 rad/sec, the bandwidth expansion ratio is β, and $T = 2\pi/16$ sec to permit adequately fast sampling of the FM baseband process.

A commonly used steady-state performance display consists of plots of ξ_λ^{-1}, the inverse of the "evaluated" mean-square message error, versus CNR $= 2\beta/r$, the carrier-to-noise ratio in the message bandwidth. For our curves, ξ_λ is evaluated as the average over 40 sample paths of the quantity

$$\frac{1}{2000} \sum_{k=100}^{2100} [\lambda_k - \hat{\lambda}_{k/k}]^2$$

A set of such steady-state performance curves for $\beta = 100$ is presented in Fig. 8.4-1. The curve $m = 1$ corresponds to the performance of the decoupled

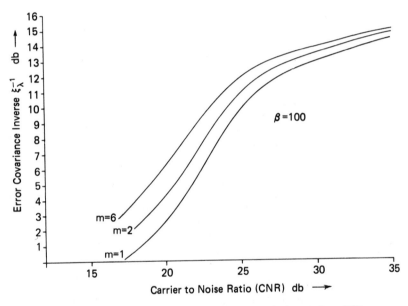

Fig. 8.4-1 Performance of FM demodulators—m is the number of filters in the filter bank.

quasi-optimum demodulator as presented in an earlier section. In the region of high CNR, the performance of the demodulators is the same for all m. However, in the region of low CNR, the performance for larger m is improved over that for smaller m. In such a region, ξ_k^{-1} for $m = 2$ is roughly 1 db better than that for $m = 1$, while for $m = 6$, the improvement is about 3 db.

The demodulators also have improved transient performance in the high CNR region, as illustrated in Fig. 8.4-2.

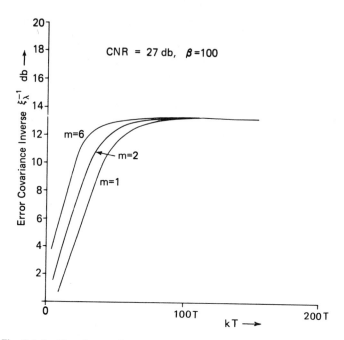

Fig. 8.4-2 Transient performance for FM demodulators—*m* is the number of parallel filters.

Main Points of the Section

We have demonstrated that significant performance improvement can be achieved by using gaussian sum nonlinear estimators involving a bank of extended Kalman filters rather than the simple extended Kalman filter. Unfortunately, the cost associated with increase in filter complexity is considerable, and as we have demonstrated in the FM example, ingenuity may be required in any application to achieve a useful tradeoff between performance and complexity.

It is of theoretical significance that optimal nonlinear estimation in high noise can be achieved in the limit as the number of filters in the filter bank becomes infinite. It is also of interest to view the simple extended Kalman filter as a suboptimal version of the optimal gaussian sum filter.

Finally, we have demonstrated in this chapter that an understanding of the optimal linear filter goes a long way in penetrating some of the deeper problems associated with nonlinear filtering.

Problem 4.1 (Gaussian Sum Approximation). Let x be a scalar quantity, and suppose $p(x)$ is prescribed. Choose $\epsilon > 0$. Select α such that

$$\int_{|x| > \alpha} p(x)\, dx < \epsilon$$

Let $x_0 = -\alpha, x_1, x_2, \ldots, x_N, \alpha$ be points in $[-\alpha, \alpha]$ uniformly spaced with $x_i - x_{i-1} = \Delta$. Set

$$p_A(x) = \sum_i \Delta p(x_i) \gamma [x - x_i, k\Delta^2]$$

where k is a fixed quantity. Show that for suitably small Δ,

$$\int_{|x| < \alpha} |p(x) - p_A(x)| \, dx < \epsilon$$

The quantity k can be adjusted to minimize $\int |p(x) - p_A(x)| \, dx$, which is $0(\epsilon)$.

Problem 4.2. Carry out the manipulations required for the derivation of (4.5).

Problem 4.3. In step $\{1\}$ of the gaussian sum algorithm described in the text, show that for the FM demodulation example the covariance constraint $\bar{B}_{ik} < \epsilon I$ for some small $\epsilon > 0$ can be relaxed to requiring that the phase variance $\bar{B}_{ik}^{22}$ (the 2-2 component of $\bar{B}_{ik}$) should satisfy $\bar{B}_{ik}^{22} < \epsilon$ for some small $\epsilon > 0$. Demonstrate this by using the observation that the message is a nonlinear function of the phase θ but not the frequency λ.

REFERENCES

[1] BUCY, R. S., and K. D. SENNE, "Digital Synthesis of Nonlinear Filters," *Automatica*, Vol. 7, No. 3, May 1971, pp. 287–289.

[2] DE FIGUEIREDO, R. J. P., and Y. G. JAN, "Spline Filters," *Proc. 2nd Symp. on Nonlinear Estimation Theory and its Applications*, San Diego, 1971, pp. 127–141.

[3] WILLSKY, A. S., "Fourier Series and Estimation on the Circle with Applications to Synchronous Communication, Part I (Analysis) and Part II (Implementation)", *IEEE Trans. Inform. Theory*, Vol. IT-20, No. 5, September 1974, pp. 577–590.

[4] TENNEY, R. R., R. S. HEBBERT, and N. R. SANDELL, JR., "Tracking Filter for Maneuvering Sources," *IEEE Trans. Automatic Control*, Vol. AC-22, No. 2, pp. 246–251, April 1977.

[5] WEISS, H., and J. B. MOORE, "Dither in Extended Kalman Filters," Optimization Days, Montreal, May 1977, also submitted for publication.

[6] JAZWINSKI, A. H., *Stochastic Processes and Filtering Theory*, Academic Press, Inc., New York, 1970.

[7] SAGE, A. P., and J. L. MELSA, *Estimation Theory with Applications to Communications and Control*, McGraw-Hill Book Company, New York, 1971.

[8] POLK, O. R., and S. C. GUPTA, "Quasi-optimum Digital Phase-locked Loops," *IEEE Trans. Communications*, Vol. COM-21, No. 1, January 1973, pp. 75–82.

[9] MCBRIDE, A. L., "On Optimum Sample-data FM Demodulation," *IEEE Trans. Communications*, Vol. COM-21, No. 1, January 1973, pp. 40–50.

[10] TAM, P. K., and J. B. MOORE, "Improved Demodulation of Sampled-FM Signals in High Noise," *IEEE Trans. Communications*, Vol. COM-25, No. 9, September 1977, pp. 1052–1053.

[11] BUCY, R. S., C. HECHT, and K. D. SENNE, "An Application of Bayes' Law Estimation to Nonlinear Phase Demodulation," *Proc. 3rd Symp. on Nonlinear Estimation Theory and its Applications*, San Diego, September 1972, pp. 23–35.

[12] GILMAN, A. S., and I. B. RHODES, "Cone-bounded Nonlinearities and Mean Square Bounds—Estimation Upper Bounds," *IEEE Trans. Automatic Control*, Vol. AC-18, No. 6, June 1973, pp. 260–265.

[13] DESOER, C. A., and M. VIDYASAGAR, *Feedback Systems: Input & Output Properties*, Academic Press, Inc., New York, 1975.

[14] SORENSON, H. W., and D. L. ALSPACH, "Recursive Bayesian Estimation Using Gaussian Sums," *Automatica*, Vol. 7, No. 4, July 1971, pp. 465–479.

[15] ALSPACH, D. L., and H. W. SORENSON, "Nonlinear Bayesian Estimation Using Gaussian Sum Approximations," *IEEE Trans. Automatic Control*, Vol. AC-17, No. 4, August 1972, pp. 439–448.

[16] TAM, P. K., and J. B. MOORE, "A Gaussian Sum Approach to Phase and Frequency Estimation," *IEEE Trans. Communications*, Vol. COM-25, No. 9, September, 1977, pp. 435–942.

[17] LO, J., "Finite-dimensional Sensor Orbits and Optimal Nonlinear Filtering," *IEEE Trans. Inform. Theory*, Vol. IT-18, No. 5, September, 1972, pp. 583–589.

[18] TAUB, H., and D. L. SCHILLING, *Principles of Communication Systems*, McGraw-Hill Book Company, New York, 1971.

CHAPTER 9

INNOVATIONS REPRESENTATIONS, SPECTRAL FACTORIZATION, WIENER AND LEVINSON FILTERING

9.1 INTRODUCTION

The first central idea of this chapter, developed in Sec. 9.2, is that the one Kalman filter can be optimal for many different signal models, although its performance may vary between these models; i.e. the filter gain is the same for a collection of different models, while the error covariance is not.

The collection of signal models with the same Kalman filter have a common property: their output covariance is the same. Equivalently then, the first idea of the chapter is that the Kalman filter is determined by the covariance of the measurement process rather than by the detail of the signal model. This idea seems to have first been put forward in [1]. More recent developments are contained in [2–5], the last reference considering the discrete-time problem.

Once the many-to-one nature of the signal model to Kalman filter mapping is understood, the question arises as to whether there is one particular model, in some way special as compared with the others, among the collection of signal models associated with the one Kalman filter. Indeed there is; this model is the *innovations model*, so-called because its input white noise process is identical with the innovations process of the associated filter. Perhaps the most crucial property of the innovations model is that it is causally invertible, in the sense that the input noise process to the model can be computed from the output process in a causal fashion. There are, however, many other very important properties and alternative characterizations; for example, it is immediately computable from the Kalman filter, being in a sense an inverse to the Kalman filter.

It turns out that the problem of computing an innovations representation from a covariance parallels, in case the covariance is stationary, a classical problem of *minimum phase spectral factorization*. In this classical problem, one is given a power density spectrum and one seeks a transfer function matrix with certain properties, including the property that a linear system excited by white noise and with the sought-after transfer function matrix has output spectrum equal to that prescribed. Many solutions are known to this classical problem [6–18].

In this chapter, we also examine two classical approaches to filtering: Wiener filtering and Levinson filtering. Wiener filtering theory [19–21] introduced the idea of statistically representing signals, and was in many ways a precursor of Kalman filtering. To understand it, the concept of spectral factorization is required. Levinson filtering theory [22, 23] aimed at simplifying the computational aspects of Wiener theory, and has since found wide application. Both theories require stationarity of the underlying processes.

In the remainder of this section, we discuss the problem of signal estimation, showing that covariance data alone suffices to determine the estimate. Though signal estimation is not the same as state estimation, this perhaps suggests the reasonableness of the claim that the Kalman filter qua state estimator is definable using covariance data.

We shall also illustrate the fact that there is an infinity of different signal models with the same output covariance. In conjunction with the later proof that this covariance alone determines the Kalman filter, this shows that the mapping {signal model} $\longrightarrow$ {Kalman filter} is certainly a many-to-one map.

Signal Estimation Using Covariance Data

Let us suppose that there is a zero mean signal process $\{y_k\}$, zero mean, white noise process $\{v_k\}$, and measurement process $\{z_k = y_k + v_k\}$, all jointly gaussian. Let v_k be independent of y_l for $l \leq k$.

Further, suppose that $E[z_k z_l']$ is known for all k and l. Let us observe certain properties of formulas for one-step prediction estimates and the associated errors.

From the projection theorem as pointed out in Chap. 5, we have

$$E[z_k | Z_{k-1}] = E[z_k Z_{k-1}']\{E[Z_{k-1} Z_{k-1}']\}^{-1} Z_{k-1} \qquad (1.1)$$

Every entry of $E[z_k Z_{k-1}']$ and $E[Z_{k-1} Z_{k-1}']$ is known, being part of the covariance data. Therefore,

$$\hat{y}_{k/k-1} = \hat{z}_{k/k-1} = M_{k-1}Z_{k-1} \tag{1.2}$$

where M_{k-1} is computable from the output covariance data only. What of the estimation error? We have

$$E\{[z_k - \hat{z}_{k/k-1}][z_k - \hat{z}_{k/k-1}]'\} = E[z_k z'_k]$$
$$- E[z_k Z'_{k-1}]\{E[Z_{k-1}Z'_{k-1}]\}^{-1}E[Z_{k-1}z'_k] \tag{1.3}$$

and this quantity is known from the output covariance data. On the other hand, the error covariance associated with the signal estimate requires more than the *output* covariance data. For we have

$$E\{[y_k - \hat{y}_{k/k-1}][y_k - \hat{y}_{k/k-1}]'\} = E[y_k y'_k]$$
$$- E[y_k Z'_{k-1}]\{E[Z_{k-1}Z'_{k-1}]\}^{-1}E[Z_{k-1}y'_k] \tag{1.4}$$

Now $E[y_k Z'_{k-1}] = E[(z_k - v_k)Z'_{k-1}] = E[z_k Z'_{k-1}]$, so that this quantity is known from output covariance data only. On the other hand, $E[y_k y'_k] = E[z_k z'_k] - E[v_k v'_k]$, and we evidently need to know $E[v_k v'_k]$ *in addition to* $E[z_k z'_l]$ for all k and l.

For state estimation, as we shall see in the next section, the estimate is determinable from the output covariance data alone, as for the signal estimate. However, the state estimate error covariance is not determinable from the output covariance data above, again as for the signal estimate error covariance.

The use of (1.1) in practice could be very burdensome, since the inversion of matrices of ever-increasing dimension is required. If it is known that $y_k = H'_k x_k$, with x_k the state of a finite-dimensional process, and if a finite-dimensional system generating $\hat{x}_{k/k-1}$ can be found, then $\hat{y}_{k/k-1} = H'_k \hat{x}_{k/k-1}$ can be found much more easily than via (1.1). However, without an underlying finite dimensionality, there is really no escape from (1.1) or an equivalent.

An alternative to (1.1), incidentally, is an estimate involving the innovations. Thus in lieu of (1.1), one has

$$\hat{z}_{k/k-1} = E[z_k \tilde{Z}'_{k-1}]\{E[\tilde{Z}_{k-1}\tilde{Z}'_{k-1}]\}^{-1}\tilde{Z}_{k-1} \tag{1.5}$$

which, because of the whiteness of the innovations sequence $\tilde{z}_k$, becomes

$$\hat{z}_{k/k-1} = \sum_{l<k} E[z_k \tilde{z}'_l]\{E[\tilde{z}_l \tilde{z}'_l]\}^{-1}\tilde{z}_l \tag{1.6}$$

Though this formula appears to have eliminated the problem of computing $\{E[Z_{k-1}Z'_{k-1}]\}^{-1}$, it introduces another problem, that of computing $E[z_k \tilde{z}'_l]$, $\tilde{z}_l$, and $E[\tilde{z}_l \tilde{z}'_l]$ for all k and l. This issue is taken up in Prob. 1.1.

Signal Models with the Same Output Covariance

In this subsection, we illustrate the existence of many signal models with the same output covariance. Consider

$$x_{k+1} = \tfrac{1}{2}x_k + w_k \qquad z_k = x_k + v_k$$

We shall show that with an initial time in the infinitely remote past, we can obtain the same output power spectrum for an infinity of different noise statistics. In view of the fact that a power spectrum is a z-transform of a stationary covariance, this is equivalent to showing that many different signal models have the same output covariance.

Accordingly, in addition to taking an initial time in the infinitely remote past, we take $[w_k \quad v_k]$ white, with

$$E\left\{\begin{bmatrix} w_k \\ v_k \end{bmatrix} [w_l \quad v_l]\right\} = \begin{bmatrix} \alpha & \frac{2}{3}(1-\alpha) \\ \frac{2}{3}(1-\alpha) & 3 - \frac{4\alpha}{3} \end{bmatrix} \delta_{kl}$$

Here α is a parameter chosen to make the covariance matrix nonnegative definite, but is otherwise unrestricted. Note that the set of allowable α is nonempty, since $\alpha = 1$ is a member.

The transfer function matrix linking $\left\{\begin{bmatrix} w_k \\ v_k \end{bmatrix}\right\}$ to $\{z_k\}$ is $\begin{bmatrix} \frac{1}{z - \frac{1}{2}} & 1 \end{bmatrix}$. Therefore, the power spectrum of $\{z_k\}$ is

$$\begin{bmatrix} \frac{1}{z - \frac{1}{2}} & 1 \end{bmatrix} \begin{bmatrix} \alpha & \frac{2}{3}(1-\alpha) \\ \frac{2}{3}(1-\alpha) & 3 - \frac{4\alpha}{3} \end{bmatrix} \begin{bmatrix} \frac{1}{z^{-1} - \frac{1}{2}} \\ 1 \end{bmatrix} = \frac{\frac{37}{12} - \frac{5}{6}z - \frac{5}{6}z^{-1}}{(z - \frac{1}{2})(z^{-1} - \frac{1}{2})}$$

which is independent of α. Therefore, an infinity of signal models can produce the one output power spectrum.

Main Points of the Section

It will be shown that the output covariance of a signal model is sufficient to define a Kalman filter, that there is a special signal model among the class of all those with the same output covariance, and that when the covariance is stationary, this special model is related to the classical concept of spectral factorization.

For signal estimation, output covariance data alone is enough to define the filter, even with no finite dimensionality, though more data is needed to compute the associated error variance.

There can exist an infinity of signal models with the same output covariance.

Problem 1.1. Suppose that $z_0, z_1, \ldots$ is a scalar output process. Set $E[z_k z_l] = r_{kl}$ and form the semi-infinite matrix R with kl entry as r_{kl}. Show that it is possible to recursively define a factorization of R as $T'ST$, where S is diagonal and T is upper triangular with unity elements on the diagonal. Interpret the formulas (1.1) and (1.5) using this factorization.

Problem 1.2. Suppose $z_k = y_k + v_k$ with $\{y_k\}$, $\{v_k\}$ jointly gaussian and of zero mean, with $\{v_k\}$ white and v_k independent of y_l for $l \leq k$. Determine what data are necessary for computing predicted and smoothed estimates of y_k and what data are required for computing the associated error variances. Does it make a difference if $\{y_k\}$ and $\{v_k\}$ are independent?

Problem 1.3. Consider the signal models defined in the last subsection. Compute the steady-state Kalman filter for different values of α, and check that the same filter is obtained, but that its performance depends on α.

9.2 KALMAN FILTER DESIGN FROM COVARIANCE DATA

Our main task in this section is to illustrate that the Kalman filter can be determined simply from the output covariance of the signal model. Accordingly, we shall suppose that the signal model is

$$x_{k+1} = F_k x_k + G_k w_k \tag{2.1}$$

$$z_k = y_k + v_k = H'_k x_k + v_k \tag{2.2}$$

where $\{w_k\}$ and $\{v_k\}$ are zero mean, jointly gaussian, white processes with

$$E\left\{\begin{bmatrix} w_k \\ v_k \end{bmatrix}[w'_l \quad v'_l]\right\} = \begin{bmatrix} Q_k & S_k \\ S'_k & R_k \end{bmatrix}\delta_{kl} \tag{2.3}$$

Also, x_0 is a $N(\bar{x}_0, P_0)$ random variable, independent of $\{w_k\}$ and $\{v_k\}$. For simplicity, let us take $\bar{x}_0 = 0$.

We shall compute the output covariance of this model and then show that the Kalman filter gain as computed from the model is also computable from the covariance.

Covariance of $\{z_k\}$

In Chap. 2, we calculated $E[z_k z'_l]$ in case $S_k = 0$. Coping with nonzero S_k is simple provided one uses the easily established fact that $E[x_k v'_l] = \Phi_{k,l+1} G_l S_l$ for $k > l$ and is zero otherwise; here, $\Phi_{k,l}$ is the transition matrix associated with F_k. With $P_k = E[x_k x'_k]$ and given recursively by

$$P_{k+1} = F_k P_k F'_k + G_k Q_k G'_k \tag{2.4}$$

one has

$$E[z_k z'_k] = L_k \qquad E[z_k z'_l] = H'_k \Phi_{k,l+1} M_l \qquad (k > l) \tag{2.5}$$

where

$$L_k = H'_k P_k H_k + R_k \qquad M_l = F_l P_l H_l + G_l S_l \tag{2.6}$$

Of course, for $k < l$, we have

$$E[z_k z'_l] = \{E[z_l z'_k]\}' = [H'_l \Phi_{l,k+1} M_k]' = M'_k \Phi'_{l,k+1} H_l$$

Knowing the Covariance of $\{z_k\}$

Earlier we have intimated that knowledge of the covariance of $\{z_k\}$ is sufficient to determine the Kalman filter. Let us make more precise the notion of knowledge of the covariance of $\{z_k\}$.

First, we can conceive of knowing the values taken by $E[z_k z_l']$ for all k and l, without knowing separately the values of F_k, etc. In the stationary case, one has $E[z_k z_l'] = C_{k-l}$ for some C_{k-l}.

This knowledge is all we need in order to draw certain conclusions about the calculation of signal estimates as opposed to state estimates, as we discussed in Sec. 9.1.

Second, we can conceive of knowing the values taken by $E[z_k z_l']$ for all k and l by knowing certain matrix functions L_k, A_k, and B_l for which $E[z_k z_k'] = L_k$ and $E[z_k z_l'] = A_k' B_l$ for $k > l$. This means that one knows something about the finite-dimensional structure of the $\{z_k\}$ process, but one does not know F_k, H_k, and M_k individually. Using this knowledge, we can show that signal estimates may be calculated by means of a finite-dimensional filter.

Third, we can conceive of knowing the quantities F_k, H_k, M_k, and L_k in (2.5). This knowledge, as it turns out, is sufficient to calculate the Kalman filter as a state filter for all signal models of the form of (2.1) through (2.3). Since state estimation can only take place when a coordinate basis has been fixed, the fact that one needs to know these extra quantities should be no surprise.

Actually, from the signal estimation point of view, the second and third situations are the same. For suppose that the $\{A_k\}$ and $\{B_k\}$ sequences are known. Define $F_k = I$, $H_k = A_k$, and $M_k = B_k$; solve the state filtering problem associated with these parameters. Then a signal estimate is immediately obtainable from the state estimate of the filter.

State Estimation Using Covariance Data

When the signal model of (2.1) through (2.3) is known, we compute the Kalman filter error covariance and gain matrix as follows:

$$\Sigma_{k+1/k} = F_k \Sigma_{k/k-1} F_k' - (F_k \Sigma_{k/k-1} H_k + G_k S_k)(H_k' \Sigma_{k/k-1} H_k + R_k)^{-1}$$
$$\times (F_k \Sigma_{k/k-1} H_k + G_k S_k)' + G_k Q_k G_k' \tag{2.7}$$

with $\Sigma_{0/-1} = P_0$, and

$$K_k = (F_k \Sigma_{k/k-1} H_k + G_k S_k)(H_k' \Sigma_{k/k-1} H_k + R_k)^{-1} \tag{2.8}$$

The relation with the covariance data is obtained in the following way.

THEOREM 2.1 [5]. Consider the signal model of (2.1) through (2.3), with output covariance as defined in (2.5) and Kalman filter gain and error covariance as defined in (2.7) and (2.8). Consider also the equation

$$T_{k+1} = F_k T_k F_k' + (F_k T_k H_k - M_k)(L_k - H_k' T_k H_k)^{-1}(F_k T_k H_k - M_k)' \tag{2.9}$$

initialized by $T_0 = 0$. Then

$$K_k = -(F_k T_k H_k - M_k)(L_k - H_k' T_k H_k)^{-1} \tag{2.10}$$

and

$$P_k = \Sigma_{k/k-1} + T_k \tag{2.11}$$

The crucial points to note are that both T_k and K_k are determined from output covariance data only. On the other hand, since P_k is not determined by output covariance data only, $\Sigma_{k/k-1}$ is not determined by output covariance data only.

The proof of the theorem can be easily obtained by induction, or see [5]. Several miscellaneous points follow.

1. As we know, $0 \leq \Sigma_{k/k-1} \leq P_k$. It follows from (2.11) that $0 \leq T_k \leq P_k$. The matrix T_k can be identified with a covariance in the following way. Because $x_k - \hat{x}_{k/k-1}$ is orthogonal to Z_{k-1}, it is orthogonal to $\hat{x}_{k/k-1}$, and accordingly

$$E[x_k x_k'] = E\{[(x_k - \hat{x}_{k/k-1}) + \hat{x}_{k/k-1}][(x_k - \hat{x}_{k/k-1}) + \hat{x}_{k/k-1}]'\}$$
$$= E[(x_k - \hat{x}_{k/k-1})(x_k - \hat{x}_{k/k-1})'] + E[\hat{x}_{k/k-1}\hat{x}_{k/k-1}']$$

The definitions of P_k and $\Sigma_{k/k-1}$ and comparison with (2.11) show, then, that

$$T_k = E[\hat{x}_{k/k-1}\hat{x}_{k/k-1}'] \tag{2.12}$$

Thus T_k is the state covariance of the Kalman filter.

2. The quantity $H_k' \Sigma_{k/k-1} H_k$ is the error covariance associated with the estimate $\hat{y}_{k/k-1}$ of y_k. We noted in the last section that this quantity had to depend only on $E[z_k z_l']$ for all k and l and on $E[v_k v_l'] = R_k$. This can be checked from the fact that

$$H_k' \Sigma_{k/k-1} H_k = (H_k' P_k H_k + R_k) - H_k' T_k H_k - R_k$$

The first term is $E[z_k z_k']$, the second is computable from covariance data, and the third is known.

3. One might conjecture results similar to those above for smoothing. Actually, it is true that optimal smoothed estimates of y_k depend on $E[y_k y_l']$, $E[y_k v_l']$, and $E[v_k v_l']$ separately; i.e., knowledge of $E[z_k z_l']$ alone is not sufficient to define a smoothed estimate, while to obtain an optimal smoothed estimate of x_k, the full signal model needs to be known. See Prob. 2.3. Of course, optimal smoothed estimates of z_k can be obtained from knowledge of $E[z_k z_l']$. (Why?)

4. Equation (2.11) shows that larger P_k are associated with larger $\Sigma_{k/k-1}$ for signal models with the same output covariance. Especially in the

stationary case, one can give interesting interpretations (some involving frequency domain ideas) of the ordering properties of the P_k of different signal models with the same output covariance (see [24, 25]).

Main Points of the Section

The Kalman filter gain is computable from the signal model output covariance via (2.9) and (2.10), while filter performance depends on the particular signal model via the equation $P_k = \Sigma_{k/k-1} + T_k$. Here, T_k is the state covariance of the Kalman filter and depends only on the signal model output covariance.

Problem 2.1. Suppose that $\{z_k\}$ is a stationary process, generated by an asymptotically stable signal model of the form of (2.1) through (2.3) with all matrices constant. Show that if F, H are known and $E[z_k z_l']$ is known for all k and l, and if $[F, H]$ is completely observable, the quantity M in (2.6) is computable.

Problem 2.2. Consider the matrix T_k defined recursively by (2.9). Show that T_{k+1} can be characterized by

$$T_{k+1} = \min \left\{ X \middle| \begin{bmatrix} X & M_k \\ M_k' & L_k \end{bmatrix} - \begin{bmatrix} F_k \\ H_k' \end{bmatrix} T_k [F_k' \quad H_k] \geq 0 \right\}$$

[*Hint:* Show that with the nonnegativity assumption, the nonnegative matrix is congruent to the direct sum of $X - F_k T_k F_k' - (F_k T_k H_k - M_k)(L_k - H_k' T_k H_k)^{-1}$ $(F_k T_k H_k - M_k)'$ and $L_k - H_k' T_k H_k$.]

Problem 2.3. Compute $E[x_k z_l']$ for $l > k$ for the signal model of (2.1) through (2.3). Using the formula

$$E[x_k | Z_{k+m}] = E[x_k Z_{k+m}'] \{ E[Z_{k+m} Z_{k+m}'] \}^{-1} Z_{k+m}$$

argue that $\hat{x}_{k/k+m}$ depends on the particular signal model and is not determined solely by knowledge of F_k, H_k, M_k, and L_k in the formulas for $E[z_k z_l']$. Discuss also the quantity $E[y_k | Z_{k+m}]$.

9.3 INNOVATIONS REPRESENTATIONS WITH FINITE INITIAL TIME

As we now know, there is an infinity of signal models corresponding to the one Kalman filter. For this section, we identify one such model and identify some of its key properties.

The model is termed the *innovations model*. Here are some of its important properties.

1. It is determinable from the covariance data only and is unique.
2. The input to the innovations model can be determined from its output.

3. The Kalman filter can estimate the state of the innovations model with zero error, and the Kalman filter innovations sequence is identical with the input noise sequence of the innovations model (hence the name).

The first property has great significance. In many practical situations, a signal model of a process may not be known, yet covariance data is obtainable by experiment. In principle then, this data can be used to construct one possible signal model for the process.

The results in this section all apply in nonstationary situations and with finite initial time. In the next section, we take up the special aspects associated with stationary processes, and we allow an initial time in the infinitely remote past.

One way of obtaining the innovations model is as a sort of inverse of a Kalman filter. We shall introduce the model this way.

The Kalman Filter as a Whitening Filter

We usually think of the output of a Kalman filter as a state estimate $\hat{x}_{k/k-1}$. Instead let us think of the output as the innovations sequence $\{\tilde{z}_k\}$. Thus the filter equations (with input z_k, state $\hat{x}_{k/k-1}$, and output $\tilde{z}_k$) are

$$\hat{x}_{k+1/k} = (F_k - K_k H_k')\hat{x}_{k/k-1} + K_k z_k \qquad \hat{x}_{0/-1} = 0 \qquad (3.1a)$$

$$\tilde{z}_k = -H_k'\hat{x}_{k/k-1} + z_k \qquad\qquad (3.1b)$$

(We assume operation is over $[0, \infty)$ and $\bar{x}_0 = 0$ in the usual notation.) Recall that the innovations sequence $\{\tilde{z}_k\}$ is white. Then (3.1) define a system with the following property: *The input is a non-white process* $\{z_k\}$, *the system is computable from* $E[z_k z_i']$, *and the output is a white process* $\{\tilde{z}_k\}$. Such a system is termed a whitening filter.

Inverse of the Whitening Filter

Now let us turn the whitening filter idea round. We seek a system with input $\tilde{z}_k$ and output z_k, i.e., an inverse to (3.1). This is easy to obtain. For when z_k in (3.1a) is replaced by $\tilde{z}_k + H_k'\hat{x}_{k/k-1}$ using (3.1b), we have

$$\hat{x}_{k+1/k} = F_k\hat{x}_{k/k-1} + K_k\tilde{z}_k \qquad \hat{x}_{0/-1} = 0 \qquad (3.2a)$$

while (3.1b) also implies

$$z_k = H_k'\hat{x}_{k/k-1} + \tilde{z}_k \qquad\qquad (3.2b)$$

What we are saying is the following: Suppose that a Kalman filter is known and that the innovations sequence $\{\tilde{z}_k\}$ of the filter is available. Then from $\{\tilde{z}_k\}$, we can construct $\{z_k\}$ by taking $\{\tilde{z}_k\}$ as the input to a certain finite-dimensional system, the output of which will be $\{z_k\}$.

To obtain the innovations model, we need the following idea.

DEFINITION 3.1. Suppose there is given a process $\{a_k\}$. Whenever we can define a process* $\{\bar{a}_k\}$ whose statistics are the same as those of $\{a_k\}$, we say that $\{\bar{a}_k\}$ is a realization of $\{a_k\}$. If a set of statistics are given and we can define a process $\{\bar{a}_k\}$ whose statistics agree with those prescribed, we say that $\{\bar{a}_k\}$ is a process realizing the prescribed statistics.

Equation (3.2) shows that there is a linear system which, if the input is a special white noise process $\{\tilde{z}_k\}$, has output $\{z_k\}$. What if the input is simply a white noise process $\{\bar{v}_k\}$ that has the same statistics as $\{\tilde{z}_k\}$, i.e., is a realization of $\{\tilde{z}_k\}$? Then we should expect the corresponding output process $\{\bar{z}_k\}$ to no longer be identical with $\{z_k\}$, but to have the same statistics as $\{z_k\}$, i.e., to be a realization of $\{z_k\}$.

This thinking leads to the following definition.

DEFINITION 3.2 (Innovations Model). Let there be a zero mean gaussian process $\{z_k\}$ resulting from some signal model of the form (2.1) through (2.3). Suppose that $E[z_k z_k'] = L_k$ and $E[z_k z_l'] = H_k' \Phi_{k,l+1} M_l$ for $k > l$ in the usual notation. Define the sequences $\{T_k\}$, $\{\Omega_k\}$, $\{K_k\}$ by†

$$T_{k+1} = F_k T_k F_k' + (F_k T_k H_k - M_k)(L_k - H_k' T_k H_k)^{-1}(F_k T_k H_k - M_k)'$$
$$T_0 = 0 \tag{3.3a}$$

$$\Omega_k = L_k - H_k' T_k H_k \tag{3.3b}$$

$$K_k = -(F_k T_k H_k - M_k)\Omega_k^{-1} \tag{3.3c}$$

The innovations model for $\{z_k\}$, defined for $k \geq 0$, is the linear system

$$\bar{x}_{k+1} = F_k \bar{x}_k + K_k \bar{v}_k \qquad \bar{x}_0 = 0 \tag{3.4a}$$

$$\bar{z}_k = H_k' \bar{x}_k + \bar{v}_k \tag{3.4b}$$

with $\{\bar{v}_k\}$ a zero mean gaussian sequence with

$$E[\bar{v}_k \bar{v}_l'] = \Omega_k \delta_{kl} \tag{3.5}$$

It is important to note that the innovations model falls within the class of signal models introduced at the start of the last section.

The argument prior to the above definition and comparison of (3.2) and (3.4) suggests that $\{\bar{z}_k\}$ must be a realization of $\{z_k\}$. Let us formally state this fact; we leave the straightforward proof to the reader.

---

*The overbar does not denote a mean, but is used simply to distinguish two related quantities.

†Equations (2.1) through (2.3) actually only force $L_k - H_k' T_k H_k$ to be nonnegative definite. If the inverse in (3.3a) fails to exist, a pseudo-inverse should be used. Though the theory will cover this situation, we shall assume existence of the inverse to keep life simple.

THEOREM 3.1. The output process $\{\bar{z}_k\}$ of the innovations model is a realization for $\{z_k\}$, and

$$E[\bar{x}_k \bar{x}_k'] = T_k \tag{3.6}$$

In the definition of the innovations model, we gave formulas allowing its computation from covariance data. The crucial quantities are of course K_k and Ω_k; these are the Kalman filter gain and filter innovations covariance. So the innovations model is also immediately computable from the Kalman filter. Again, the filter quantities K_k and Ω_k are computable directly from a prescribed signal model; i.e. one does not have to compute first the output covariance of the signal model, then from it K_k and Ω_k; it follows that the innovations model is computable from any signal model. In summary, the innovations model associated with a given covariance $E[z_k z_l']$ is computable from one of the following.

1. The covariance itself
2. The Kalman filter corresponding to the covariance
3. A signal model generating the covariance

The first observation means that the construction of an innovations model provides a solution to the *covariance factorization* problem (sometimes termed the spectral factorization problem, though this terminology is more properly reserved for situations in which $\{z_k\}$ is stationary and the initial time is in the infinitely remote past). The covariance factorization problem is, of course, to pass from a prescribed covariance to a linear system with white noise input with output covariance equal to that prescribed. In this section, we are restricting attention to a finite initial time, but in the next section, we allow an initial time in the infinitely remote past, which in turn allows us to capture some of the classical ideas of spectral factorization (see [1–19]). Nonstationary covariance factorization is discussed in [5, 26–33], with [5] providing many of the ideas discussed in this section. References [26–33] consider the continuous-time problem, with [30–33] focusing on state-variable methods for tackling it.

Point 3 above raises an interesting question. Suppose there is prescribed a signal model

$$x_{k+1} = F_k x_k + K_k v_k \tag{3.7a}$$

$$z_k = H_k' x_k + v_k \tag{3.7b}$$

with v_k a zero mean, white gaussian process with

$$E[v_k v_l'] = \Omega_k \delta_{kl} \tag{3.8}$$

Is it the innovations model for the $\{z_k\}$ process? Equivalently, would the gain of the associated Kalman filter equal the quantity K_k in (3.7)?

The answer is yes, provided that Ω_k is nonsingular.

THEOREM 3.2 (Uniqueness of K_k, Ω_k). Consider the signal model defined by (3.7) and (3.8) with $x_0 = 0$ and Ω_k nonsingular. Then it is an innovations model in the sense of Definition 3.2.

Proof. To prove the claim, the strategy is to compute the Kalman filter gain and innovations covariance and check that these quantities are the same as K_k and Ω_k. From (3.9), $x_{k+1} = F_k x_k + K_k(z_k - H'_k x_k)$ with $x_0 = 0$. This equation shows that x_k is computable from z_l for $l < k$, i.e., $E[x_k | Z_{k-1}] = x_k$. Thus the equation can be rewritten as

$$\hat{x}_{k+1/k} = F_k \hat{x}_{k/k-1} + K_k(z_k - H'_k \hat{x}_{k/k-1})$$

Also

$$\tilde{z}_k = z_k - H'_k \hat{x}_{k/k-1} = z_k - H'_k x_k = v_k$$

So $E[\tilde{z}_k \tilde{z}'_k] = E[v_k v'_k] = \Omega_k$. This completes the proof.

Further Relations between the Innovations Model and the Kalman Filter

In the course of proving Theorem 3.2, we constructed the Kalman filter gain and error covariance for a signal model which proved to be an innovations model. The error covariance turns out to be zero since $x_k = E[x_k | Z_{k-1}]$. We also showed that the filter innovations process was identical with the input of the innovations model. In summary:

THEOREM 3.3. Consider a signal model of the form of (3.7) and (3.8), with $x_0 = 0$ and Ω_k nonsingular. The associated Kalman filter is

$$\hat{x}_{k+1/k} = F_k \hat{x}_{k/k-1} + K_k(z_k - H'_k \hat{x}_{k/k-1}) \tag{3.9}$$

and one has $\hat{x}_{k+1/k} = x_{k+1}$, or zero error covariance, and $\tilde{z}_k = z_k - H'_k \hat{x}_{k/k-1} = v_k$.

This result is also consistent with the fact that for an arbitrary signal model, one has $P_k = \Sigma_{k/k-1} + T_k$ in terms of the earlier notation, while, according to (3.6), one has $P_k = T_k$ for the innovations model. This means that for the innovations model $\Sigma_{k/k-1} = 0$.

Causal Invertibility Property of Innovations Representation

The innovations model (3.4) has the property that it is causally invertible, i.e., $\bar{v}_k$ is computable from $\bar{z}_l$ for $l \leq k$. This comes about in the following way. First, $\bar{v}_k$ appears in the output equation (3.4b), which allows us to rewrite (3.4a) as

$$\bar{x}_{k+1} = F\bar{x}_k + K_k(\bar{z}_k - H'_k \bar{x}_k)$$

Second, because $\bar{x}_0 = 0$, $\bar{x}_k$ can be determined from $\bar{z}_l$ for $l < k$, and thus, from (3.4b), $\bar{v}_k$ can be determined from $\bar{z}_l$ for $l \leq k$.

This causal invertibility property can be taken as the defining property of an innovations representation; i.e., we are making the following claim.

THEOREM 3.4. Consider a signal model of the form

$$x_{k+1} = F_k x_k + G_k w_k \qquad (3.10a)$$

$$z_k = H'_k x_k + v_k \qquad (3.10b)$$

with $x_0, \{w_k\}, \{v_k\}$ jointly gaussian, x_0 is $N(0, P_0)$ and independent of $\{w_k\}$ and $\{v_k\}$. Also, $\{w_k\}$ and $\{v_k\}$ are zero mean and with covariance

$$E\left\{ \begin{bmatrix} w_k \\ v_k \end{bmatrix} [w'_l \quad v'_l] \right\} = \begin{bmatrix} Q_k & S_k \\ S'_k & R_k \end{bmatrix} \delta_{kl} \qquad (3.10c)$$

Suppose that the driving noise $[w'_k \quad v'_k]'$ is causally computable from $\{z_k\}$. Then the model must be of the form of (3.4) and (3.5), save that possibly $P_0 \neq 0$ but $H'_k \Phi_{k,0} P_0 \Phi'_{k,0} H_k = 0$ for all k.

Proof. Since $z_0 = H'_0 x_0 + v_0$ and v_0 is computable from z_0, one must have no uncertainty in $H'_0 x_0$, i.e., $H'_0 P_0 H_0 = 0$. Then $v_0 = z_0$. Since w_0 is computable from z_0, one must then have $w_0 = L_0 v_0$ for some L_0. Since $z_1 = H'_1 x_1 + v_1 = H'_1 F_0 x_0 + H'_1 G_0 w_0 + v_1$ and v_1 is computable from z_0 and z_1, there must be no uncertainty about $H'_1 F_0 x_0$, i.e., $H'_1 F_0 P_0 F'_0 H_1 = 0$. Then v_1 is known exactly. Then $w_1 = L_1 v_1$, because w_1 is known exactly from z_0 and z_1 or equivalently v_0 and v_1; one could not have $w_1 = L_{11} v_1 + L_{10} v_0$ with $L_{10} \neq 0$ without violating the whiteness property.

More generally, we conclude that $H'_k \Phi_{k,0} P_0 \Phi'_{k,0} H_k = 0$ for all k and $w_k = L_k v_k$. Setting $K_k = G_k L_k$, the model of (3.4) and (3.5) is recovered. This proves the theorem.

The condition $H'_k \Phi_{k,0} P_0 \Phi'_{k,0} H_k = 0$ for all k has an obvious interpretation; none of the initial uncertainty in x_0 is allowed to show up in $\{z_k\}$ for $k \geq 0$. This means that the statistics of $\{z_k\}$ would be unaltered if x_0 were changed from being $N(0, P_0)$ to $N(0, 0)$. So the difference in a causally invertible model with $P_0 \neq 0$ and the true innovations model where $P_0 = 0$ is trivial. For this reason, one identifies the notion of innovations model and causally invertible model.

Other Types of Innovations Representation

So far, the innovations representations dealt with have been state-variable models. In the remainder of this section, we examine other types of representations—those associated with infinite-dimensional processes, and those associated with ARMA representations.

Innovations Representations Lacking Finite-Dimensional Content

Let $\{z_k\}$ be a process defined for $k \geq 0$ with $E[z_k z'_l]$ prescribed, but not necessarily associated with a finite-dimensional system. An innovations representation is a writing of z_k as

$$z_k = \sum_{l=0}^{k} g_{kl} \bar{v}_l \qquad (3.11)$$

with $\{\bar{v}_k\}$ zero mean, white and gaussian, and with $g_{kk} = I$ for all k. One can show that $\{\bar{v}_k\}$ must be identical with $\tilde{z}_k = z_k - \hat{z}_{k/k-1}$. The causal invertibility is easily checked.

Such a representation is essentially defined in Prob. 1.1 in the scalar case. In the vector case, let $\Re$ be the $(m + 1) \times (m + 1)$ block matrix whose kl block entry is $E[z_k z_l']$. One writes

$$\Re = T'ST \qquad (3.12)$$

where T is block upper triangular with identity blocks on the diagonal, and S is block diagonal. The entries of T define the g_{kl}, and the uniqueness of the factorization corresponds to the uniqueness of the innovations representation. See Prob. 3.4.

Innovations Representations for ARMA Processes

An ARMA process is defined by a vector difference equation of the form

$$z_k + A_{1k}z_{k-1} + \cdots + A_{nk}z_{k-n} = B_{0k}v_k + \cdots + B_{mk}v_{k-m} \qquad (3.13)$$

More often than not, the A_{ik} and B_{jk} do not depend on k; the process $\{v_k\}$ is zero mean, white and gaussian, and $\{z_k\}$ is the output process. If (3.13) is defined for $k \geq 0$, some form of initialization, deterministic or random, is needed.

An innovations representation of the process is provided by

$$z_k + A_{1k}z_{k-1} + \cdots + A_{nk}z_{k-n} = C_{0k}\bar{v}_k + \cdots + C_{mk}\bar{v}_{k-m} \qquad (3.14)$$

where C_{ik} are coefficients determined in a way described below, $\{\bar{v}_k\}$ is a zero mean, white gaussian process, and is causally computable from $\{z_k\}$. Initial conditions are $z_{-1} = z_{-2} = \cdots = z_{-n} = 0$ and $\bar{v}_{-1} = \cdots = \bar{v}_{-m} = 0$.

We obtain the results by setting up an equivalent state-space model to (3.13), finding a corresponding state-space innovations model, and obtaining (3.14) from this model.

The key to doing this is the following lemma:

LEMMA 3.2. For $m \leq n$ the state-variable equations

$$x_{k+1} = \begin{bmatrix} -A_{1,k+1} & I & \cdots & 0 & 0 \\ -A_{2,k+2} & & \cdots & & \\ \cdot & & \cdots & & \\ \cdot & & \cdots & & \\ \cdot & & \cdots & I & 0 \\ -A_{n-1,k+(n-1)} & & \cdots & 0 & I \\ -A_{n,k+n} & & \cdots & 0 & 0 \end{bmatrix} x_k + \begin{bmatrix} B_{1,k+1} - A_{1,k+1}B_{0k} \\ B_{2,k+2} - A_{2,k+2}B_{0k} \\ \cdot \\ \cdot \\ \cdot \\ \cdot \\ \cdot \end{bmatrix} v_k$$

$$(3.15a)$$

$$z_k = [I \quad 0 \quad \cdots \quad 0 \quad 0]x_k + B_{0k}v_k \qquad (3.15b)$$

imply (3.13).

Proof. We use a superscript to denote a subvector of x_k. We have

$$z_k = x_k^1 + B_{0k}v_k \quad \text{[from (3.15b)]}$$

$$= -A_{1k}x_{k-1}^1 + x_{k-1}^2 + (B_{1k} - A_{1k}B_{0,k-1})v_{k-1}$$
$$\quad + B_{0k}v_k \quad \text{[from (3.15a)]}$$

$$= -A_{1k}z_{k-1} + A_{1k}B_{0,k-1}v_{k-1} + x_{k-1}^2 + (B_{1k} - A_{1k}B_{0,k-1})v_{k-1}$$
$$\quad + B_{0k}v_k \quad \text{[from (3.15b)]}$$

$$= -A_{1k}z_{k-1} + x_{k-1}^2 + B_{0k}v_k + B_{1k}v_{k-1} \quad \text{[by rearrangement]}$$

$$= -A_{1k}z_{k-1} - A_{2k}x_{k-2}^1 + x_{k-2}^3 + (B_{2k} - A_{2k}B_{0,k-2})v_{k-2}$$
$$\quad + B_{0k}v_k + B_{1k}v_{k-1} \quad \text{[by (3.15a)]}$$

$$= -A_{1k}z_{k-1} - A_{2k}z_{k-2} + A_{2k}B_{0,k-2}v_{k-2} + x_{k-2}^3$$
$$\quad + (B_{2k} - A_{2k}B_{0,k-2})v_{k-2} + B_{0k}v_k + B_{1k}v_{k-1} \quad \text{[by (3.15b)]}$$

$$= -A_{1k}z_{k-1} - A_{2k}z_{k-2} + x_{k-2}^3 + B_{0k}v_k + B_{1k}v_{k-1}$$
$$\quad + B_{2k}v_{k-2} \quad \text{[by rearrangement]}$$

The general pattern should then be clear, and the lemma is proved.

The lemma shows how to connect (3.13) to a state-variable equation in case $m \leq n$. If $m > n$, we add further terms to the left side of (3.13), viz., $A_{n+1,k}z_{k-(n+1)} + \cdots + A_m z_{k-m}$, with $A_{n+1,k} = A_{n+2,k} = \cdots = A_{m,k} = 0$. Then we can use the lemma again.

The Kalman filter for the state-variable model of (3.15) is readily derived, as is the innovations model, which has the form of (3.15) save that the matrix multiplying v_k in (3.15a) is replaced by the Kalman gain matrix, $\{v_k\}$ is replaced by a different white process $\{\bar{v}_k\}$, and $x_0 = 0$. Then (3.14) follows from this state-space innovations model. As initial conditions for (3.14), we take $z_{-1} = z_{-2} = \cdots = z_{-n} = 0$ and $\bar{v}_{-1} = \cdots = \bar{v}_{-m} = 0$ to reflect the fact that $x_0 = 0$.

Main Points of the Section

Among the class of state-variable signal models with the same output covariance, one stands out—the innovations model. Its important properties are as follows.

1. It is computable from either the output covariance, or the Kalman filter, or an arbitrary signal model.
2. It is essentially unique.
3. The Kalman filter applied to an innovations model estimates the innovations model states with zero error, and the filter innovations process is identical with the innovations model input process.

4. The innovations model solves the covariance factorization problem, while the Kalman filter solves the whitening filter problem.
5. The innovations model is causally invertible, and any signal model which is causally invertible is virtually an innovations model; causal invertibility is the property that the input is causally obtainable from the output.

One can also define innovations representations where no finite dimensionality is involved, and innovations ARMA models can be associated with an arbitrary ARMA model.

Problem 3.1. Show that among the class of all signal models of the form of (2.1) through (2.3) with the same output covariance, the innovations model has the least state covariance.

Problem 3.2. Let T_k be as in Definition 3.2, and suppose that the signal model of the form of (2.1) through (2.3), realizing the $\{z_k\}$ process, has $E[x_k x_k'] = T_k$. Show that the model must be an innovations model as in (3.4) and (3.5), save that K_k is only determined to within addition of a matrix whose rows are in $\mathfrak{N}[\Omega_k]$.

Problem 3.3. Consider the causally invertible model defined in Theorem 3.4. Suppose that F, H are a constant, completely observable pair. Show that $P_0 = 0$.

Problem 3.4. Consider the equation $\mathfrak{R} = T'ST$, where $\mathfrak{R}$ is an $(m + 1) \times (m + 1)$ block matrix, T is upper triangular with identity blocks on the diagonal, and S is diag $[\Omega_0, \Omega_1, \ldots, \Omega_m]$. Let the $k - l$ block entry of T be $T_{kl} = g'_{lk}$. Show that if $z_k = \sum_{l=0}^{k} g_{kl} \bar{v}_l$ with $E[\bar{v}_l \bar{v}_k'] = \Omega_k \delta_{kl}$, then $E[z_k z_l']$ is the $k - l$ block entry of $\mathfrak{R}$. Show also that if $T_1' S_1 T_1 = T_2' S_2 T_2$, with T_i, S_i possessing the properties listed above, then $T_1 = T_2$, $S_1 = S_2$. [*Hint for second part:* $(T_2')^{-1} T_1'$ is lower triangular and equals $S_2 T_2 T_1^{-1} S_1^{-1}$, which is upper triangular.]

9.4 STATIONARY INNOVATIONS REPRESENTATIONS AND SPECTRAL FACTORIZATION

In this section, we concentrate on innovations representations for stationary processes. The section divides naturally into four parts:

1. Review of classical frequency domain based ideas of spectral factorization, with foreshadowing of connection to the innovations ideas.
2. Discussion of state-variable innovations representations commencing at a finite initial time for stationary processes; such representations turn out to be time varying, but asymptotically time invariant.

3. Discussion of state-variable innovations representations commencing in the infinitely remote past for stationary processes. Such representations are time invariant, and are connected with the classical ideas.
4. Discussion of other approaches than those using state-variable models: ARMA representations and the Wold decomposition.

Time-invariant Innovations Representations—Frequency Domain Properties

Suppose that $\{z_k\}$ is the output process of a linear, asymptotically stable, finite-dimensional system driven by zero mean, stationary white noise $\{v_k\}$ commencing in the infinitely remote past. Then $\{z_k\}$ is a stationary process, and a calculation of Chap. 4 effectively led to the result.

$$\Phi_{ZZ}(z) = W(z)\Omega W'(z^{-1}) \qquad (4.1)$$

Here $W(z)$ is the transfer function matrix of the system in question, Ω is the covariance of v_k, i.e., $\Omega = E[v_k v_k']$, and $\Phi_{ZZ}(z)$ is the power spectrum matrix of $\{z_k\}$. The power spectrum is related to the covariance in the standard way:

$$\Phi_{ZZ}(z) = \sum_{k=-\infty}^{\infty} E[z_k z_0']z^{-k} \qquad (4.2)$$

Passing from $W(z)$ and Ω to $\Phi_{ZZ}(z)$ is straightforward. The converse problem of spectral factorization is harder. From $\Phi_{ZZ}(z)$, one is required to construct $W(z)$ and Ω satisfying (4.1), with Ω nonnegative definite symmetric and $W(z)$ rational with all poles in $|z| < 1$ [so that $W(z)$ corresponds to an asymptotically stable system].

Throughout this section, we restrict attention to the case when z_k is a *full-rank* process, as now defined. (This restriction is frequently a very reasonable one, and is standard in the literature.)

DEFINITION 4.1. $\{z_k\}$ is a full-rank process if there exists no signal model with output covariance $E[z_k z_l']$ which is driven by a white noise process $\{v_k\}$ with the dimension of v_k less than that of z_k.

In terms of (4.1), we see that a process will not be full rank if there exists a decomposition with Ω of smaller dimension than $\Phi_{ZZ}(z)$; and in that case, $\Phi_{ZZ}(z)$ will be singular for all values of z. The converse is also true; i.e., if $\Phi_{ZZ}(z)$ is singular for all z, $\{z_k\}$ is not a full-rank process, but we shall omit any proof of this result. If $\{z_k\}$ is a full-rank process, $\Phi_{ZZ}(z)$ will be nonsingular for almost all z, but not usually for all z. Note also that any scalar process is automatically full rank.

It is clear that any solution of the spectral factorization problem defines a realization of the $\{z_k\}$ process: one simply drives the system with transfer function matrix $W(z)$ by a white noise process $\{v_k\}$ with $E[v_k v_k'] = \Omega$. The question then arises as to what the innovations representation is. The answer is described in the following theorem for full-rank processes.

THEOREM 4.1. Suppose that $\Phi_{ZZ}(z)$ is constructed in the manner described above and is of full rank for almost all z. Then there is a factorization of $\Phi_{ZZ}(z)$ as

$$\Phi_{ZZ}(z) = \bar{W}(z)\bar{\Omega}\bar{W}'(z^{-1}) \qquad (4.3)$$

where $\bar{W}(z)$ is a square, real, rational, transfer function matrix, all poles lie in $|z| < 1$, $\lim_{z\to\infty} \bar{W}(z) = I$, $\bar{W}^{-1}(z)$ is analytic in $|z| > 1$ [or, equivalently, $\bar{W}(z)$ has constant rank in $|z| > 1$], and $\bar{\Omega}$ is positive definite symmetric. Moreover, the factorization is unique and defines an innovations representation.

We shall offer a proof here of only the uniqueness part. (The remainder of the proof will be filled in by our state-variable treatment of innovations representations.) Following the uniqueness proof, we offer a number of remarks concerning the theorems.

*Proof of Uniqueness.** Suppose that

$$\bar{W}_1(z)\bar{\Omega}_1\bar{W}'_1(z^{-1}) = \bar{W}_2(z)\bar{\Omega}_2\bar{W}'_2(z^{-1})$$

where $\bar{W}_i(z)$ and $\bar{\Omega}_i$ are as described in the theorem statement. Then

$$\bar{\Omega}_1\bar{W}'_1(z^{-1})[\bar{W}'_2(z^{-1})]^{-1} = [\bar{W}_1(z)]^{-1}\bar{W}_2(z)\bar{\Omega}_2$$

The assumptions on the $\bar{W}_i$ imply that the right side is analytic in $|z| > 1$, the left side in $|z| < 1$. Therefore,

$$U(z) = [\bar{W}_1(z)]^{-1}\bar{W}_2(z)\bar{\Omega}_2^{1/2}$$

is analytic everywhere except possibly on $|z| = 1$. One checks easily that $U(z)U'(z^{-1}) = \bar{\Omega}_1$. Now on $|z| = 1$, $U'(z^{-1}) = U'^*(z)$ since $U(z)$ is real, rational, and $z^{-1} = z^*$. Therefore

$$\text{trace } \bar{\Omega}_1 = \text{trace } [U'^*(z)U(z)] = \sum_{i,j} |u_{ij}(z)|^2$$

So clearly, no element of $U(z)$ can have a pole on $|z| = 1$. Hence $U(z)$ is analytic everywhere. Letting $z \to \infty$ yields $\lim_{z\to\infty} U(z) = \bar{\Omega}_2^{1/2}$ and so it is also bounded. Hence $U(z)$ is a constant by Liouville's theorem. It is immediate then that $\bar{W}_1(z) = \bar{W}_2(z)$ and $\bar{\Omega}_1 = \bar{\Omega}_2$.

Remarks on Theorem 4.1

1. If $\Phi_{ZZ}(z)$ is nonsingular for all $|z| = 1$, as is often the case, the claim of the theorem can be strengthened in that $\bar{W}(z)$ has constant rank in $|z| \geq 1$, rather than $|z| > 1$.

*This proof may be omitted without loss of continuity.

2. The innovations property is linked with the fact that $\bar{W}^{-1}(z)$ is analytic in $|z| > 1$. At least if the analyticity extends to $|z| \geq 1$, this means that $\bar{W}^{-1}(z)$ is a causal inverse for $\bar{W}(z)$. The state variable interpretation is as follows. If $\bar{W}(z) = I + H'(zI - F)^{-1}K$, then $\bar{W}^{-1}(z) = I - H'(zI - F + KH')^{-1}K$ and if $|\lambda_i(F - KH')| < 1$, $\bar{W}^{-1}(z)$ has the required analyticity property.

3. Theorems very like Theorem 4.1 can be found in the literature [6–15]. However, instead of requiring $\Phi_{ZZ}(z)$ to be constructed in the manner described, it is usual simply to postulate the following properties for $\Phi_{ZZ}(z)$:
   (a) $\Phi_{ZZ}(z)$ is analytic on $|z| = 1$, is rational, and has full rank almost everywhere.
   (b) $\Phi_{ZZ}(z) = \Phi'_{ZZ}(z^{-1})$.
   (c) $\Phi_{ZZ}(z)$ is nonnegative definite hermitian on $|z| = 1$ (in fact positive definite hermitian almost everywhere on $|z| = 1$).
   If $\Phi_{ZZ}(z)$ is defined via (4.1), where $W(z)$ and Ω have the properties specified earlier, it has the three properties just listed. But the important point is that if these properties are simply assumed for $\Phi_{ZZ}(z)$, the theorem is still true. In this chapter, we shall not prove this last claim (although it can be proved using state variable ideas [18], as well as by more classical procedures).

4. It is possible to prove results for the case when $\{z_k\}$ is not a full-rank process. They are a good deal more complicated.

5. Classical treatments often give the name *minimum phase* to the $\bar{W}(z)$ of the theorem statement. This arises because in the scalar case, the phase of $\bar{W}(e^{j\omega})$ for any real ω is less than that of any other spectral factor. Another name is *minimum delay*. The heuristic reasoning for this name is well described in [34]. See also [25] for the connection with state-variable ideas.

"Knowing" the Power Spectral Matrix

In the classical treatments of spectral factorization referred to above, it is usual to assume that $\Phi_{ZZ}(z)$ is given as a matrix of rational functions of z. However, in treatments of the classical spectral factorization problem via state-variable ideas, it is necessary to assume that one knows matrices F, H, L, and M such that

$$\Phi_{ZZ}(z) = L + H'(zI - F)^{-1}M + M'(z^{-1}I - F')^{-1}H \qquad (4.4)$$

and $|\lambda_i(F)| < 1$. Let us describe how these might be found from $\Phi_{ZZ}(z)$, expressed as a matrix of rational functions. Broadly, there are two approaches.

1. One carries out a partial fraction expansion of $\Phi_{ZZ}(z)$. [This necessitates factoring of the least common denominator of all elements of

$\Phi_{zz}(z)$.] There results

$$\Phi_{zz}(z) = A + \sum_i \frac{B_i}{z - z_i} + \sum_i \frac{C_i}{z - z_i^{-1}}$$

where $|z_i| < 1$, and we have assumed for convenience that Φ_{zz} is bounded as $z \to \infty$ and has no repeated poles. Rewrite $C_i(z - z_i^{-1})^{-1}$ as $-C_i z_i - C_i z_i^2 (z^{-1} - z_i)^{-1}$, which yields

$$\Phi_{zz}(z) = L + \sum_i \frac{B_i}{z - z_i} + \sum_i \frac{B_i'^*}{z^{-1} - z_i}$$

[The $B_i'^*$ arise because $\Phi_{zz}(z) = \Phi_{zz}'(z^{-1})$.] Then one obtains a state-variable realization of $\sum_i B_i(z - z_i)^{-1}$ (see Appendix C); i.e., one finds F, H, and M such that $H'(zI - F)^{-1}M = \sum_i B_i(z - z_i)^{-1}$. One has $|\lambda_i(F)| < 1$ as a result.

2. One obtains a Laurent series expansion of $\Phi_{zz}(z)$ convergent in an annulus $a < |z| < a^{-1}$:

$$\Phi_{zz}(z) = \sum_{i=-\infty}^{+\infty} C_i z^{-i} \qquad (4.5)$$

The quantity C_i is precisely $E[z_i z_0']$; therefore, if covariance data is known, C_i is available at once. Otherwise, one has

$$C_i = \frac{1}{2\pi j} \int_{|z|=1} \Phi_{zz}(z) z^{i-1} \, dz \qquad (4.6)$$

the integral being taken in an counterclockwise direction. Having (4.5), one then finds F, H, and M such that $H'(zI - F)^{-1}M = \sum_{i\geq 1} C_i z^{-i}$ (see Appendix C). The integral in (4.6) can be computed via Cauchy residue theory (but this is almost like using the partial fraction method), or by using numerical values of $\Phi_{zz}(z)$ on $|z| = 1$, or, approximately, by a discrete Fourier transform type of calculation.

Innovations Representations for Stationary Processes with Finite Initial Time

Suppose that $\{z_k\}$ is a stationary, full-rank process as described above. Then for some quantities F, H, M, and L, we have

$$\begin{aligned} E[z_k z_0'] &= L & k &= 0 \\ &= H'F^{k-1}M & k &> 0 \end{aligned} \qquad (4.7)$$

or equivalently

$$\Phi_{zz}(z) = L + H'(zI - F)^{-1}M + M'(z^{-1}I - F')^{-1}H \qquad (4.8)$$

Also, $|\lambda_i(F)| < 1$.

The data F, H, M, and L might be all that is available. Alternatively, we might know a signal model with $\{z_k\}$ as the output process. Let us work, however, just with the covariance data.

As we know, the innovations representation with initial time $k = 0$ is given by

$$\bar{x}_{k+1} = F\bar{x}_k + K_k\bar{v}_k \qquad \bar{x}_0 = 0 \qquad (4.9a)$$

$$\bar{z}_k = H'\bar{x}_k + \bar{v}_k \qquad (4.9b)$$

where $\{\bar{v}_k\}$ is zero mean, white, and gaussian with $E[\bar{v}_k\bar{v}'_k] = \Omega_k$, and where K_k and Ω_k are defined as follows. Let

$$T_{k+1} = FT_kF' + (FT_kH - M)(L - H'T_kH)^{-1}(FT_kH - M)' \quad (4.10)$$

$$T_0 = 0$$

$$\Omega_k = L - H'T_kH \qquad (4.11a)$$

$$K_k = -(FT_kH - M)\Omega_k^{-1} \qquad (4.11b)$$

We see immediately that though z_k is stationary, *the innovations model is not time invariant*. However, we do have the following important result.

THEOREM 4.2. With definitions as above, the innovations model is asymptotically time invariant.

To prove this theorem, we shall first study the sequence T_k.

LEMMA 4.1. The sequence T_k converges monotonically to a limit T, satisfying (4.10) with T_k and T_{k+1} replaced by T.

*Proof.** We first show the monotonicity of T_k. The signal model (4.9) has the property that $E[\bar{x}_1\bar{x}'_1] = T_1$. Therefore, the following signal model, obtained by studying (4.9) on $[1, \infty)$ and then shifting the time origin, has the same output covariance as (4.9):

$$x_{k+1} = Fx_k + K_{k+1}v_k$$

$$z_k = H'x_k + v_k$$

with $E[v_kv'_k] = \Omega_{k+1}$, $E[x_0x'_0] = T_1$, $E[x_0] = 0$. Now (4.9) is the innovations representation for this model, so that $P_k - T_k \geq 0$, where $P_k = E[x_kx'_k]$. (See Theorem 2.1.) On the other hand,

$$E[x_kx'_k] = E[\bar{x}_{k+1}\bar{x}'_{k+1}] = T_{k+1}$$

by Theorem 3.1, so that $T_{k+1} - T_k \geq 0$. (An algebraic proof of monotonicity is contained in the problems.)

Convergence of T_k follows from the monotonicity and the existence of an upper bound on T_k provided by the state covariance of the signal model generating $\{z_k\}$. (See Theorem 2.1.)

---

*The proof may be omitted without loss of continuity.

The proof of Theorem 4.2 is now virtually immediate. Using (4.11), we have

$$\Omega = \lim_{k \to \infty} \Omega_k = L - H'TH \tag{4.12a}$$

$$K = \lim_{k \to \infty} K_k = -(FTH - M)\Omega^{-1} \tag{4.12b}$$

Innovations Representations for Stationary Processes with Initial Time in the Infinitely Remote Past

The first matter we have to explore is the definition of an innovations representation with initial time in the infinitely remote past. Actually, we proceed in a very straightforward fashion. We consider Definition 3.2, which applies to a finite initial time. We let the initial time approach $-\infty$ and see if the resulting quantities, e.g., Ω_k, K_k, have limits. If they do, we associate these with the innovations representation with initial time in the infinitely remote past.

Now suppose the initial time in Definition 3.2 is changed from 0 to k_0; if F, H, M, and L are constant, it follows easily that $T_{k,k_0} = T_{k-k_0,0} = T_{k-k_0}$; here, the second subscript denotes the initial time, the first subscript the running time. Likewise, $K_{k,k_0} = K_{k-k_0}$ and $\Omega_{k,k_0} = \Omega_{k-k_0}$. Letting $k_0 \to -\infty$ is equivalent to letting $k \to \infty$. In the light then of (4.12), we have the following.

DEFINITION 4.2. Let z_k be a stationary process with covariance as given in (4.7). Let K, Ω defined in (4.12) be the steady-state values of the matrices defining the innovations model with finite initial time. Then the innovations model with initial time in the infinitely remote past is defined by

$$\bar{x}_{k+1} = F\bar{x}_k + K\bar{v}_k \tag{4.13a}$$

$$\bar{z}_k = H'\bar{x}_k + \bar{v}_k \tag{4.13b}$$

and

$$E[\bar{v}_k \bar{v}'_l] = \Omega\delta_{kl} \qquad E[\bar{v}_k] = 0 \tag{4.13c}$$

Observe that for this model, $E[\bar{x}_k \bar{x}'_k] = T$, the limiting solution of (4.10). Of course, it is time invariant, and this is a helpful property in practice.

Now that the model has been defined, we need to check to what extent the properties applicable with finite initial time carry over to an initial time in the infinitely remote past. We start by studying the issue of causal invertibility, which, it should be recalled, is a property possessed by all innovations models, and only by models which are, to all intents and purposes, innovations models.

DEFINITION 4.3. A signal model

$$\bar{x}_{k+1} = F\bar{x}_k + G\bar{w}_k \tag{4.14a}$$

$$\bar{z}_k = H'\bar{x}_k + \bar{v}_k \tag{4.14b}$$

$$E\left\{\begin{bmatrix} \bar{w}_k \\ \bar{v}_k \end{bmatrix} [\bar{w}'_l \ \ \bar{v}'_l]\right\} = \begin{bmatrix} Q & S \\ S' & R \end{bmatrix} \delta_{kl}, \qquad E\begin{bmatrix} \bar{w}_k \\ \bar{v}_k \end{bmatrix} = 0 \tag{4.14c}$$

with $|\lambda_i(F)| < 1$, is causally invertible if from $\bar{z}_{k_0}, \bar{z}_{k_0+1}, \ldots, \bar{z}_k$ one can construct quantities $\bar{w}_{k,k_0}$ and $\bar{v}_{k,k_0}$ such that as $k_0 \rightarrow -\infty$, $\bar{w}_{k,k_0} \rightarrow \bar{w}_k$ and $\bar{v}_{k,k_0} \rightarrow \bar{v}_k$, convergence being in mean square.

We then have the following theorem.

THEOREM 4.3. The innovations model (4.13) is causally invertible. Any causally invertible model is an innovations model.

Proof.* First suppose that (4.13) is an innovations model. Consider the Kalman filter for (4.13) assumed to commence operation at time k_0. Then this filter will simply be a time-shifted version of the inverse of (4.9). (Why?) The filter is

$$\hat{x}_{k+1/k} = F\hat{x}_{k/k-1} + K_{k-k_0}(\bar{z}_k - H'\hat{x}_{k/k-1})$$

with $\hat{x}_{k_0/k_0-1} = 0$. Let $\Sigma_{k/k-1}$ denote the error covariance of the filter. Then we have the relation

$$E[\bar{x}_k \bar{x}'_k] = \Sigma_{k/k-1} + E[\hat{x}_{k/k-1} \hat{x}'_{k/k-1}]$$

or $T = \Sigma_{k/k-1} + T_{k-k_0}$, where T_k is defined by (4.10). Letting $k_0 \rightarrow -\infty$ shows that $\Sigma_{k/k-1} \rightarrow 0$, i.e., $\hat{x}_{k/k-1} \rightarrow \bar{x}_k$. Define $\bar{v}_{k,k_0} = \bar{z}_k - H'\hat{x}_{k/k-1}$, this quantity being computable from the $\bar{z}_k$ sequence restricted to $k \geq k_0$. Then $\bar{v}_{k,k_0} \rightarrow \bar{z}_k - H'\bar{x}_k = \bar{v}_k$ when $k_0 \rightarrow -\infty$, as required.

Second, suppose that (4.14) is causally invertible. We shall argue first that $E[\bar{x}_k \bar{x}'_k]$ must be T. For any model of the form of (4.14), with k fixed and k_1 arbitrary but $k_1 < k$, we have

$$\bar{x}_k - F^{k-k_1}\bar{x}_{k_1} = \sum_{k_1}^{k-1} F^{k-j-1}G\bar{w}_j$$

For arbitrary $\epsilon > 0$, choose k_1 so that

$$E[F^{k-k_1}\bar{x}_{k_1}\bar{x}'_{k_1}(F')^{k-k_1}] < \epsilon I$$

This is possible because $|\lambda_i(F)| < 1$.

Choose k_0 such that $z_{k_0}, z_{k_0+1}, \ldots, z_k$ define an estimate of $\sum_{k_1}^{k-1} F^{k-j-1}G\bar{w}_j$ with error covariance less than ϵI. This is possible because

---

*The proof may be omitted without loss of continuity.

of the causal invertibility property and the fact that only a finite number of terms occur in the summation. Call the estimate a. Observe then that

$$E[\|\tilde{x}_k - a\|^2] = E\left[\left\|F^{k-k_1}\tilde{x}_{k_1} + \sum_{k_1}^{k-1} F^{k-j-1}G\bar{w}_j - a\right\|^2\right]$$

Since $\|c + d\|^2 \leq 2\|c\|^2 + 2\|d\|^2$ for all vectors c, d, we have

$$E[\|\tilde{x}_k - a\|^2] \leq 2E[\|F^{k-k_1}\tilde{x}_{k_1}\|^2] + 2E\left[\left\|\sum_{k_1}^{k-1} F^{k-j-1}G\bar{w}_j - a\right\|^2\right] < 4\epsilon n$$

where $n = \dim \tilde{x}_k$. Now $E[\tilde{x}_k \,|\, z_{k_0}, \ldots, z_k]$ is the minimum variance estimate of $\tilde{x}_k$ given $z_{k_0}, \ldots, z_k$, and so the associated mean square error is also overbounded by $4\epsilon n$.

Imagine a Kalman filter for (4.14) with initial time of the filter equal to k_0. Let $\Sigma_{k/k-1}$ denote the error covariance. Let P denote the state covariance of (4.14). Then the by now frequently used connection between these quantities yields

$$T_{k-k_0} = P - \Sigma_{k/k-1}$$

As $k_0 \rightarrow \dot{-}\infty$, we have $\Sigma_{k/k-1} \rightarrow 0$ as a result of the bound developed above, and so $E[\tilde{x}_k\tilde{x}_k'] = P = T$.

This almost completes the proof. We know that $E[z_k z_0'] = H'P^{k-1}M$, where $M = FPH + GS$. Also from the innovations model it follows that $M = FTH + K\Omega$. Since $P = T$, we have $GS = K\Omega$. Also $E[z_0 z_0'] = L = H'PH + R$ from the signal model, and $L = H'TH + \Omega$ from the innovations model. Therefore, since $P = T$, $R = \Omega$. Next, from the signal model, $P - FPF' = GQG'$ while from the innovations model, since $P = T$, $P - FPF' = K\Omega K'$. Summing up,

$$\begin{bmatrix} GQG' & GS \\ S'G' & R \end{bmatrix} = \begin{bmatrix} K\Omega K' & K\Omega \\ \Omega K' & \Omega \end{bmatrix} = \begin{bmatrix} K \\ I \end{bmatrix}\Omega[K' \quad I]$$

That (4.14) is the innovations model is then immediate.

Uniqueness: Models "Like" the Innovations Model

Consider any signal model of the form

$$x_{k+1} = Fx_k + Kv_k \tag{4.15a}$$

$$z_k = H'x_k + v_k \tag{4.15b}$$

$$E[v_k v_l'] = \Omega\delta_{kl} \qquad E[v_k] = 0 \tag{4.15c}$$

where $|\lambda_i(F)| < 1$ and the initial time is in the infinitely remote past. Must it be an innovations model for the $\{z_k\}$ process? As we know, with minor qualification in the finite initial time case, the answer is affirmative. A major upset, however, occurs in the infinitely remote initial time case. What we shall prove below is that (4.15) is an innovations model if and only if $|\lambda_i(F - KH')|$ ≤ 1.

THEOREM 4.4. If (4.15) defines an innovations model, then $|\lambda_i(F - KH')|$ ≤ 1. Conversely, if $|\lambda_i(F - KH)| \leq 1$, then (4.15) is an innovations model if $[F, H]$ is completely observable, and has the same transfer function as the innovations model otherwise.

Proof.[*] Consider the signal model

$$x_{k+1} = Fx_k + Kv_k + w_k \qquad (4.16a)$$

$$z_k = H'x_k + v_k \qquad (4.16b)$$

with $\{w_k\}$ a white noise sequence independent of $\{v_k\}$ and with $E[w_k w_k']$ $= \epsilon I$. Let P_ϵ denote $E[x_k x_k']$ and Σ_ϵ the steady-state value of the filter error covariance associated with this model, and let L_ϵ, M_ϵ, Ω_ϵ, K_ϵ, and T_ϵ be defined in the obvious manner. The conditions are fulfilled to ensure that Σ_ϵ exists and that the associated steady-state Kalman filter is asymptotically stable, i.e., $|\lambda_i(F - K_\epsilon H')| < 1$. One way to see this latter fact is to derive it from the steady-state equation

$$\Sigma_\epsilon = (F - K_\epsilon H')\Sigma_\epsilon(F - K_\epsilon H')' + (K - K_\epsilon)\Omega(K - K_\epsilon)' + \epsilon I$$

say by use of the discrete-time lemma of Lyapunov.

Now let us consider the effect of letting $\epsilon \to 0$. It is trivial to establish that $P_\epsilon \to T$, the state covariance of (4.15), and that $M_\epsilon \to M$. Now Σ_ϵ is monotone increasing with ϵ increasing. (One way to see this is to show that the solutions of the *transient* error covariance equation are monotone with ϵ.) Therefore, $\lim_{\epsilon \downarrow 0} \Sigma_\epsilon$ exists; call the limit $\bar{\Sigma}$. Since $P_\epsilon - \Sigma_\epsilon = T_\epsilon$ must satisfy

$$T_\epsilon = FT_\epsilon F' + (FT_\epsilon H - M_\epsilon)(L_\epsilon - H'T_\epsilon H)^{-1}(FT_\epsilon H - M_\epsilon)'$$

and $\lim_{\epsilon \downarrow 0} T_\epsilon = \bar{T}$ exists, we obtain

$$\bar{T} = F\bar{T}F' + (F\bar{T}H - M)(L - H'\bar{T}H)^{-1}(F\bar{T}H - M)' \qquad (4.17)$$

Because $\lim_{\epsilon \downarrow 0} P_\epsilon = T$ and $\lim_{\epsilon \downarrow 0} \Sigma_\epsilon \geq 0$, $\bar{T} \leq T$. Let us show that $\bar{T} = T$. The signal model

$$x_{k+1} = Fx_k - (F\bar{T}H - M)\bar{\Omega}^{-1}v_k$$

$$z_k = H'x_k + v_k$$

is easily checked to be a realization of $\{z_k\}$ because (4.17) holds. Therefore its state covariance is underbounded by the state covariance of the innovations representation, i.e., $\bar{T} \geq T$. Hence $\bar{T} = T$.

---

[*]The proof may be omitted at a first reading.

Now $K_\epsilon = -(FT_\epsilon H - M)\Omega_\epsilon^{-1}$ and $|\lambda_i(F - K_\epsilon H')| < 1$. Taking limits yields

$$\bar{K} = -(F\bar{T}H - M)\bar{\Omega}^{-1} = -(FTH - M)\Omega^{-1} = K$$

Also, $|\lambda_i(F - \bar{K}H')| \leq 1$. This proves the first part of the theorem.

To prove the second part, let $W(z)$ denote the transfer function matrix $I + H'(zI - F)^{-1}K$. Then because (4.15) is a realization of a certain power spectrum $\Phi_{zz}(z)$, we have

$$W(z)\Omega W'(z^{-1}) = \Phi_{zz}(z)$$

Now $W(z)$ is analytic in $|z| \leq 1$, $\lim_{z \to \infty} W(z) = I$, and because

$$W^{-1}(z) = I - H'[zI - (F - KH')]^{-1}K$$

$W(z)$ is of constant rank in $|z| > 1$. [Use the fact that $|\lambda_i(F - KH')| \leq 1$.] By Theorem 4.1, for a given $\Phi_{zz}(z)$ there is only one $W(z)$ with these properties. Since the innovations representation is guaranteed to have a transfer function with these properties, any representation with these properties must have the same transfer function matrix as the innovations representation and must have the same input noise covariance as the innovations representation. If $[F, H]$ is completely observable,

$$H'(zI - F)^{-1}K = H'(zI - F)^{-1}K_1$$

implies $K = K_1$, so that (4.15) is then our innovations model. This proves the theorem.

The above theorem provides an important link to the result of Theorem 4.1. As noted in the proof of the theorem, the innovations model transfer function matrix, $I + H'(zI - F)^{-1}K$, is of constant rank in $|z| > 1$ because $|\lambda_i(F - KH')| \leq 1$. Thus we have:

THEOREM 4.5. The transfer function matrix of the innovations representation is minimum phase.

Let us make a number of other observations.

1. The existence part of Theorem 4.1 was not proved earlier. The results of Theorems 4.2 through Theorem 4.4 allow the filling in of this proof, in that a constructive procedure has been given for the minimum phase spectral factor; i.e., the spectral factorization problem has been solved.
2. There is no suggestion that the formal inverse of the innovations representation, viz.,

$$x_{k+1} = (F - KH')x_k + Kz_k \tag{4.18a}$$

$$v_k = z_k - H'x_k \tag{4.18b}$$

is necessarily stable, let alone asymptotically stable. This is because $F - KH'$ may have repeated eigenvalues on the unit circle (see the problems).

3. It turns out that if $\Phi_{ZZ}(z)$ is positive definite everywhere on $|z| = 1$, many of the preceding ideas can be strengthened. This we now do.

Power Spectra Positive Definite on $|z| = 1$

Frequently, a power spectrum matrix will be positive definite on $|z| = 1$. For example, if $\{z_k\}$ is the output of a system with both white measurement and input noise, and if these noises are independent, $\Phi_{ZZ}(z)$ will be positive definite. For with $z_k = y_k + v_k$, the independence of $\{y_k\}$ and $\{v_k\}$ means that $\Phi_{ZZ} = \Phi_{YY} + \Phi_{VV}$. Since Φ_{YY} is nonnegative definite and Φ_{VV} is constant and positive definite on $|z| = 1$, Φ_{ZZ} is positive definite on $|z| = 1$. The main result is as follows:

THEOREM 4.6. Let $\{z_k\}$ be the output process of a signal model of the form of (4.14), with $|\lambda_i(F)| < 1$ and initial time in the infinitely remote past. (Causal invertibility of the model is not assumed.) The following conditions are equivalent:

1. $\Phi_{ZZ}(z)$ is positive definite on $|z| = 1$.
2. If (4.15) is the innovations model, then $|\lambda_i(F - KH')| < 1$.
3. The system

$$\hat{x}_{k+1/k} = (F - KH')\hat{x}_{k/k-1} + Kz_k \qquad (4.19)$$

with initial time in the infinitely remote past is a Kalman filter for (4.15) with zero error covariance and is exponentially stable.
4. The system of (4.18) will reconstruct $\{v_k\}$ causally from $\{z_k\}$ (see Fig. 9.4-1); conversely if the system (4.15) follows the system (4.18), the output of (4.15) reconstructs the input of (4.18).

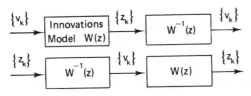

Fig. 9.4-1 Invertibility of innovations model when power spectrum is positive definite on unit circle.

The reader may find it helpful to review the factorization result of Sec. 4.6 at this point.

Proof.* $1 \Rightarrow 2$. $\Phi_{ZZ}(z) = W(z)\Omega W'(z^{-1})$, where $W(z) = I + H'(zI - F)^{-1}K$. Since $\Phi_{ZZ}(z)$ is positive definite on $|z| = 1$, $W^{-1}(z)$ is analytic on

---

*The proof may be omitted at first reading.

$|z| = 1$. Because (4.5) is an innovations model, $W^{-1}(z)$ is analytic in $|z| > 1$. Since

$$W(z) = I - H'(zI - \overline{F - KH'})^{-1}K$$

this means all observable and controllable modes of $F - KH'$ are asymptotically stable. All other modes are asymptotically stable, since unobservable and/or uncontrollable modes of $F - KH'$ are also modes of F, and $|\lambda_i(F)| < 1$.

$2 \Rightarrow 3$. Equation (4.19) is obtained as the limit as $k_0 \rightarrow -\infty$ of the Kalman filter for finite k_0. Since $|\lambda_i(F - KH')| < 1$, the equation defines a filter with initial time in the infinitely remote past. Also, from (4.15) we have $x_{k+1} = (F - KH')x_k + Kz_k$, and with an initial time in the infinitely remote past it follows from this and (4.19) that $\hat{x}_{k+1/k} = x_k$.

$3 \Rightarrow 4$. We just argued that x_k in (4.15) also obeys (4.18a). Also (4.18b) is trivial from (4.15b). The converse is equally easy.

$4 \Rightarrow 1$. The hypothesis demands that $|\lambda_i(F - KH')| < 1$. Modification of the argument that $1 \Rightarrow 2$ then establishes that 1 holds. This proves the theorem.

We remark that without the knowledge that $|\lambda_i(F - KH')| < 1$, one cannot use (4.19) as a Kalman filter. Thus there are covariances which have innovations representations with infinitely remote initial time, but not an associated Kalman filter for an infinitely remote initial time.

We also remark that though it may be the case that there exists no Kalman filter with infinitely remote initial time because $|\lambda_i(F - KH')| < 1$ fails, there is always a Kalman filter with finite initial time. It is *always* asymptotically stable, and exponentially asymptotically stable in case $|\lambda_i(F - KH')| < 1$. A guided derivation is called for in the problems.

In case one desires a constant filter when $\Phi_{ZZ}(z)$ is not positive definite on $|z| = 1$, a suboptimal, constant, asymptotically stable filter can be obtained which performs close to the optimum (see the problems).

Other Types of Innovations Representations for Stationary Processes

The innovations representations described above for stationary processes involve state-variable equations. We mention two other types of representations here which have found frequent use.

Vector ARMA representation. Consider a process defined by

$$z_k + A_1 z_{k-1} + \cdots + A_n z_{k-n} = B_0 v_k + \cdots + B_m v_{k-m} \qquad (4.20)$$

where the A_i, B_j are coefficient matrices and $\{v_k\}$ is a zero mean, white gaussian process. In case the initial time is $-\infty$, to ensure stationarity of $\{z_k\}$, we require

$$\det [Iz^n + A_1 z^{n-1} + \cdots + A_n] \neq 0 \qquad |z| \geq 1 \qquad (4.21)$$

In case (4.20) is to be an innovations representation of a full-rank process valid for an infinitely remote initial time, one requires B_0 to be nonsingular (in which case by slight redefinition of v_k it can be taken as the identity matrix) and also that

$$\det [Iz^m + B_1 z^{m-1} + \cdots + B_m] \neq 0 \qquad |z| > 1 \qquad (4.22)$$

In case this does not hold, (4.20) is, of course, still a representation, and there exists an associated innovations representation which is unique.

If the initial time associated with the innovations representation is finite, then (4.20) cannot be an innovations representation; the innovations representation involves time-varying parameter matrices which are asymptotically constant; it is of the form

$$\bar{z}_k + A_1 \bar{z}_{k-1} + \cdots + A_n \bar{z}_{k-n} = C_{0,k} \bar{v}_k + \cdots + C_{m,k} \bar{v}_{k-m} \qquad (4.23)$$

with $C_{i,k} \to C_i$ as $k \to \infty$. These results can be found in, for example, [35, 36]. They follow from the state-variable result of this section, and the ARMA process/state-variable connection of the previous section. See Prob. 4.7.

Wold decomposition. Suppose a stationary process $\{z_k\}$ has a power spectrum $\Phi_{zz}(z)$ which is analytic on $|z| = 1$. Then Wold's theorem states that there exists a stationary white process $\{v_k\}$ causally obtainable from $\{z_k\}$ and such that

$$z_k = v_k + \Gamma_1 v_{k-1} + \Gamma_2 v_{k-2} + \cdots \qquad (4.24)$$

with $\sum_k \|\Gamma_k\|^2 < \infty$. This is an innovations representation, since

$$E[z_k | Z_{k-1}] = E[z_k | V_{k-1}] = \Gamma_1 v_{k-1} + \Gamma_2 v_{k-2} + \cdots$$

so that $\bar{z}_k = v_k$. One can also conceive of representations given by (4.24) with v_k stationary and white, but without the causal invertibility property. These ideas are discussed in, for example, [37, 38], with [38] providing a discussion in the context of Kalman filtering.

Provided one works with second order statistics only, these ideas extend to nongaussian processes and wide-sense stationary processes.

Main Points of the Section

The various results are all for full-rank processes, which class includes all scalar processes.

Given a power spectral matrix $\Phi_{ZZ}(z)$, one is interested in finding $W(z)$ and Ω such that

$$\Phi_{ZZ}(z) = W(z) \Omega W'(z^{-1})$$

If $W(z)$ is analytic in $|z| \leq 1$, $\lim_{z \to \infty} W(z) = I$, and of constant rank in $|z| > 1$, $W(z)$ is unique and is termed minimum phase.

Given a stationary output covariance of a signal model, one can define an innovations representation with finite initial time which is time varying but asymptotically time invariant. In the steady state, it is also an innovations representation associated with an initial time in the infinitely remote past and is causally invertible. In order that

$$x_{k+1} = Fx_k + Kv_k \qquad z_k = H'x_k + v_k \qquad E[v_k v_l'] = \Omega \delta_{kl} \qquad E[v_k] = 0$$

with $|\lambda_i(F)| < 1$ and initial time in the infinitely remote past, be an innovations representation, it is necessary and virtually sufficient that $|\lambda_i(F - KH')| \leq 1$. The transfer function matrix of an innovations representation is minimum phase.

When the spectral factorization problem is described in state-variable terms, the constructive technique for obtaining the innovations representation provides a classically important solution of the spectral factorization problem.

If $\Phi_{ZZ}(z)$ is nonsingular everywhere on $|z| = 1$, one has $|\lambda_i(F - KH')| < 1$, and the formal inverse of the innovations model is the Kalman filter.

Similar conclusions apply to vector ARMA representations. The vector ARMA model

$$z_k + A_1 z_{k-1} + \cdots + A_n z_{k-n} = v_k + B_1 v_{k-1} + \cdots + B_m v_{k-m}$$

is an innovations model with initial time in the infinitely remote past if and only if

$$\det [Iz^n + A_1 z^{n-1} + \cdots + A_n] \neq 0 \quad \text{for } |z| \geq 1$$

and

$$\det [Iz^m + B_1 z^{m-1} + \cdots + B_m] \neq 0 \quad \text{for } |z| > 1$$

Problem 4.1. (Monotonicity of T_k). Show that

$$T_{k+1} = \min \left\{ X \mid X = X', \begin{bmatrix} X & M \\ M' & L \end{bmatrix} - \begin{bmatrix} F \\ H' \end{bmatrix} T_k [F' \quad H] \geq 0 \right\}$$

for an arbitrary initial condition of (4.10), so long as $L - H'T_k H \geq 0$. Conclude that if for two initial conditions T_0^i ($i = 1, 2,$) one has $T_0^1 \geq T_0^2$, then $T_k^1 \geq T_k^2$. Take $T_0^2 = 0$; take $T_0^1 = T_1^2$. Conclude that $T_k^1 = T_{k+1}^2$ and establish the monotonicity.

Problem 4.2. Show that of all time-invariant representations of the one covariance, the innovations representation has the smallest state covariance. Consider the steady-state equation

$$X = FXF' + (FXH - M)(L - H'XH)^{-1}(FXH - M)'$$

in which the symbols have the usual meanings and $L - H'XH > 0$. Show that if there are multiple solutions, the smallest is T, the state covariance of the time-invariant innovations representation.

Problem 4.3. Consider the full-rank spectral factorization problem as one of passing from a quadruple $\{F, H, M, L\}$ to $\{F, H, K, \Omega\}$. Suppose that $[F, M]$ is completely reachable. Show that $[F, K]$ is completely reachable. (*Hint:* Using the steady-state equation for T, show that if $F'w = \lambda w$ and $K'w = 0$ for some scalar λ and nonzero w, then $Tw = 0$. Conclude that $M'w = 0$. Suppose also that

$$F = \begin{bmatrix} F_{11} & F_{12} \\ 0 & F_{22} \end{bmatrix}, \qquad M = \begin{bmatrix} M_1 \\ 0 \end{bmatrix}$$

with $[F_{11}, M_1]$ completely reachable. Let w be such that $w'F_{22} = \lambda w'$. Show that $[0 \quad w']T_k = 0$ for all k and that $[0 \quad w']K = 0$. Thus if $[F, K]$ is completely reachable, $[F, M]$ must be completely reachable.)

Problem 4.4.
(a) Consider the following two systems, defined with initial time in the infinitely remote past:

$$x_{k+1} = \tfrac{1}{2}x_k + v_k \qquad\qquad x_{k+1} = \tfrac{1}{2}x_k + \tfrac{5}{2}v_k$$

$$z_k = x_k + v_k \qquad\qquad\qquad z_k = x_k + v_k$$

$$E[v_k v_l] = \delta_{kl} \qquad\qquad\qquad E[v_k v_l] = \tfrac{1}{4}\delta_{kl}$$

Check that $E[z_k z_l]$ is the same for both systems, and show that the first system is an innovations representation but the second is not.
(b) Show that

$$x_{k+1} = \begin{bmatrix} \tfrac{1}{2} & 1 \\ 0 & \tfrac{1}{2} \end{bmatrix} x_k + \begin{bmatrix} 1 & 0 \\ 0 & 1 \end{bmatrix} v_k, \qquad z_k = \begin{bmatrix} -\tfrac{1}{2} & 0 \\ 0 & -\tfrac{1}{2} \end{bmatrix} x_k + v_k$$

is an innovations representation with formal inverse that is unstable.

Problem 4.5. (Asymptotic Stability of Kalman Filter). Consider a stationary signal model with innovations representation

$$\bar{x}_{k+1} = F\bar{x}_k + K_k \bar{v}_k \qquad \bar{z}_k = H'\bar{x}_k + \bar{v}_k$$

Then the error covariance equation is

$$\Sigma_{k+1/k} = (F - K_k H')\Sigma_{k/k-1}(F - K_k H')' + \text{nonnegative quantities}$$

and $\Sigma_{0/-1} \geq T$. Suppose first that $[F, M]$ is completely reachable; show that T is nonsingular, and consequently that the Kalman filter is asymptotically stable (but not necessarily exponentially stable). Can you extend the result to the case of $[F, M]$ not completely reachable. (*Hint:* Use the ideas of Prob. 4.3.) When is the filter exponentially stable?

Problem 4.6. (Suboptimal Filter). Consider a Kalman filtering problem in which $\Phi_{zz}(z)$ is singular for some $|z| = 1$. Design a filter on the assumption that there is additional white measurement noise of covariance ϵI, independent of the actual input and measurement noise processes. Show that the resulting filter will be asymptotically stable and definable with infinitely remote initial time. Show also that as $\epsilon \to 0$, it approaches the quasi-optimal performance obtainable by taking the Kalman filter for the original problem with very negative, but finite, initial time. (*Hint:* Recall the material of Chap. 6 on modeling errors.)

Problem 4.7. Model the ARMA process (4.20), with $B_0 = I$ and $m \leq n$ assumed, by

$$
x_{k+1} = \begin{bmatrix} -A_1 & I & \cdots & 0 & 0 \\ \cdot & & & \cdot & \cdot \\ \cdot & & & \cdot & \cdot \\ \cdot & & & \cdot & \cdot \\ -A_{n-2} & \cdot & \cdots & I & \cdot \\ -A_{n-1} & \cdot & \cdots & 0 & I \\ -A_n & \cdot & \cdots & 0 & 0 \end{bmatrix} x_k + \begin{bmatrix} B_1 - A_1 \\ B_2 - A_2 \\ \cdot \\ \cdot \\ \cdot \\ \cdot \\ \cdot \end{bmatrix} v_k
$$

$$ z_k = [I \quad 0 \quad \cdots \quad 0 \quad 0] x_k + v_k $$

Show that for $\{z_k\}$ to be stationary with infinitely remote initial time, one requires

$$ \det [Iz^n + A_1 z^{n-1} + \cdots + A_n] \neq 0 \quad \text{in} \quad |z| \geq 1 $$

and that for the state-variable model [and therefore (4.20)] to be an innovations representation,

$$ \det [Iz^m + B_1 z^{m-1} + \cdots + B_m] \neq 0 $$

Show that a vector ARMA innovations representation with finite initial time has the form of (4.23), with $C_{i,k} \longrightarrow C_i$ as $k \longrightarrow \infty$. Discuss the situation of $\Phi_{zz}(z)$ nonsingular for all $|z| = 1$.

Problem 4.8. Consider the vector ARMA representation of (4.20) with

$$ \det [z^n I + A_1 z^{n-1} + \cdots + A_n] \neq 0 \quad \text{for} \quad |z| \geq 1 $$

Let $w_k = z_k + A_1 z_{k-1} + \cdots + A_n z_{k-n}$; thus $\{w_k\}$ is a moving average process, since

$$ w_k = v_k + B_1 v_{k-1} + \cdots + B_n v_{k-n} $$

Show that $w_k - E[w_k \,|\, W_{k-1}] = z_k - E[z_k \,|\, Z_{k-1}]$. (*Hint:* First form an innovations representation for $\{z_k\}$ and argue that it is, in effect, an innovations representation for $\{w_k\}$.)

Problem 4.9. Let $\Phi_{zz}(z)$ be a rational power spectral matrix. Explain how to find a linear time-invariant system with white noise input defined for $k \geq 0$ such that the output spectrum is $\Phi_{zz}(z)$. (Random initial conditions are permitted.)

9.5 WIENER FILTERING

In this section, we state a basic Wiener filtering problem, we describe its solution with the aid of some of the foregoing ideas, and we make the connection with Kalman filtering ideas. In contrast to Wiener's work [19], we work in discrete time rather than in continuous time and we omit discussion of many of the interesting facets of the Wiener theory. For fuller discussion, see for example [19, 20]. Our main purpose is simply to make contact between the classical and more modern results.

Wiener Filtering Problem

Suppose that there is a signal process $\{y_k\}$ and noise process $\{n_k\}$ with measurement process $\{z_k = y_k + n_k\}$. Suppose that $\{y_k\}$ and $\{n_k\}$ are independent, zero mean, and stationary, but that $\{n_k\}$ is *not* necessarily white. The initial time is in the infinitely remote past.

Figure 9.5-1 depicts the situation. In the figure, $W_1(z)$ and $W_2(z)$ are real, rational, transfer function matrices with all poles in $|z| < 1$, and $\{w_k\}$ and $\{v_k\}$ are unit variance, zero mean, independent, white noise processes.

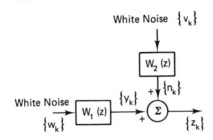

Fig. 9.5-1 Prototype situation for Wiener filtering.

We consider the problem of forming $E[y_k | Z_{k+m}]$ for zero, positive, and negative integer m. (These are filtering, smoothing, and prediction problems respectively.) We make the assumption that $\Phi_{ZZ}(z)$ is nonsingular on $|z| = 1$. This means that there exists a $\bar{W}(z)$ which together with its inverse $\bar{W}^{-1}(z)$ is analytic in $|z| \geq 1$, with $\lim_{z \to \infty} \bar{W}(z)$ finite, and $\Phi_{ZZ}(z) = \bar{W}(z)\bar{W}^{-1}(z)$. [Thus $\bar{W}(z)$, to within a constant nonsingular matrix, is the transfer function matrix of an innovations representation of $\{z_k\}$.]

Solution Procedure for the Wiener Filtering Problem

The general strategy is to first convert the problem of estimating y_k from certain of the measurements $\{z_k\}$ to one of estimating y_k from a scaled version of the innovations, which we still label $\{\tilde{z}_k\}$. Then one computes the optimal transfer function matrix for this task. In the derivation immediately below, most calculations are done in the frequency domain; however, time domain ideas are introduced when we write down an expression for $E[y_0 | \tilde{Z}_m]$.

Accordingly, to introduce the scaled innovations into the picture, we calculate $\bar{W}(z)$ with the properties as described earlier, and consider the arrangement depicted in Fig. 9.5-2. Let m_k denote the vector $[y'_k \quad \tilde{z}'_k]'$. Via an easy calculation, we have

$$Z\{E[m_k m'_0]\} = \Phi_{MM}(z) = \begin{bmatrix} \Phi_{YY}(z) & \Phi_{YY}(z)[\bar{W}^{-1}(z^{-1})]' \\ \bar{W}^{-1}(z)\Phi_{YY}(z) & I \end{bmatrix}$$

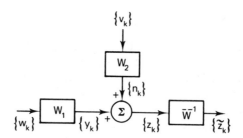

Fig. 9.5-2 Whitening filter added onto signal model.

and thus

$$Z\{E[y_k \tilde{z}_0']\} = \Phi_{YY}(z)[\bar{W}^{-1}(z^{-1})]' \tag{5.1}$$

Equation (5.1) is the key to writing down the solution of the three problems noted above. Suppose that the right side of (5.1) is $\sum\limits_{k=-\infty}^{+\infty} D_k z^{-k}$, which is a Laurent series convergent on $|z| = 1$. (The determination of such a series is discussed further in the following paragraphs.) The quantity D_k has the significance that $E[y_k \tilde{z}_0'] = E[y_0 \tilde{z}_{-k}'] = D_k$. Then since

$$E[y_0 | \tilde{Z}_0] = E[y_0 | \tilde{z}_0] \tilde{z}_0 + E[y_0 | \tilde{z}_{-1}] \tilde{z}_{-1} + \cdots$$

(using the orthogonality of the $\tilde{z}_i$), we have

$$E[y_0 | \tilde{Z}_0] = D_0 \tilde{z}_0 + D_1 \tilde{z}_{-1} + \cdots$$

which shows that the transfer function from $\{\tilde{z}_k\}$ to $\{E[y_k | \tilde{Z}_k]\}$ is $D_0 + D_1 z^{-1} + D_2 z^{-2} + \cdots$ and that from $\{z_k\}$ to $\{E[y_k | Z_k]\}$ is, accordingly,

$$[D_0 + D_1 z^{-1} + D_2 z^{-2} + \cdots] \bar{W}^{-1}(z)$$

Similarly, for $p > 0$ (corresponding to prediction)

$$E[y_0 | \tilde{Z}_{-p}] = E[y_0 | \tilde{z}_{-p}] \tilde{z}_{-p} + E[y_0 | \tilde{z}_{-(p+1)}] \tilde{z}_{-(p+1)} + \cdots$$

and so the transfer function linking $\{z_k\}$ to $E[y_{k+p} | Z_k]$ is

$$[D_p + D_{p+1} z^{-1} + \cdots] \bar{W}^{-1}(z)$$

Smoothing is obviously tackled in the same way.

We noted above that we would comment on the determination of the D_k. Basically, one simply does a partial fraction expansion of $\Phi_{YY}(z)[\bar{W}^{-1}(z^{-1})]'$ to write this quantity as

$$\sum_i \frac{A_i}{z - z_i} + B_0 + \sum_j \frac{C_i}{z^{-1} - z_j^{-1}}$$

where $|z_i| < 1, |z_j| > 1$. Then one expands each term of the first summation in powers of z^{-1} and each term of the second summation in powers of z. This yields the desired Laurent series. When $W_1(z)$ or $W_2(z)$ in Fig. 9.5-1 are known, the process can be easier; that this is so will be exemplified below in connecting the above ideas to the Kalman filter ideas.

Rapprochement with Kalman Filtering

Let us suppose that $\{y_k\}$ is the output of a linear system of transfer function matrix $H'(zI - F)^{-1}GQ^{1/2}$ driven by white noise of unit variance. Suppose that $W_2(z)$ in Fig. 9.5-2 is the constant $R^{1/2}$, so that $\{n_k\}$ is a white noise sequence of covariance $R\delta_{kl}$. Then

$$\Phi_{ZZ}(z) = R + H'(zI - F)^{-1}GQG'(z^{-1}I - F')^{-1}H$$

and the Kalman filter is given by

$$\hat{x}_{k+1/k} = (F - KH')\hat{x}_{k/k-1} + Kz_k$$

Here

$$K = F\bar{\Sigma}H(H'\bar{\Sigma}H + R)^{-1}$$

where $\bar{\Sigma}$ satisfies the steady-state covariance equation:

$$\bar{\Sigma} = F[\bar{\Sigma} - \bar{\Sigma}H(H'\bar{\Sigma}H + R)^{-1}H'\bar{\Sigma}]F' + GQG'$$

The transfer function matrix linking $\{z_k\}$ to $\{E[y_k \,|\, Z_{k-1}]\}$ is

$$W_f(z) = H'[zI - (F - KH')]^{-1}K \tag{5.2}$$

Let us check that the same result follows from the Wiener filtering approach. With $\{n_k\}$ a white process, we have

$$\Phi_{ZZ}(z) = \Phi_{YY}(z) + R = \bar{W}(z)\bar{W}'(z^{-1})$$

Therefore post-multiplying by $[\bar{W}^{-1}(z^{-1})]'$, we have

$$\Phi_{YY}(z)[\bar{W}^{-1}(z^{-1})]' = -R[\bar{W}^{-1}(z^{-1})]' + \bar{W}(z) \tag{5.3}$$

From the material of the previous section, we know that

$$\bar{W}(z) = [I + H'(zI - F)^{-1}K][H'\bar{\Sigma}H + R]^{1/2}$$

Now $\bar{W}(z)$ has all its poles in $|z| < 1$. Also, it has all its zeros in $|z| < 1$ by its minimum phase character and the fact that $\Phi_{ZZ}(z)$ is positive definite on $|z| = 1$. Therefore, $V(z) = \bar{W}^{-1}(z^{-1})$ has all its poles in $|z| > 1$. This means that when the left side of (5.3) is expanded as a Laurent series $\sum_{k=-\infty}^{+\infty} D_k z^{-k}$ convergent on $|z| = 1$, we must have

$$\sum_{k\geq 1} D_k z^{-k} = \bar{W}(z) - (H'\bar{\Sigma}H + R)^{1/2}$$
$$= H'(zI - F)^{-1}K(H'\bar{\Sigma}H + R)^{1/2}$$

Accordingly, the transfer function linking $\{z_k\}$ to $E[y_k \,|\, Z_{k-1}]$ is

$$\sum_{k\geq 1} D_k z^{-k}\bar{W}^{-1}(z) = H'(zI - F)^{-1}K[I + H'(zI - F)^{-1}K]^{-1}$$
$$= H'[zI - (F - KH')]^{-1}K$$

This is identical with the transfer function matrix computed in (5.2) via the Kalman theory.

In a similar way, we can recover prediction and smoothing results.

Review of Some Differences between the Kalman and Wiener Theories

Having considered the common ground between Kalman and Wiener filtering, let us consider ways in which they are dissimilar. The obvious differences between the Kalman and Wiener theories are listed below.

1. The Kalman theory allows consideration of nonstationary processes, including a finite initial time; the Wiener theory does not.
2. The Wiener theory does not draw great distinction between colored and white measurement noise. The Kalman theory in the first instance demands white measurement noise, but extension of the theory to the colored noise case is possible by modeling colored noise as the output of a linear system driven by white noise. This point will be discussed in a later chapter.
3. The Kalman theory is essentially concerned with finite-dimensional systems. The Wiener theory permits infinite-dimensional systems, although the task of spectral factorization becomes much more difficult, and is still central to application of the theory.

Main Points of the Section

Wiener and Kalman filtering theory make contact when the measurement noise is white, the signal model is finite dimensional, and all processes are stationary.

The two theories then (naturally) lead to the same result.

Problem 5.1. Relate the optimum prediction filters for prediction intervals greater than 1 for the Wiener and Kalman theories.

Problem 5.2. Show that the rationality of $W_1(z)$ and $W_2(z)$ in Fig. 9.5-1 leads to rationality of the transfer function matrix linking $\{z_k\}$ to $\{E[y_k | Z_{k+m}]\}$ for arbitrary fixed m.

9.6 LEVINSON FILTERS

In this section, we look at another classical filtering problem associated with stationary processes. The original ideas are due to Levinson [22]. In an attempt to lighten the computational burden associated with Wiener filtering,

Levinson (working in discrete time) suggested that prediction estimates should be derived from a finite window of past data, rather than all past data. Since that time, the ideas following from Levinson's suggestion and the associated theory have found wide applications by statisticians and in geophysical data processing (See, e.g., [34].)

We begin by stating the Levinson problem. Next we indicate its solution first for scalar processes, then for vector processes.

The Levinson Problem

Let $\{z_k\}$ be a stationary time series, and suppose that one has available the segment $z_0, z_1, \ldots, z_{N-1}$ and wishes to estimate z_N. Thus one seeks coefficients $A_{N,N-i}$ such that

$$\hat{z}_{N/N-1} = -\sum_{i=0}^{N-1} A_{N,N-i} z_i \tag{6.1}$$

By the orthogonality principle,

$$E\{[z_N - \hat{z}_{N/N-1}]z_j'\} = 0 \qquad \text{for } j = 0, 1, \ldots, N-1$$

so that

$$E[z_N z_j'] = -\sum_{i=0}^{N-1} A_{N,N-i} E[z_i z_j'] \qquad j = 0, 1, \ldots, N-1$$

or

$$C_{N-j} = -\sum_{i=0}^{N-1} A_{N,N-i} C_{i-j} \tag{6.2}$$

where $C_j = E[z_j z_0']$. The associated mean square error is

$$\Pi_N = E\{[z_N - \hat{z}_{N/N-1}][z_N - \hat{z}_{N/N-1}]'\} = E\{[z_N - \hat{z}_{N/N-1}]z_N'\}$$

$$= E[z_N z_N'] + \sum_{i=0}^{N-1} A_{N,N-i} E[z_i z_N']$$

$$= C_0 + \sum_{i=0}^{N-1} A_{N,N-i} C_{i-N} \tag{6.3}$$

Since the more measurements there are, the better our estimate will be, Π_N is seen to be monotone decreasing with N. Its value can be successively computed for $N = 1, 2, \ldots$ to decide whether or not more data should be collected in order to predict values of the $\{z_k\}$ sequence, the idea being that we would compare Π_N for each N with a preassigned desired value of mean square estimation error. (Of course, it may be that the desired value could not be achieved no matter how large we took N.) This means that we desire a procedure for calculating successively for each N the quantities $A_{N,N-i}$ and Π_N. Finding such a procedure is the Levinson problem.

Levinson's own solution to the problem just mentioned was for scalar $\{z_k\}$. The vector case is more complicated, essentially because it lacks a self-duality inherent in the scalar case. Discussion of the vector case can be found in [23, 34] and is dealt with a little later.

What we are concerned about is the following equation obtained by combining (6.2) and (6.3) and by using the fact that, in the scalar case, $E[z_0 z_j] = E[z_j z_0]$:

$$[1 \quad a_{N,1} \quad \cdots \quad a_{N,N}] \begin{bmatrix} c_0 & c_1 & \cdots & c_N \\ c_1 & c_0 & \cdots & c_{N-1} \\ \cdot & \cdot & & \cdot \\ \cdot & \cdot & & \cdot \\ \cdot & \cdot & & \cdot \\ c_N & c_{N-1} & \cdots & c_0 \end{bmatrix} = [\Pi_N \quad 0 \quad \cdots \quad 0] \quad (6.4)$$

(We have switched to lower-case quantities to emphasise the scalar nature of the problem.) Suppose the values $a_{N,i}$ and Π_N have been found and we seek the quantities $a_{N+1,i}$ and Π_{N+1}. Let us define a quantity α_N by

$$[1 \quad a_{N,1} \quad \cdots \quad a_{N,N} \quad 0] \begin{bmatrix} c_0 & c_1 & \cdots & c_{N+1} \\ c_1 & c_0 & \cdots & c_N \\ \cdot & \cdot & & \cdot \\ \cdot & \cdot & & \cdot \\ \cdot & \cdot & & \cdot \\ c_{N+1} & c_N & \cdots & c_0 \end{bmatrix} = [\Pi_N \quad 0 \quad \cdots \quad 0 \quad \alpha_N]$$

$$(6.5)$$

Note that if

$$\alpha_N = c_{N+1} + \sum_{i=1}^{N-1} a_{N,i} c_{N+1-i}$$

were zero, we should immediately have the desired quantities $a_{N+1,i}$ and Π_{N+1}. In general, α_N is not zero.

Now observe that, essentially because of the Toeplitz structure of the matrix on the left side of (6.5), we have from (6.5) that

$$[0 \quad a_{N,N} \quad \cdots \quad a_{N,1} \quad 1] \begin{bmatrix} c_0 & c_1 & \cdots & c_{N+1} \\ c_1 & c_0 & \cdots & c_N \\ \cdot & \cdot & & \cdot \\ \cdot & \cdot & & \cdot \\ \cdot & \cdot & & \cdot \\ c_{N+1} & c_N & \cdots & c_0 \end{bmatrix} = [\alpha_N \quad 0 \quad \cdots \quad 0 \quad \Pi_N]$$

$$(6.6)$$

Now add $-\alpha_N/\Pi_N$ times (6.6) to (6.5). The effect is to obtain the row vector on the right side to have all zero entries except for the first, while the row vector on the left side has leading entry 1:

$$\left[1 \quad a_{N,1} - \frac{\alpha_N}{\Pi_N} a_{N,N} \quad \cdots \quad a_{N,N} - \frac{\alpha_N}{\Pi_N} a_{N,1} \quad -\frac{\alpha_N}{\Pi_N}\right] \begin{bmatrix} c_0 & c_1 & \cdots & c_{N+1} \\ c_1 & c_0 & \cdots & c_N \\ \cdot & \cdot & & \cdot \\ \cdot & \cdot & & \cdot \\ \cdot & \cdot & & \cdot \\ c_{N+1} & c_N & \cdots & c_0 \end{bmatrix}$$

$$= \left[\Pi_N - \frac{\alpha_N^2}{\Pi_N} \quad 0 \quad \cdots \quad 0\right] \quad (6.7)$$

This equation is precisely an updated version of (6.4). In other words, we have the following recursions:

$$\alpha_N = c_{N+1} + \sum_{i=1}^{N-1} a_{N,i} c_{N+1-i} \qquad (6.8a)$$

$$a_{N+1,i} = a_{N,i} - \frac{\alpha_N}{\Pi_N} a_{N,N+1-i} \qquad i = 1, \ldots, N \qquad (6.8b)$$

$$a_{N+1,N+1} = -\frac{\alpha_N}{\Pi_N} \qquad (6.8c)$$

$$\Pi_{N+1} = \Pi_N - \frac{\alpha_N^2}{\Pi_N} \qquad (6.8d)$$

Initialization is provided by

$$a_{11} = -\frac{c_1}{c_0} \qquad \Pi_1 = c_0 - \frac{c_1^2}{c_0} \qquad (6.8e)$$

Notice from (6.8d) that $\Pi_{N+1} \leq \Pi_N$, as expected. Of course, if $\Pi_N = 0$, the algorithm stops. In that case, N measurements will predict the next value of $\{z_k\}$ with no error.

Solution to the Levinson Problem—Vector $\{z_k\}$ Processes

If one attempts to mimic the above derivation for the case of vector $\{z_k\}$, one runs into difficulty on account of the fact that $E[z_1 z_0'] \neq E[z_0 z_1']$, i.e., $C_1 \neq C_{-1}$; rather $C_1 = C'_{-1}$. The cancellation used to obtain (6.7) cannot be executed. However, it turns out that if we work with a Levinson algorithm associated with a *backward* estimation problem *as well*, then we can obtain the recursion. The backward estimation problem is to estimate z_0 given measurements $z_1, \ldots, z_N$.

Suppose that

$$E[z_0 | z_1, \ldots, z_N] = -\sum_{i=1}^{N} B_{N,i} z_i$$

From the orthogonality principle, we obtain

$$C_{-j} = -\sum_{i=1}^{N} B_{N,i} C_{i-j} \qquad j = 1, \ldots, N \qquad (6.9)$$

with the associated mean square error as

$$\Gamma_N = C_0 + \sum_{i=1}^{N} B_{N,i} C_i \qquad (6.10)$$

This means that from (6.2) and (6.3),

$$[I \quad A_{N,1} \quad \cdots A_{N,N}] \begin{bmatrix} C_0 & C_1 & \cdots & C_N \\ C_{-1} & C_0 & \cdots & C_{N-1} \\ \cdot & \cdot & & \cdot \\ \cdot & \cdot & & \cdot \\ \cdot & \cdot & & \cdot \\ C_{-N} & C_{-N+1} & \cdots & C_0 \end{bmatrix} = [\Pi_N \quad 0 \quad \cdots \quad 0] \quad (6.11)$$

and from (6.9) and (6.10),

$$[B_{N,N} \quad B_{N,N-1} \quad \cdots \quad I] \begin{bmatrix} C_0 & C_1 & \cdots & C_N \\ C_{-1} & C_0 & \cdots & C_{N-1} \\ \cdot & \cdot & & \cdot \\ \cdot & \cdot & & \cdot \\ \cdot & \cdot & & \cdot \\ C_{-N} & C_{-N+1} & \cdots & C_0 \end{bmatrix} = [0 \quad \cdots \quad 0 \quad \Gamma_N]$$

$$(6.12)$$

We seek recursions simultaneously for the quantities $A_{N,i}$, Π_N, $B_{N,i}$, and Γ_N, and this time we are more successful in our attempt to mimic the scalar $\{z_k\}$ process derivation. From (6.11) and (6.12), we have

$$[I \quad A_{N,1} \quad \cdots \quad A_{N,N} \quad 0] \begin{bmatrix} C_0 & C_1 & \cdots & C_{N+1} \\ C_{-1} & C_0 & \cdots & C_N \\ \cdot & \cdot & & \cdot \\ \cdot & \cdot & & \cdot \\ \cdot & \cdot & & \cdot \\ C_{-N-1} & C_{-N} & \cdots & C_0 \end{bmatrix} = [\Pi_N \quad 0 \quad \cdots \quad \alpha_N]$$

and

$$[0 \quad B_{N,N} \quad \cdots \quad B_{N,1} \quad I] \begin{bmatrix} C_0 & C_1 & \cdots & C_{N+1} \\ C_{-1} & C_0 & \cdots & C_N \\ \cdot & \cdot & & \cdot \\ \cdot & \cdot & & \cdot \\ \cdot & \cdot & & \cdot \\ C_{-N-1} & C_{-N} & \cdots & C_0 \end{bmatrix}$$
$$= [\beta_N \quad 0 \quad \cdots \quad 0 \quad \Gamma_N]$$

where α_N, β_N are easily derived, the precise formulas being given below. Premultiplying the second of these equations by $\alpha_N \Gamma_N^{-1}$ and subtracting it from the first gives the equation for the $A_{N+1,i}$, and premultiplying the first equation by $\beta_N \Pi_N^{-1}$ and subtracting it from the second gives the equation for the $B_{N+1,i}$. The various equations are as follows:

$$\alpha_N = C_{N+1} + \sum_{i=1}^{N} A_{N,i} C_{N+1-i} \tag{6.13a}$$

$$\beta_N = C_{-N-1} + \sum_{i=1}^{N} B_{N,i} C_{-N-1+i} \tag{6.13b}$$

$$[I \quad A_{N+1,1} \quad \cdots \quad A_{N+1,N+1}] = [I \quad A_{N,1} \quad \cdots \quad A_{N,N} \quad 0)$$
$$- \alpha_N \Gamma_N^{-1} [0 \quad B_{N,N} \quad \cdots \quad B_{N,1} \quad I] \tag{6.13c}$$

$$[B_{N+1,N+1} \quad \cdots \quad B_{N+1,1} \quad I] = [0 \quad B_{N,N} \quad \cdots \quad B_{N,1} \quad I]$$
$$- \beta_N \Pi_N^{-1} [I \quad A_{N,1} \quad \cdots \quad A_{N,N} \quad 0] \tag{6.13d}$$

$$\Pi_{N+1} = \Pi_N - \alpha_N \Gamma_N^{-1} \beta_N \tag{6.13e}$$

$$\Gamma_{N+1} = \Gamma_N - \beta_N \Pi_N^{-1} \alpha_N \tag{6.13f}$$

Initialization of these equations is left to the reader. The last equations have an apparent asymmetry about them. However, one can show that

$$\alpha_N = \beta_N' \tag{6.14}$$

This is an identity attributed to Burg in [38]. A derivation is called for in the problems.

A number of interesting interpretations and remarks relating to the Levinson problem and its solution (6.13) can be found in [38].

Main Points of the Section

The coefficients yielding one-step prediction estimates in a stationary random sequence using a finite window of data can be recursively determined by considering windows of increasing length. For vector sequences, one must also introduce "backward" prediction estimates to obtain the desired recursions.

Problem 6.1. Why does it follow from (6.4) that $\Pi_N \geq 0$?

Problem 6.2. In the scalar $\{z_k\}$ case, how are the coefficients $B_{N,i}$ related to the $A_{N,i}$?

Problem 6.3. Show that with α_N defined by (6.13a), one has

$$E\{[z_{N+1} - E[z_{N+1} | z_1, z_2, \ldots, z_N]] z_0'\} = \alpha_N$$

and

$$E\{[z_0 - E[z_0 | z_1, z_2, \ldots, z_N]] z_{N+1}'\} = \beta_N$$

Show then that

$$\alpha_N' = \beta_N = E\{[z_0 - E[z_0 | z_1, \ldots, z_N]][z_{N+1}' - E[z_{N+1} | z_1, \ldots, z_N]']\}$$

Problem 6.4. Consider the scalar problem and set

$$F_N = \begin{bmatrix} 0 & 1 & 0 & \cdots & 0 \\ 0 & 0 & 1 & \cdots & 0 \\ \cdot & \cdot & \cdot & & \cdot \\ \cdot & \cdot & \cdot & & \cdot \\ \cdot & \cdot & \cdot & & 1 \\ -a_{NN} & -a_{N,N-1} & -a_{N,N-2} & \cdots & -a_{N1} \end{bmatrix}$$

$$P_N = \begin{bmatrix} c_0 & c_1 & \cdots & c_{N-1} \\ c_1 & c_0 & \cdots & c_{N-2} \\ \cdot & \cdot & & \cdot \\ \cdot & \cdot & & \cdot \\ \cdot & \cdot & & \cdot \\ c_{N-1} & c_{N-2} & \cdots & c_0 \end{bmatrix}$$

Show that

$$P_N - F_N P_N F_N' = -\begin{bmatrix} 0 & 0 & \cdots & 0 & 0 \\ 0 & 0 & \cdots & 0 & 0 \\ \cdot & \cdot & & \cdot & \cdot \\ \cdot & \cdot & & \cdot & \cdot \\ \cdot & \cdot & & \cdot & \cdot \\ 0 & 0 & \cdots & 0 & 0 \\ 0 & 0 & \cdots & 0 & \Pi_N \end{bmatrix}$$

Conclude that if $P_N > 0$ and $\Pi_N > 0$, all roots of $z^N + a_{N1}z^{N-1} + \cdots a_{NN}$ lie inside $|z| < 1$. Can this be extended to the vector $\{z_k\}$ case?

REFERENCES

[1] ANDERSON, B. D. O., and J. B. MOORE, "Solution of a Time-varying Wiener Filtering Problem," *Electronics Letters*, Vol. 3, No. 12, December 1967, pp. 562–563.

[2] ANDERSON, B. D. O., and J. B. MOORE, "State Estimation via the Whitening Filter", *Proc. Joint Automatic Control Conf.*, Ann Arbor, Michigan, June 1968, pp. 123–129.

[3] ANDERSON, B. D. O., and J. B. MOORE, "The Kalman-Bucy Filter as a True Time-varying Wiener Filter," *IEEE Trans. Systems, Man and Cybernetics*, Vol. SMC-1, No. 2, April 1971, pp. 119–127.

[4] KAILATH, T., and R. A. GEESEY, "An Innovations Approach to Least-squares Estimation: Part IV, Recursive Estimation Given Lumped Covariance Functions," *IEEE Trans. Automatic Control*, Vol. AC-16, No. 6, December 1971, pp. 720–727.

[5] SON, L. H., and B. D. O. ANDERSON, "Design of Kalman Filters Using Signal Model Output Statistics," *Proc. IEE*, Vol. 120, No. 2, February 1973, pp. 312–318.

[6] WIENER, N., and L. MASANI, "The Prediction Theory of Multivariate Stochastic Processes, Parts 1 and 2," *Acta Mathematica*, Vol. 98, June 1958.

[7] YOULA, D. C., "On the Factorization of Rational Matrices," *IRE Trans. Inform. Theory*, Vol. IT-7, No. 3, July 1961, pp. 172–189.

[8] DAVIS, M. C., "Factoring the Spectral Matrix," *IEEE Trans. Automatic Control*, Vol. AC-8, No. 4, October 1963, pp. 296–305.

[9] WONG, E., and J. B. THOMAS, "On the Multidimensional Prediction and Filtering Problem and the Factorization of Spectral Matrices," *J. Franklin Inst.*, Vol. 272, No. 2, August 1961, pp. 87–99.

[10] POPOV, V. M., "Hyperstability and Optimality of Automatic Systems with Several Control Functions," *Revue Roumaine des Sciences Technique, Electrotechn. et Energ.*, Vol. 9, No. 4, 1964, pp. 629–690.

[11] RIDDLE, A. C., and B. D. O. ANDERSON, "Spectral Factorization, Computational Aspects," *IEEE Trans. Automatic Control*, Vol. AC-11, No. 4, October 1966, pp. 764–765.

[12] ANDERSON, B. D. O., "An Algebraic Solution to the Spectral Factorization Problem," *IEEE Trans. Automatic Control*, Vol. AC-12, No. 4, August 1967, pp. 410–414.

[13] CSÁKI, F., and P. FISCHER, "On the Spectrum Factorization," *Acta Technica Academiae Scientarum Hungaricae*, Vol. 58, 1967, pp. 145–168.

[14] STRINTZIS, M. G., "A Solution to the Matrix Factorization Problem," *IEEE Trans. Inform. Theory*, Vol. IT-18, No. 2, March 1972, pp. 225–232.

[15] BAUER, F. L., "Ein direktes Iterationsverfahren zur Hurwitzzerlegung eines Polynoms," *Archiv der Elektrischen Übertragung*, Vol. 9, 1955, pp. 285–290.

[16] TUEL, W. G., "Computer Algorithm for Spectral Factorization of Rational Matrices," *IBM J. Research and Development*, March 1968, pp. 163–170.

[17] TUNNICLIFFE WILSON, G., "The Factorization of Matricial Spectral Densities," *SIAM J. Applied Math.*, Vol. 23, No. 4, December 1972, pp. 420–426.

[18] ANDERSON, B. D. O., K. L. HITZ, and N. DIEM, "Recursive Algorithms for Spectral Factorization," *IEEE Trans. Circuit and Systems*, Vol. CAS-21, No. 6, November 1974, pp. 742–750.

[19] WIENER, N., *Extrapolation, Interpolation and Smoothing of Stationary Time Series*, The M.I.T. Press, Cambridge, Mass., 1949.

[20] BODE, H. W., and C. E. SHANNON, "A Simplified Derivation of Linear Least Square Smoothing and Prediction Theory," *Proc. IRE*, Vol. 38, No. 4, April 1950, pp. 417–425.

[21] WAINSTEIN, L. A., and V. D. ZUBAKOV, *Extraction of Signals from Noise* (trans. R. A. Silverman), Prentice-Hall, Inc., Englewood Cliffs, N.J., 1962.

[22] LEVINSON, N., "The Wiener rms (root-mean-square) Error Criterion in Filter Design and Prediction," *J. Math. Phys.*, Vol. 25, January 1947, pp. 261–278.

[23] WHITTLE, P., *Prediction and Regulation*, Van Nostrand Rheinhold Company, New York, 1963.

[24] ANDERSON, B. D. O., and T. KAILATH, "The Choice of Signal-process Models in Kalman-Bucy Filtering," *J. Mathematical Analysis and Applications*, Vol. 35, No. 3, September 1971, pp. 659–668.

[25] ANDERSON, B. D. O., "Algebraic Properties of Minimal Degree Spectral Factors," *Automatica*, Vol. 9, No. 4, July 1973, pp. 491–500.

[26] BATKOV, A., "Generalization of the Shaping Filter Method to Include Nonstationary Random Processes," *Automation and Remote Control*, Vol. 20, No. 8, August 1959, pp. 1049–1062.

[27] DARLINGTON, S., "Nonstationary Smoothing and Prediction Using Network Theory Concepts," *Transactions of the 1959 International Symposium on Circuit and Information Theory*, Los Angeles, pp. 1–13.

[28] STEAR, E. B., "Shaping Filters for Stochastic Processes," in *Modern Control Systems Theory*, McGraw-Hill Book Company, New York, 1965.

[29] ZADEH, L. A., "Time-varying Networks, I," *Proc. IRE*, Vol. 49, No. 10, October 1961, pp. 1488–1503.

[30] ANDERSON, B. D. O., "Time-varying Spectral Factorization," Rept. No. SEL-66-107, (TR 6560-8), *Stanford Electronics Lab.*, Stanford, Calif., October 1966.

[31] ANDERSON, B. D. O., J. B. MOORE, and S. G. LOO, "Spectral Factorization of Time-varying Covariance Functions," *IEEE Trans. Inform. Theory*, Vol. IT-15, No. 5, September 1969, pp. 550–557.

[32] MOORE, J. B., and B. D. O. ANDERSON, "Spectral Factorization of Time-varying Covariance Functions: The Singular Case," *Mathematical System Theory*, Vol 4, No. 1, 1970, pp. 10–23.

[33] ANDERSON, B. D. O., and P. J. MOYLAN, "Spectral Factorization of a Finite-dimensional Nonstationary Matrix Covariance," *IEEE Trans. Automatic Control*, Vol. AC-19, No. 6, December 1974, pp. 680–692.

[34] ROBINSON, E. A., *Multi-channel Time-series Analysis with Digital Computer Programs*, Holden-Day, Inc., San Francisco, 1967.

[35] RISSANEN, J., and L. BARBOSA, "Properties of Infinite Covariance Matrices and Stability of Optimum Predictors," *Inform. Sciences*, Vol. 1, 1969, pp. 221–236.

[36] RISSANEN, J., and P. E. CAINES, "Consistency of Maximum Likelihood Estimators for Multivariate Gaussian Processes with Rational Spectrum," *Control Systems Report 7424*, Department of Electrical Engineering, University of Toronto, December 1974.

[37] DOOB, J. L., *Stochastic Processes*, John Wiley & Sons, Inc., New York, 1953.

[38] KAILATH, T., "A View of Three Decades of Linear Filtering Theory," *IEEE Trans. Inform. Theory*, Vol. IT-20, No. 2, March 1974, pp. 146–181.

CHAPTER 10

PARAMETER IDENTIFICATION
AND ADAPTIVE ESTIMATION

10.1 ADAPTIVE ESTIMATION VIA
PARALLEL PROCESSING

When signal models and noise statistics are known, we have seen that linear minimum variance estimation is easily obtainable. In some engineering applications, however, the underlying processes are either too complex or too obscure for scientific laws and analytical studies to yield a precise signal model. Of course, when there are signal model uncertainties, the guidelines arising from the error analysis of Chap. 6 may be helpful, but these do not assist very much when there is little or no knowledge of the signal model.

More often than not a signal model must be calculated off-line from test input-output data or on-line from the measurement data itself. With test-data and off-line calculations, it may be possible using time-series analysis techniques [1] to estimate quantities such as the signal model output covariance. Then the techniques of the previous chapter allow construction of a signal model. These techniques are almost completely restricted to situations in which the signal model parameters do not vary in any way with time. However, a process usually has parameters which vary slowly in some random manner, and for on-line filtering of such processes it is clearly preferable to employ schemes for adapting the filter on-line to the signal model parameter variations.

We now take up the question of how to carry out filtering when the signal models (assumed linear) are unknown, or are known except for an unknown parameter matrix θ which may be slowly time varying. Estimation for such models is termed *adaptive estimation* and is invariably carried out in some suboptimal fashion, since the simultaneous optimal estimation of states and parameters (also viewed as states) is usually a highly nonlinear filtering problem too formidable to solve directly without the introduction of simplifying assumptions.

In this chapter, we demonstrate for adaptive estimation a notion already encountered earlier in discussing nonlinear filtering, namely that in situations where *optimal* filtering is out of the question because of its complexity, optimal linear filter results for modified models can still be applied to achieve useful *near optimal* nonlinear filtering algorithms.

In this section, we first introduce the simplifying assumption that the unknown parameters belong to a discrete set; we can then achieve optimal parallel processing adaptive schemes. These schemes also extend to useful near optimal filtering when the unknown parameters belong to a continuous range. In the following section even simpler suboptimal schemes based on extended Kalman filter ideas and least squares parameter identification are studied. These may work very well for some signal models, but for others there could well be difficulties in preventing divergence.

Parallel processing techniques have been applied by a number of authors [2–4] to the adaptive estimation problem, and, in fact, adaptive estimators requiring many Kalman filters can be implemented using minicomputers. In essence, this approach to the adaptive estimation problem is as follows. Assume that the unknown parameter vector θ is discrete or suitably quantized to a finite number of grid points $\{\theta_1, \ldots, \theta_N\}$, with known or assumed a priori probability for each θ_i. The conditional mean estimator includes a parallel bank of N Kalman filters, each driven by the noisy signal measurement, and with the ith filter a standard Kalman filter designed on the assumption that $\theta = \theta_i$ and yielding conditional state estimates $\hat{x}_{k/k-1,\theta_i}$. The conditional mean estimate $\hat{x}_{k/k-1}$ is given by a weighted sum of the conditional state estimates $\hat{x}_{k/k-1,\theta_i}$. The weighting coefficient of the state of the ith Kalman filter is the a posteriori probability that $\theta = \theta_i$, which is updated recursively using the noisy signal measurements and the state of the ith Kalman filter. (Figure 10.1-1 on p. 274 illustrates the adaptive estimation scheme.)

We now derive a recursive method for updating a posteriori parameter probabilities. The resulting recursions are crucial to the parallel processing algorithm for state estimation next described. Certain analytical results concerning such algorithms are then derived (see also [5–7].)

Suppose there is a signal model expressed in terms of an unknown parameter vector θ. Further assume that θ is known to belong to the discrete set $\{\theta_1 \quad \theta_2 \quad \dots \quad \theta_N\}$; then the application of Bayes' rule yields the relationships

$$
\begin{aligned}
p(\theta_i | Z_k) &= \frac{p(Z_k, \theta_i)}{p(Z_k)} \\
&= \frac{p(Z_k | \theta_i) p(\theta_i)}{\sum\limits_{i=1}^{N} p(Z_k | \theta_i) p(\theta_i)}
\end{aligned} \tag{1.1}
$$

Here, as in earlier chapters, Z_k denotes the sequence of measurements $z_0, z_1,$ $\dots, z_k$, while $p(\theta_i | Z_k)$ is shorthand for $p(\theta = \theta_i | Z_k)$. A lower-case p is used interchangeably to denote a probability or probability density. The $p(\theta_i | Z_k)$ are termed *a posteriori probabilities* (or, less precisely, conditional probabilities) and $p(Z_k | \theta_i)$ are termed *likelihood functions*. It is for the recursive calculation of these quantities that the Kalman filter, or, more precisely, a bank of conditional Kalman filters, comes into its own. We have, following [2],

$$
\begin{aligned}
p(\theta_i | Z_k) &= \frac{p(z_k, Z_{k-1}, \theta_i)}{p(z_k, Z_{k-1})} \\
&= \frac{p(z_k, \theta_i | Z_{k-1}) p(Z_{k-1})}{p(z_k | Z_{k-1}) p(Z_{k-1})} \\
&= \frac{p(z_k, \theta_i | Z_{k-1})}{p(z_k | Z_{k-1})} \\
&= \frac{p(z_k | Z_{k-1}, \theta_i) p(\theta_i | Z_{k-1})}{\sum\limits_{i=1}^{N} p(z_k | Z_{k-1}, \theta_i) p(\theta_i | Z_{k-1})}
\end{aligned} \tag{1.2}
$$

Actually, the denominator of both (1.1) and (1.2) is just a normalizing constant, and should θ_i belong to a continuous range, (1.2) is still valid save that the summation is replaced by an integral.

The calculation of $p(z_k | Z_{k-1}, \theta_i)$ is crucial, but is readily implemented for gaussian signal models; in this case, $p(z_k | Z_{k-1}, \theta_i)$ is gaussian with mean $\hat{z}_{k|\theta_i}$ and covariance $E[\tilde{z}_{k|\theta_i} \tilde{z}'_{k|\theta_i}]$, which we denote by $\Omega_{k|\theta_i}$. That is, for a p-vector z_k,

$$
p(z_k | Z_{k-1}, \theta_i) = (2\pi)^{-p/2} |\Omega_{k|\theta_i}^{-1}|^{1/2} \exp\left\{ -\tfrac{1}{2} \tilde{z}'_{k|\theta_i} \Omega_{k|\theta_i}^{-1} \tilde{z}_{k|\theta_i} \right\} \tag{1.3}
$$

and clearly $p(\theta_i | Z_k)$ can be calculated recursively from

$$
p(\theta_i | Z_k) = c |\Omega_{k|\theta_i}^{-1}|^{1/2} \exp\left\{ -\tfrac{1}{2} \tilde{z}'_{k|\theta_i} \Omega_{k|\theta_i}^{-1} \tilde{z}_{k|\theta_i} \right\} p(\theta_i | Z_{k-1}) \tag{1.4}
$$

where c is a normalizing constant independent of θ_i, chosen to ensure that $\sum\limits_{i=1}^{N} p(\theta_i | Z_k) = 1$. The quantities $\Omega_{k|\theta_i}$ and $\tilde{z}_{k|\theta_i}$ are of course available from the conditional Kalman filter covariance equations and the filter equations, respectively.

Comments on the a Posteriori Probability Update Equation (1.4)

The quantity $\Omega_{k|\theta_i}$ can be computed in advance of filter operations, while naturally $\tilde{z}_{k|\theta_i}$ can only be computed on-line; for each i, a Kalman filter tuned to θ_i is constructed, and the $\{z_k\}$ sequence drives all these Kalman filters. The innovations sequence of the ith filter is denoted $\tilde{z}_{k|\theta_i}$, even if the signal model has $\theta \neq \theta_i$. The sequence $\{\tilde{z}_{k|\theta_i}\}$ will be white with covariance $\Omega_{k|\theta_i}$ so long as the signal model parameter θ is θ_i; otherwise, the sequence $\{\tilde{z}_{k|\theta_i}\}$ will in general not be white, and in general will not have covariance $\Omega_{k|\theta_i}$.

In a sense, the notation $\{\tilde{z}_{k|\theta_i}\}$ is ambiguous; thus it can denote the innovations sequence of the Kalman filter tuned to θ_i only if the signal model θ is θ_i, while it can also denote the innovations sequence of the Kalman filter tuned to θ_i irrespective of the signal model θ. Evidently in Eq. (1.4), $\{\tilde{z}_{k|\theta_i}\}$ has this second meaning, while $\Omega_{k|\theta_i}$ is the covariance of $\tilde{z}_{k|\theta_i}$ only when the signal model θ is θ_i.

Two other minor points should also be made. First, it can be simpler to update $\ln p(\theta_i|Z_k)$ as opposed to $p(\theta_i|Z_k)$, recursion formulas being immediately derivable. (The convergence results following will illustrate this point.) Second, if $\Omega_{k|\theta_i} = \Omega_{k|\theta_j}$ and $\tilde{z}_{k|\theta_i} = \tilde{z}_{k|\theta_j}$ (the second equality is equivalent to the θ_i and θ_j Kalman filters being the same), then $p(\theta_i) = p(\theta_j)$ implies $p(\theta_i|Z_k) = p(\theta_j|Z_k)$ for all k and there is no possible way of distinguishing θ_i and θ_j on the basis of the measurement data. More generally, even if $p(\theta_i) \neq p(\theta_j)$ but the Kalman filters tuned to θ_i and θ_j are the same and the innovations sequences have the same covariance, the measurements will add nothing to the a priori data which is useful for distinguishing between θ_i and θ_j as possible true values of θ.

A Convergence Result

One would hope that the use of (1.4) in a particular situation would imply that if the true value of θ were, say, θ_1, then $p(\theta_1|Z_k) \rightarrow 1$ as $k \rightarrow \infty$ and $p(\theta_j|Z_k) \rightarrow 0$ as $k \rightarrow \infty$ for $j \neq 1$. Indeed, results of this type hold. We prove one such result below, assuming ergodicity. The proof depends on the following lemma.

LEMMA 1.1. Let A, B be two $p \times p$ positive definite matrices. Then

$$p + \ln \left(\frac{|A|}{|B|}\right) - \text{tr}\,[B^{-1}A] \leq 0 \tag{1.5}$$

with equality if and only if $A = B$.

Proof: We use the easily verified fact that for all $x > 0$,

$$1 + \ln x - x \leq 0$$

with equality only at $x = 1$. Let $\lambda_1, \ldots, \lambda_p$ be the eigenvalues of $B^{-1}A$; since this matrix is similar to $B^{-1/2}AB^{-1/2}$, which is positive definite, the λ_i are all positive. Therefore,

$$1 + \ln \lambda_i - \lambda_i \leq 0$$

with equality if and only if $\lambda_i = 1$, and summing yields

$$p + \ln \Pi\lambda_i - \Sigma\lambda_i \leq 0$$

Equation (1.5) is immediate, using standard matrix theory results. Equality holds if and only if the eigenvalues of $B^{-1}A$ are all 1, i.e., $B^{-1}A = I$.

The results to follow are all valid in case the sequences $\tilde{z}_{k|\theta_i}$ are asymptotically wide sense stationary. However, we shall assume ergodicity in order to keep the notation simpler. Thus $\Omega_{k|\theta_i}$, the covariance of $\tilde{z}_{k|\theta_i}$ when the signal model θ is the same as θ_i, is constant: we shall write it as Ω_i. Further, $E[\tilde{z}_{k|\theta_i}\tilde{z}'_{k|\theta_i}]$ for $\theta_i \neq \theta$ will also be independent of k.

Let us now fix $i \neq 1$ and set $L_k = [p(\theta_i | Z_k)][p(\theta_1 | Z_k)]^{-1}$. From (1.4),

$$L_k = \frac{|\Omega_i^{-1}|^{1/2} \exp\left[-\frac{1}{2}\tilde{z}'_{k|\theta_i}\Omega_i^{-1}\tilde{z}_{k|\theta i}\right]}{|\Omega_1^{-1}|^{1/2} \exp\left[-\frac{1}{2}\tilde{z}'_{k|\theta_1}\Omega_1^{-1}\tilde{z}_{k|\theta_1}\right]}L_{k-1} \tag{1.6}$$

and so

$$\ln \frac{L_{k+n-1}}{L_{k-1}} = \frac{1}{2}n \ln \frac{|\Omega_1|}{|\Omega_i|} - \frac{1}{2}\,\text{tr}\left[\sum_{j=k}^{k+n-1} \tilde{z}_{j|\theta_i}\tilde{z}'_{j|\theta_i}\Omega_i^{-1}\right]$$
$$+ \frac{1}{2}\,\text{tr}\left[\sum_{j=k}^{k+n-1} \tilde{z}_{j|\theta_1}\tilde{z}'_{j|\theta_1}\Omega_1^{-1}\right] \tag{1.7}$$

Let us now assume that θ_1 is the true parameter and introduce what will prove to be a convergence criterion.

CONVERGENCE CONDITION. For $\theta_i \neq \theta_j$, either $\tilde{z}_{k|\theta_i} - \tilde{z}_{k|\theta_j}$ fails to approach zero as $k \longrightarrow \infty$, or $\Omega_i \neq \Omega_j$, or both.

Evidently, this is a sort of distinguishability criterion. If $\tilde{z}_{k|\theta_i} - \tilde{z}_{k|\theta_j} = 0$ (as $k \longrightarrow \infty$) and $\Omega_i = \Omega_j$, there can be no basis for deciding whether $\theta = \theta_i$ or $\theta = \theta_j$.

Now with θ_1 the true parameter, ergodicity yields, as $n \longrightarrow \infty$,

$$\frac{1}{n}\sum_{j=k}^{k+n-1} \tilde{z}_{j|\theta_1}\tilde{z}'_{j|\theta_1}\Omega_1^{-1} \longrightarrow I$$

Further, since Ω_1 is the error covariance in estimating z_j using the optimal filter, it underbounds the error covariance in estimating z_j using any suboptimal filter. In particular, it underbounds the error covariance in using the suboptimal filter designed under the assumption that $\theta = \theta_i$, where $i \neq 1$. Therefore

$$\lim_{n \to \infty} \frac{1}{n}\sum_{j=k}^{k+n-1} \tilde{z}_{j|\theta_i}\tilde{z}'_{j|\theta_i} \geq \Omega_1 \tag{1.8}$$

Let us now observe that strict equality can only occur if $\Omega_i \neq \Omega_1$. For if strict equality occurs, this implies that the limiting error covariance associated with use of both the θ_i and θ_1 filters on the signal model with $\theta = \theta_1$ are the same; therefore, the θ_i filter in the limit is optimal. Because of the uniqueness of the linear minimum variance estimate, this means that $\tilde{z}_{j|\theta_i} \to \tilde{z}_{j|\theta_1}$ as $j \to \infty$. By the Convergence Condition, we then have a contradiction. [A situation such as this could occur if θ_1 corresponded to a signal model with parameters F, G, H, Q, and R, and θ_i to F, G, H, αQ, and αR, with α a constant. The Kalman filter is the same for each, but the innovations covariances Ω_1 and Ω_i are different. If the true signal model is $\theta = \theta_1$, then the filter will yield

$$E[\tilde{z}_{j|\theta_i} \tilde{z}'_{j|\theta_i}] = E[\tilde{z}_{j|\theta_1} \tilde{z}'_{j|\theta_1}] = \Omega_1$$

Of course, $E[\tilde{z}_{j|\theta_i} \tilde{z}'_{j|\theta_i}] \neq \Omega_i$ because the signal model driving the Kalman filter producing $\tilde{z}_{j|\theta_i}$ is not that defined by $\theta = \theta_i$.]

Now use (1.8) in (1.6) to conclude that under the convergence condition, as $n \to \infty$,

$$2n^{-1} \ln \frac{L_{k+n-1}}{L_{k-1}} \longrightarrow \ln \frac{|\Omega_1|}{|\Omega_i|} - \text{tr } [\Omega_i^{-1} \Omega_1] + \text{tr } I - \text{tr } [\Omega_i^{-1} C]$$

where $\hspace{9cm}$ (1.9)

$$C = \lim_{n \to \infty} n^{-1} \sum_{j=k}^{k+n-1} \tilde{z}_{j|\theta_i} \tilde{z}'_{j|\theta_i} - \Omega_1$$

By Lemma 1.1, the first three terms on the right side yield a nonpositive quantity, negative if $\Omega_i \neq \Omega_1$. By the remarks preceding (1.9), we see that tr $[\Omega_i^{-1} C]$ is nonnegative, and positive if $\Omega_i = \Omega_1$. Therefore the right side of (1.9) is negative.

Thus for some positive α,

$$2n^{-1} \ln \frac{L_{k+n-1}}{L_{k-1}} \longrightarrow -\alpha$$

or

$$L_{k+n-1} \longrightarrow K \exp\left(-\frac{n\alpha}{2}\right) L_{k-1}$$

for some constant K. We conclude that

$$\frac{p(\theta_i | Z_k)}{p(\theta_1 | Z_k)} \longrightarrow 0 \qquad i \neq 1$$

and thus $p(\theta_i | Z_k) \to 0$, $i \neq 1$, and $p(\theta_1 | Z_k) \to 1$. Further, convergence is exponentially fast. Several remarks should be made.

1. If one is not interested in the recursive values of the $p(\theta_i | Z_k)$, but simply in identifying the correct value of θ, we see from (1.8) that

$$\lim_{n \to \infty} \frac{1}{n} \sum_{j=k}^{k+n-1} \tilde{z}_{j|\theta_1} \tilde{z}'_{j|\theta_1} \leq \lim_{n \to \infty} \frac{1}{n} \sum_{j=k}^{k+n-1} \tilde{z}_{j|\theta_i} \tilde{z}'_{j|\theta_i} \qquad (1.10)$$

and if equality does not hold, θ is immediately identified as θ_1. If equality does hold, for $i = 2$ say, then θ_1 is determined by the fact that the two quantities in (1.10) equal Ω_1, while $\Omega_1 \neq \Omega_2$.

2. We have refrained from specifying the precise mode of convergence. This can be linked to the ergodicity hypothesis; if the z_k are gaussian, the convergence is almost sure.
3. As noted earlier, the ergodicity assumptions can be relaxed to asymptotic stationarity, so that initial condition effects are not important. Further relaxation again is, however, possible to consider some nonstationary situations (see [7]).
4. The above convergence results are useful for *fault detection*. Each θ_i could represent a possible fault condition in a signal model [10].
5. For the nongaussian case, the algorithms are still useful, but the interpretations in terms of probabilities breaks down. The schemes of this section for the more general case are known as prediction error schemes since they involve calculation with $z_k - \hat{z}_{k/k-1,\theta_i}$, where now $\hat{z}_{k/k-1,\theta_i}$ denotes a one step ahead prediction estimate, rather than a conditional mean estimate. See also [7].

Parallel Processing State Estimation

Under the assumption that θ is in the set $\{\theta_1, \theta_2, \ldots, \theta_N\}$, the conditional mean state estimate can be expressed in terms of the conditional estimates $\hat{x}_{k/k-1, \theta_i}$ and the conditional probabilities $p(\theta_i | Z_k)$ as

$$\hat{x}_{k/k-1} = \sum_{i=1}^{N} \hat{x}_{k/k-1, \theta_i} p(\theta_i | Z_k) \tag{1.11a}$$

One can also form an estimate

$$\hat{x}_{k/k-1}^{\text{MAP}} = \hat{x}_{k/k-1, \theta_{\text{MAP}}} \tag{1.11b}$$

with $\hat{\theta}_{\text{MAP}}$ chosen such that $p(\hat{\theta}_{\text{MAP}} | Z_k) \geq p(\theta_i | Z_k)$ for $i = 1, 2, \ldots N$; i.e., we use a maximum a posteriori estimate of θ.

Recall that the conditional estimates $\hat{x}_{k/k-1,\theta_i}$ are calculated using conditional Kalman filters with θ set as θ_i, while the conditional probabilities $p(\theta_i | Z_k)$ are best calculated recursively as in (1.4) using the conditional innovations quantities $\tilde{z}_{k|\theta_i}$ with covariances $\Omega_{k|\theta_i}$ given from the conditional Kalman filter equations. See Fig. 10.1-1 for a block diagram of the adaptive estimator.

Time-varying Parameter Case

So far we have presented adaptive estimation schemes in which the unknown parameters are constant (or asymptotically constant in the above example). For the case when the unknown parameters are in fact time varying, various modifications to the parallel processing scheme are possible. One approach is to use exponential data weighting. This applies both to the filters themselves and the update of $p(\theta_i | Z_k)$ via (1.4). A second approach

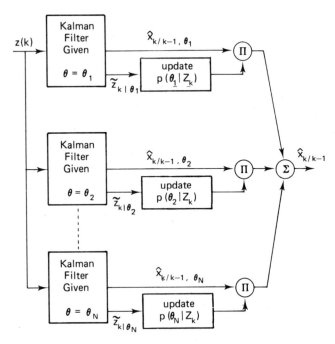

Fig. 10.1-1 Adaptive estimator for time-invariant θ.

requires reinitializing, which in effect amounts to throwing away old data. More specifically, one should reset any $p(\theta_i|Z_k)$ which are zero to a nonzero value (implying that some which are nonzero must be reduced), and one also resets the states of the Kalman filters in the filter bank. The logical state at which to reset them all is the current conditional mean or MAP state estimate. Obviously the choice of the weighting factor in the first approach and frequency of reset in the second approach should be related to the rate of time variation of the unknown parameters. Note that at one extreme, the filters in the bank can be reinitialized at every time instant, in which case the algorithm simplifies considerably.

Unknown Parameter Drawn from an Infinite Set

To this point, the unknown parameter has been assumed to belong to a finite set. But what if this is not the case, and instead θ is, say, contained in some closed, bounded region? One can proceed by selecting a discrete set of points $\theta_1, \ldots, \theta_N$ in the region and acting as if $\theta = \theta_i$ for some unknown i. Clearly an approximation is introduced thereby; intuitively, one can see that the greater is the integer N, or the denser is the covering of the allowable region by discrete points, the more accurate the approximation will be.

Let us leave aside for the moment the problem of how the θ_i might be selected and consider what happens if one assigns some a priori value of

$p(\theta = \theta_i)$, and then implements (1.4) to update $p(\theta_i | Z_k)$. Of course, $p(\theta = \theta_i)$ cannot really be the probability that $\theta = \theta_i$, but we might think of it as a pseudo-probability. A reasonable value for it would be

$$p(|\theta - \theta_i| < |\theta - \theta_j|, \ \forall j \neq i)$$

However, these values are forgotten in the computation of a posteriori probabilities, and so are not especially critical. As the following result shows, under reasonable assumptions one of the quantities $p(\theta_i | Z_k)$—say $p(\theta_1 | Z_k)$ — converges to 1 and the remainder converge to 0. If θ_1 is indeed picked out in this way, one cannot of course conclude that the true value of θ is θ_1; but one can conclude that, in a certain sense, the true value of θ is closer to θ_1 than to θ_i for $i = 2, 3, \ldots, N$.

THEOREM 1.1. With notation as above, let the true value of θ be θ_0 and let $\tilde{z}_{k|\theta_i}$ for $i = 1, 2, \ldots, N$ be the innovations sequence of the Kalman filter tuned to θ_i and driven by the signal model output. Let $\Omega_{k|\theta_i}$ denote the design covariance of the filter innovations, i.e., the value of $E[\tilde{z}_{k|\theta_i}\tilde{z}'_{k|\theta_i}]$ should the signal model have $\theta = \theta_i$. Suppose that $\tilde{z}_{k|\theta_i}$ is asymptotically ergodic in the autocorrelation function; suppose that $\Omega_{k|\theta_i} \longrightarrow \Omega_i$ as $k \longrightarrow \infty$ with $\Omega_i > 0$; and denote the actual limiting covariance of the filter innovations, viz.,

$$\lim_{n\to\infty} n^{-1} \sum_{j=k}^{k+n-1} \tilde{z}_{j|\theta_i} \tilde{z}'_{j|\theta_i}$$

by Σ_i. Suppose that a priori pseudo-probabilities $p(\theta_1), \ldots, p(\theta_N)$ are assigned, with (1.4) providing the recursive update for these pseudo-probabilities. Define

$$\beta_i = \ln |\Omega_i| + \text{tr}\,(\Omega_i^{-1}\Sigma_i) \tag{1.12}$$

and assume that for some i, say $i = I$, and all $j \neq I$, one has

$$\beta_I < \beta_j \tag{1.13}$$

Then $p(\theta_I | Z_k) \longrightarrow 1$ as $k \longrightarrow \infty$ and $p(\theta_j | Z_k) \longrightarrow 0$ as $k \longrightarrow \infty$ for $j \neq I$, convergence being exponentially fast.

Proof. Set $L_k^i = [p(\theta_i | Z_k)][p(\theta_0 | Z_k)]^{-1}$. As in the derivation of (1.9), we can conclude that as $n \longrightarrow \infty$,

$$2n^{-1} \ln \frac{L_{k+n-1}^i}{L_k^i} \longrightarrow \ln \frac{|\Omega_0|}{|\Omega_i|} - \text{tr}\,[\Omega_i^{-1}\Sigma_i] + \text{tr}\,I$$

so that

$$2n^{-1} \ln \left(\frac{L_{k+n-1}^i}{L_k^i} \frac{L_k^I}{L_{k+n-1}^I} \right) \longrightarrow \beta_I - \beta_j$$

i.e.,

$$\ln \frac{p(\theta_j | Z_{k+n-1})}{p(\theta_I | Z_{k+n-1})} - \ln \frac{p(\theta_j | Z_k)}{p(\theta_I | Z_k)} \longrightarrow K \frac{n(\beta_I - \beta_j)}{2}$$

for some constant K. The claimed result then follows easily.

Let us examine further the inequality (1.13), which determines which member of the set $\{\theta_1, \ldots, \theta_N\}$ is picked out by the estimation procedure. We shall indicate an interpretation via a classical idea of information theory and give a spectral description of the condition.

The *Kullback information function* [8] is defined for a finite measurement sequence as

$$J_k(\theta_s, \theta_r) = E\left\{\ln \frac{p(Z_k|\theta_s)}{p(Z_k|\theta_r)}\,\Big|\,\theta_s\right\} \tag{1.14}$$

and for an infinite measurement sequence, we have an asymptotic per sample information function

$$\bar{J}(\theta_s, \theta_r) = \lim_{k\to\infty} k^{-1}J_k(\theta_s, \theta_r) \tag{1.15}$$

A little manipulation shows that

$$\bar{J}(\theta_s, \theta_r) = \lim_{n\to\infty} n^{-1}E\left\{\ln \frac{p(\theta_s|Z_{k+n-1})}{p(\theta_r|Z_{k+n-1})}\,\Big|\,\theta_s\right\}$$

$$= -\lim_{n\to\infty} n^{-1}E\left\{\ln \frac{p(\theta_r|Z_{k+n-1})p(\theta_s|Z_k)}{p(\theta_s|Z_{k+n-1})p(\theta_r|Z_k)}\,\Big|\,\theta_s\right\}$$

With $\theta_s = \theta_0$, we have shown in proving Theorem 1.1 that

$$n^{-1}\ln \frac{L^r_{k+n-1}}{L^r_k} = n^{-1}\ln \frac{p(\theta_r|Z_{k+n-1})\,p(\theta_0|Z_k)}{p(\theta_0|Z_{k+n-1})\,p(\theta_r|Z_k)}$$

$$\longrightarrow \tfrac{1}{2}[\ln|\Omega_0| + \operatorname{tr} I - \beta_r]$$

Therefore,

$$\bar{J}(\theta_0, \theta_r) = \tfrac{1}{2}[\beta_r - \ln|\Omega_0| - \operatorname{tr} I]$$

Theorem 1.1 thus shows that convergence occurs to that member of the set $\{\theta_1, \ldots, \theta_N\}$ which is closest to θ_0 in the sense of minimizing the Kullback information measure. The Kullback information measure is, incidentally, always nonnegative, and for $\|\theta_s - \theta_r\|$ small, one has

$$\bar{J}(\theta_s, \theta_r) = \bar{J}(\theta_r, \theta_s) = \tfrac{1}{2}(\theta_s - \theta_r)'F_\theta(\theta_s - \theta_r)$$

for some positive definite F_θ. Thus the measure is locally like a metric.

The evaluation of the quantities β_i in (1.12) is not difficult. The quantity Σ_i can be found by standard procedures for computing the second order statistics of signals arising in linear systems. It can also be evaluated in terms of the power spectra $\Phi_i(z)$ and $\Phi_0(z)$ of the signal model with $\theta = \theta_i$ and the signal model with $\theta = \theta_0$, respectively as follows. Let $W_i(z)$ and $W_0(z)$ denote the transfer function matrices of the associated innovations representation. Then we have

$$\Phi_0(z) = W_0(z)\Omega_0 W_0'(z^{-1})$$

and similarly for $\Phi_i(z)$. The power spectrum of $\{\tilde{z}_{k|\theta_i}\}$ [which is the output of a linear system driven by $\{z_k\}$ with transfer function matrix $W_i^{-1}(z)$] is

$W_i^{-1}(z)\Phi_0(z)[W_i^{-1}(z^{-1})]'$. Also $\Sigma_i = E[\tilde{z}_{k|\theta_i}\,\tilde{z}'_{k|\theta_i}]$. Hence,

$$\Sigma_i = \frac{1}{2\pi j}\oint W_i^{-1}(z)\Phi_0(z)[W_i^{-1}(z^{-1})]'z^{-1}\,dz$$

and so

$$\text{tr}\,(\Omega_i^{-1}\Sigma_i) = \frac{1}{2\pi j}\oint \text{tr}\,\Phi_0(z)\{[W_i^{-1}(z^{-1})]'\Omega_i^{-1}W_i^{-1}(z)\}z^{-1}\,dz$$

$$= \frac{1}{2\pi j}\oint \text{tr}\,[\Phi_0(z)\Phi_i^{-1}(z)]z^{-1}\,dz \tag{1.16}$$

We foreshadowed earlier some comments on the selection of parameters $\theta_1,\ldots,\theta_N$ within an infinite set Θ in which lies the true value of θ. Two particular points should be made.

1. The θ_i should be evenly distributed in the allowed region Θ where the distribution is more according to the rough metric provided by the Kullback per sample asymptotic information function. In particular, one might attempt to choose the θ_i to minimize $\min_i \max_{\theta\in\Theta} J(\theta_i, \theta)$.

2. If one can choose the θ_i such that the associated Kalman filter gains are the same for subsets of the θ_i (the associated Ω_i must then be different), there will be economy in a filter bank realization. Thus if N_1 Kalman filter gains are used, and N_2 different Ω_i for each gain, $N = N_1 N_2$ values of θ_i are covered, while only N_1 filters need be implemented.

Refined Parameter Estimates

For the case when the unknown parameter θ of our signal model is in a closed bounded region, we have seen that it is possible to divide the parameter space into N decision regions and employ the detector algorithms so far described to determine in which region of the parameter space the true parameter lies. Of course for large N, the detector algorithms yield an accurate estimate of the true parameter, but at the expense of complexity. Actually, a refined parameter estimate can be obtained using a *combined detection-estimation approach*; detection determines the right local region of the parameter space, and linearization is used within it, as illustrated in Problem 1.3, to estimate the true parameter. Further details are omitted here.

Model Approximation

An entirely different application of Theorem 1.1 is to the approximation of high order models by low order models. The actual signal model may be high order, while $\theta_1,\ldots,\theta_N$ correspond to low order models; the algorithm will then identify that low order model closest (according to the Kullback information measure) to the high order model.

Main Points of the Section

For signal models with unknown parameter θ belonging to a discrete set $\{\theta_1, \theta_2, \ldots, \theta_N\}$, a bank of conditional Kalman filters yielding conditional innovation sequences $\tilde{z}_{k|\theta_i}$ can be employed to yield a posteriori parameter probabilities $p(\theta_i | Z_k)$ and, together with the conditional state estimates $\hat{x}_{k/k-1, \theta_i}$, the conditional mean or a type of maximum a posteriori state estimate. Exponential convergence of the $p(\theta_i | Z_k)$ yields the true parameter.

These ideas can be applied to the case when θ is time varying and when it belongs to an infinite set Θ. By choosing a finite set within Θ, one can identify that member of the finite set closest to the true value, with distance measured by the Kullback information measure, and readily computable in terms of various power spectra.

Problem 1.1. Let us consider the case of state estimation in the presence of unknown measurement noise environments. For simplicity, suppose that the model is time invariant. For the usual state-space model $\{F, G, H, Q, R, S = 0\}$, we assume that F, G, H, and Q are known but R is unknown. As a consequence, the Kalman gain K is unknown, as is the one-step-ahead prediction error covariance Ω. To achieve estimation results, one approach is to introduce an assumption that the gain matrix $K = \Sigma H R^{-1}$ belongs to a discrete set $\{K_1, K_2, \ldots, K_M\}$ and to obtain innovations $\tilde{z}_{k|K_i}$ for $i = 1, 2, \ldots, M$, and thereby estimates $\hat{x}_{k/k-1}$ or $\hat{x}_{k/k-1}^{\text{MAP}}$. For this purpose, knowledge of Ω is not needed. However, to obtain performance characteristics, an estimate of R or Ω must be obtained. Derive the maximum a posteriori estimate of R under the assumption that $K = K_i$. [*Hints:* Note that $\Omega_{K_i, R} = [I + H'K_i]^{-1}R$ and derive an expression for $\dfrac{\partial \ln p(Z_k | K_i, R)}{\partial R^{-1}}$ and set it to zero. This achieves a MAP estimate of R given that $K = K_i$ of

$$\hat{R}_i = \tfrac{1}{2}(A_i + A_i')$$

$$A_i = \frac{1}{k} \sum_{j=1}^{k} \tilde{z}_{j|K_i} \tilde{z}'_{j|K_i}[I - H'K_i]'$$

For the vector measurement case we need the matrix result that for matrices P and vectors a, b

$$\frac{\partial \ln |P|}{\partial P^{-1}} = -P' \quad \text{and} \quad \frac{\partial a'Pb}{\partial P} = \frac{1}{2}(ab' + ba')]$$

Problem 1.2. Demonstrate that, in calculating $p(\theta_i | Z_k)$ for the case of vector measurements, there is a computational advantage in sequentially processing the measurements (see also the relevant section of Chap. 6). This problem is solved in [9].

Problem 1.3. Consider the usual state-space signal model $\{F, G, H, Q, R, S\}$ with all matrices known except that the output matrix H' is known to belong to a neighborhood of some nominal value $\bar{H}'$. Apply extended Kalman filter algorithms to estimate $(H - \bar{H})$ as an augmented state variable. For simplicity, consider the scalar measurement case only. (Illustrates refining of parameter estimates.)

10.2 ADAPTIVE ESTIMATION VIA EXTENDED LEAST SQUARES

The parallel estimation schemes of the last section could well be too complex for implementation. A very useful class of adaptive estimators can be constructed using the following approach. Let us so restrict the signal models that

(a) if the parameters θ are assumed known, then state estimation, possibly optimal, can be readily achieved.

(b) if the states x_k are assumed measurable, then the unknown parameters can be estimated, perhaps in some optimal fashion.

Then simultaneous state and parameter estimation may be carried out if the state and parameter estimators referred to in (a) and (b) above are employed, but with the state estimator calculating $\hat{x}_k$ using parameter estimates $\{\hat{\theta}_k\}$ and the parameter estimator calculating $\hat{\theta}_k$ using state estimates $X \equiv \{\hat{x}_k\}$. Figure 10.2-1 (see next page) indicates this idea.

Let us now specialize the above approach to adaptive estimation via extended least squares. First, we foreshadow that there are many useful further specializations of the results, including schemes outlined in [11–15]. (These appear to compete favourably with other approaches, of which there are many. See, e.g., [16–21].)

Consider now the signal model*

$$x_{k+1} = Fx_k + Gy_k + Kv_k + u_k \qquad (2.1a)$$

$$y_k = \theta'x_k \qquad z_k = y_k + v_k \qquad (2.1b)$$

Here, v_k is white noise of zero mean, x_k is the state, θ is an unknown parameter, and u_k is an external input which may be a function of z_k.

Assume temporarily that x_k as well as z_k is measurable. Then certainly *least squares identification* of θ can be achieved simply by selecting θ to minimize the index $\sum_{i=0}^{k}(z_i - \theta'x_i)'(z_i - \theta'x_i)$ for each k. Denoting the minimizing θ at time k by $\hat{\theta}_k$, we have, after simple calculations, that

$$\hat{\theta}_k = \left(\sum_{i=0}^{k} x_ix_i'\right)^{-1} \sum_{i=0}^{k} x_iz_i' \qquad (2.2)$$

---

*The appearance of Gy_k in (2.1a) may seem strange. Applications will, however, justify the possibility of $G \neq 0$ on occasions.

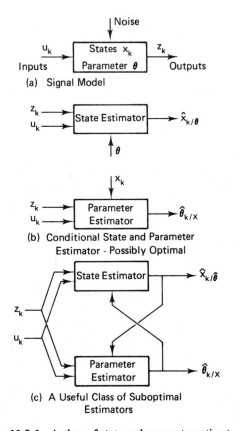

Fig. 10.2-1   A class of state and parameter estimators.

(assuming the inverse exists). This estimate may be calculated recursively from

$$P_{k+1}^{-1} = P_k^{-1} + x_k x_k' \qquad P_0^{-1} = 0 \tag{2.3a}$$

$$C_{k+1} = C_k + x_k z_k' \tag{2.3b}$$

$$\hat{\theta}_k = P_{k+1} C_{k+1} \tag{2.3c}$$

or, by application of the matrix inversion lemma (see Chap. 6),

$$\hat{\theta}_k = \hat{\theta}_{k-1} + P_{k+1} x_k (z_k' - x_k' \hat{\theta}_{k-1}) \tag{2.4a}$$

$$P_{k+1} = P_k - P_k x_k (x_k' P_k x_k)^{-1} x_k' P_k \tag{2.4b}$$

where P_0 in (2.4) is usually chosen as some suitably large positive definite matrix.

So with z_k and x_k measurable, there exist straightforward algorithms for the recursive estimation of θ. These algorithms are well understood and give almost sure convergence to the true parameter estimates under very reasonable "persistently exciting" conditions [22–24].

Actually, simpler algorithms based on stochastic approximation techniques [2] can be employed to yield consistent estimation of θ. In (2.3), P_{k+1} can be replaced by some decreasing scalar gain sequence γ_k satisfying certain conditions as discussed in [25]. Common choices of γ_k are

$$\frac{1}{k}, \quad \left[\sum_{i=0}^{k} x_i' x_i\right]^{-1} \quad \text{and} \quad \frac{1}{k}(x_k' x_k)^{-1}$$

These values for γ_k are certainly simpler to calculate than P_{k+1} above, but result in slower convergence rates. On the other hand, it is also possible to use more complicated algorithms, viz., weighted least squares rather than the unweighted version above. Such algorithms require knowledge of Ω, though if Ω is unknown, one can even consider estimating Ω and using the estimate in the weighted least squares equations. To examine these variations here would take us too far afield; we merely mention their existence for the sake of completeness.

Now if for the model (2.1) we assume knowledge of θ and measurability of z_k and u_k, but not x_k, then state estimates are readily obtained from the inverse of the signal model:

$$\hat{x}_{k+1/k,\theta} = F\hat{x}_{k/k-1,\theta} + G\hat{y}_{k/k-1,\theta} + K\hat{v}_{k/\theta} + u_k$$
$$\hat{v}_{k/\theta} = z_k - \hat{y}_{k/k-1,\theta} \tag{2.5}$$
$$\hat{y}_{k/k-1,\theta} = \hat{z}_{k/k-1,\theta} = \theta' \hat{x}_{k/k-1,\theta}$$

[Of course, if $\hat{x}_{k/k-1,\theta} = x_k$ for some $k = k_0$, then equality holds for all $k > k_0$; (2.1) is, after all, simply an innovations model, except possibly for the conditions on the initial state.]

Now we see that the earlier noted requirements (a) and (b) for achieving adaptive estimation are met. The idea of *extended least squares* is to employ $\hat{\theta}_k$ instead of θ in (2.5) and $\hat{x}_{k/k-1,\theta}$ instead of x_k in (2.2) through (2.4). With an obvious simplification of subscripts, we have the adaptive estimator equations

$$\hat{x}_{k+1} = F\hat{x}_k + G\hat{y}_k + K\hat{v}_k + u_k \tag{2.6a}$$
$$\hat{v}_k = z_k - \hat{y}_k \qquad \hat{y}_k = \hat{\theta}_k' \hat{x}_k \tag{2.6b}$$
$$\hat{\theta}_k = \hat{\theta}_{k-1} + \Lambda_{k+1} \hat{x}_k (z_k' - \hat{x}_k' \hat{\theta}_{k-1}) \tag{2.6c}$$
$$\Lambda_{k+1} = \Lambda_k - \Lambda_k \hat{x}_k (\hat{x}_k' \Lambda_k \hat{x}_k + 1)^{-1} \hat{x}_k' \Lambda_k \tag{2.6d}$$

Convergence of extended least squares algorithms. Convergence theory for the above adaptive estimator is developed in [14, 15], but in essence the convergence conditions consist of a persistently exciting condition identical to that for standard least squares but with x replaced by x_k, noise conditions, and a condition that

$$W(z) = \tfrac{1}{2}I + \theta'\{zI - [F + (G - K)\theta']\}^{-1}(G - K) \tag{2.7}$$

be strictly positive real; equivalently, $W(z)$ is real for real z, $W(z)$ has no poles in $|z| \geq 1$, and $W(e^{j\omega}) + W'(e^{-j\omega}) > 0$ for all real ω.

It is clear that this convergence theory is but a guideline, and in practice simulations or trial application of the algorithms would be advised before full implementation. It is known that when (2.7) fails to hold there could be divergence or vacillation between convergence and divergence.

The available convergence theory of other schemes as in [16–21] is not so well developed as for the schemes described here, and so there is less insight into when the schemes fail.

Let us now turn to an application of these ideas.

Adaptive Kalman Filtering

A common situation in which an adaptive filter is called for is that arising when the input and measurement noise covariances are unknown or slowly varying, while the signal model itself is known. To analyze the situation, we shall, for convenience, consider the signal model with scalar input and output

$$x_{k+1} = Fx_k + Gw_k \qquad z_k = y_k + v_k = H'x_k + v_k \qquad (2.8)$$

with

$$E\left\{\begin{bmatrix} w_k \\ v_k \end{bmatrix} [w_l \quad v_l]\right\} = \begin{bmatrix} Q & S \\ S' & R \end{bmatrix} \delta_{kl}, \qquad E\begin{bmatrix} w_k \\ v_k \end{bmatrix} = 0 \qquad (2.9)$$

Here F, G, and H' are known, while Q, S, and R are unknown. We shall be interested in obtaining a one-step predictor for $\{y_k\}$.

Now the associated innovations model will have a transfer function $H'(zI - F)^{-1}K$ for some K and an input noise covariance Ω. Both K and Ω depend not only on F, G, and H, but also on Q, R, and S. From the point of view of producing a one-step prediction of y_k, the signal model might as well be the innovations model. Moreover, in view of the scalar nature of the innovations model, we can regard it as having transfer function $K'(zI - F')^{-1}H$. Thus the adaptive filtering problem becomes one of the same form as (2.1), where the quantities F, G, K, and θ of (2.1) are replaced by F', 0, H, and K; also u_k in (2.1a) is zero. [The fact that Ω is not known makes no difference to the adaptive equations corresponding to (2.6).]

Now the adaptive algorithms derivable from (2.6) by making the appropriate replacements will converge given satisfaction of a persistently exciting condition, a noise condition, and a positive real type condition. The persistently exciting condition will hold if the state covariance in the innovations model is nonsingular; this is normally the case. Also, the noise condition will normally hold. Corresponding to (2.7), we have the requirement that the following function be positive real:

$$W(z) = \tfrac{1}{2} - K'[zI - (F' - HK')]^{-1}H \qquad (2.10)$$

Obviously, it is real for real z and has no poles in $|z| \geq 1$ should the filter be asymptotically stable, as is normal. It is a remarkable fact that in the important situations when S is known to be zero, we have $W(e^{j\omega}) + W(e^{-j\omega}) > 0$ for all real ω in a great many instances; for virtually any F, G, H, Q, for example, there exists R_0 such that with $R \geq R_0$ the positive real condition holds. Such R_0 may not have to be very large. (See Prob. 2.2.)

What now if the signal model of (2.8) and (2.9) has a vector output? One may still replace the signal model by the innovations model with transfer function $H'(zI - F)^{-1}K$ and input noise covariance Ω. However, no longer do we have $H'(zI - F)^{-1}K = K'(zI - F)^{-1}H$. We can, however, shift the "unknownness" K from the input of the innovations model to the output of another model (thus allowing use of the adaptive filtering ideas) by a technical device. We illustrate this for a 2×2 $H'(zI - F)^{-1}K$. Write

$$H'(zI - F)^{-1} = \begin{bmatrix} w_1'(z) \\ w_2'(z) \end{bmatrix}, \qquad K = [k_1 \quad k_2]$$

where $w_1(z), w_2(z), k_1$, and k_2 are all vectors. Then

$$H'(zI - F)^{-1}K = \begin{bmatrix} w_1'(z)k_1 & w_1'(z)k_2 \\ w_2'(z)k_1 & w_2'(z)k_2 \end{bmatrix}$$

$$= \begin{bmatrix} k_1' & 0 & k_2' & 0 \\ 0 & k_1' & 0 & k_2' \end{bmatrix} \begin{bmatrix} w_1 & 0 \\ w_2 & 0 \\ 0 & w_1 \\ 0 & w_2 \end{bmatrix}$$

Now the matrix θ becomes

$$\theta = \begin{bmatrix} k_1' & 0 & k_2' & 0 \\ 0 & k_1' & 0 & k_2' \end{bmatrix}$$

Because the entries of θ are constrained—some being zero, others fulfilling equality conditions—variation on (2.6) is desirable to reflect the constraints. A complication also arises (which is not present in the scalar measurement case) from the fact that Ω is a matrix and is unknown.

The interested reader is referred to [14] for techniques to handle these aspects of the problem.

Treating the Prediction Error as the Measurements

A mildly more sophisticated extended least squares algorithm with improved convergence properties can be achieved by treating the prediction error as the measurements.

The model (2.1) for the actual measurements z_k followed by the approximate whitening filter (2.6a, b) with output $\hat{v}_k$ is not convenient. Let us consider the scalar measurement case with zero input w_k so as to simply derive an alternative and more convenient model. Interchanging the order of the measurement model and whitening filter, the alternative model generating $\hat{v}_k$ is, after simplifications,

$$\psi_{k+1} = F\psi_k + K(v_k - \hat{\theta}'_k\psi_k) + G\hat{\theta}'_k\psi_k \qquad (2.11a)$$

$$\hat{v}_k = \theta'\psi_k + (v_k - \hat{\theta}'_k\psi_k) \qquad (2.11b)$$

For the model (2.11), the extended least squares ideas of this section can now be applied to yield an adaptive estimator driven from the measurements z_k as

$$\hat{\psi}_{k+1} = F\hat{\psi}_k + K(\hat{v}_k - \hat{\theta}'_k\hat{\psi}_k) + G\hat{\theta}'_k\hat{\psi}_k \qquad (2.12a)$$

$$\hat{x}_{k+1} = F\hat{x}_k + G\hat{\theta}'_k\hat{x}_k + K\hat{v}_k \qquad (2.12b)$$

$$\hat{v}_k = z_k - \hat{\theta}'_k\hat{x}_k \qquad (2.12c)$$

$$\hat{\theta}_k = \hat{\theta}_{k-1} + \Lambda_{k+1}\hat{\psi}_k(z_k - \hat{\psi}'_k\hat{\theta}_{k-1}) \qquad (2.12d)$$

$$\Lambda_{k+1} = \Lambda_k - \Lambda_k\hat{\psi}_k(\hat{\psi}'_k\Lambda_k\hat{\psi}_k + 1)^{-1}\hat{\psi}'_k\Lambda_k \qquad (2.12e)$$

This algorithm involves an additional state update equation and thus requires more computational effort, but experience shows that when it converges its convergence is more rapid than that for the standard extended least squares algorithm. To ensure convergence, it is usually necessary to test that $\lambda_i[F + (G - K)\hat{\theta}'_k] < 1$ for all i at each k, and if this is not the case, then the step size must be reduced by replacing $\Lambda_{k+1}\psi_k$ by $\frac{1}{2}\Lambda_{k+1}\psi_k, \frac{1}{4}\Lambda_{k+1}\psi_k$, et cetera, until the condition is satisfied.

The algorithm is essentially equivalent to the recursive maximum likelihood recursions of [26, 27]. The convergence analysis reported in [27] does not require a positive real condition on the model to be satisfied. It is also asymptotically equivalent to schemes achieved by application of extended Kalman filtering theory as now described.

Adaptive Estimation via Extended Kalman Filtering

Another approach to recursive state and parameter estimation, requiring additional computational effort, is to view the parameters as additional states of the signal model, and apply extended Kalman filtering algorithms to the augmented nonlinear model. The details of such an approach are a straightforward application of the extended Kalman filter theory of Sec. 8.2, and are left to the reader as an exercise.

Although this extended Kalman filter approach appears perfectly straightforward, experience has shown that with the usual state space model, it does not work well in practice. For an augmented innovations model, however, in which

$$\theta_{k+1} = \theta_k$$
$$x_{k+1} = F_k(\theta)x_k + K_k(\theta)v_k$$
$$z_k = H'_k(\theta)x_k + v_k$$

for an innovations process v_k, the linearization very naturally involves the term

$$\left.\frac{\partial K_k(\theta)}{\partial \theta} v_k\right|_{x_k = \hat{x}_k, \theta = \theta_k, v_k = \hat{v}_k}$$

For a convergence analysis, [28] has shown that the presence of this term is crucial. A simple way to make sure that $\frac{\partial K_k}{\partial \theta}(\theta)$ can be readily calculated is to include all the elements of K in θ. To ensure convergence in practice, the algorithm may require step-size reductions or other heuristics to force the poles of the filter to lie within the unit circle at each time instant as $k \to \infty$.

Main Points of the Section

For a broad class of state-space signal models with unknown parameters, the parameters can be estimated via least squares algorithms conditioned on the states being measurable and the states can be estimated conditioned on knowledge of the parameters. By a simultaneous state and parameter estimation, but with the parameter [state] estimator using state [parameter] estimates rather than the true estimates, very useful adaptive estimators can be constructed. A crucial condition for convergence of the adaptive estimator to the true optimal filter, designed given knowledge of the signal model parameters, is that a certain system derived from the signal model be positive real. The ideas can be applied to yield an adaptive Kalman filter when the noise covariance matrices are unknown but the remainder of the signal model is known. The ideas can be applied to signal models in which the prediction errors are taken as the measurements. Algorithms with improved convergence properties result at the cost of additional computational effort. Related to these algorithms are those achieved using extended Kalman filtering for an innovations signal model. For these, the unknown parameters are treated as states.

Problem 2.1. Derive the least squares identification algorithms (2.2) through (2.4). Treat also the case when

$$z_k = X_k'\theta + v_k$$

where X_k is a Z_k-measurable matrix, θ is an unknown parameter vector, and the index is

$$\sum_{i=0}^{k} (z_i - X_i'\theta)'\hat{\Omega}_k^{-1}(z_i - X_i'\theta)$$

where $\hat{\Omega}_k$ is an estimate of $\Omega = E[v_k v_k']$. Show that

$$\hat{\theta}_k = (\sum_{i=0}^{k} X_i \hat{\Omega}_k^{-1} X_i')^{-1} \sum_{i=0}^{k} X_i \hat{\Omega}_k^{-1} z_i$$

$$\hat{\theta}_k = \hat{\theta}_{k-1} + P_{k+1} X_k \hat{\Omega}_k^{-1}(z_k - X_k'\hat{\theta}_{k-1})$$

$$P_{k+1} = P_k - P_k X_k (X_k' P_k X_k + \hat{\Omega}_k)^{-1} X_k' P_k$$

[In practice, $\hat{\Omega}_k$ can be set to I or an estimate of Ω derived from the residuals $\hat{v}_k = (z_k - X'_k\hat{\theta}_k)$. Such an estimate might be

$$\hat{\Omega}_k = \frac{1}{k} \sum_{i=0}^{k} \hat{v}_i\hat{v}'_i$$

The advantages of employing different estimates $\hat{\Omega}$ is a separate study which we do not explore here.]

Problem 2.2. The usual state space signal model $\{F, G, H, Q, R\}$ leads to a model (2.5) with $K = \Sigma H(H'\Sigma H + R)^{-1}$, conclude that if F, G, H, Q are held constant and R increases, then $K \longrightarrow 0$. Thus show that if $|\lambda_i(F)| < 1$, the transfer function $W(z)$ in (2.10) will be positive real if F, G, H, Q are held constant and R is large enough.

REFERENCES

[1] Box, G. E., and G. M. Jenkins, *Time-series Analysis, Forecasting and Control*, Holden-Day, Inc., San Francisco, 1970.

[2] Sims, F. L., D. G. Lainiotis, and D. T. Magill, "Recursive Algorithm for the Calculation of the Adaptive Kalman Filter Weighting Coefficients," *IEEE Trans. Automatic Control*, Vol. AC-14, No. 2, April 1969, pp. 215–218.

[3] Lainiotis, D. G., "Optimal Adaptive Estimation: Structure and Parameter Adaptation," *IEEE Trans. Automatic Control*, Vol. AC-16, No. 2, April 1971, pp. 160–169.

[4] Tam, P., and J. B. Moore, "Adaptive Estimation Using Parallel Processing Techniques," *Computers in Electrical Engineering*, Vol. 2, Nos. 2/3, June 1975.

[5] Hawkes, R. M., and J. B. Moore, "Performance Bounds for Adaptive Estimation," *Proc. IEEE*, Vol. 64, No. 8, August 1976, pp. 1143–1151.

[6] Hawkes, R. M., and J. B. Moore, "Performance of Bayesian Parameter Estimators for Linear Signal Models," *IEEE Trans. Automatic Control*, Vol. AC-21, No. 4, August 1976, pp. 523–527.

[7] Anderson, B. D. O., J. B. Moore, and R. M. Hawkes, "Model Approximation via Prediction Error Identification," *Automatica*, to appear.

[8] Kullback, S., *Information Theory and Statistics*, John Wiley & Sons, Inc., 1959.

[9] Hawkes, R. M., and J. B. Moore, "Adaptive Estimation via Sequential Processing," *IEEE Trans. Automatic Control*, February 1975, Vol. AC-20, No. 1, February 1975, pp. 137–138.

[10] Willsky, A., "A Generalized Likelihood Ratio Approach to State Estimation in Linear Systems Subject to Abrupt Changes," *Proc. 1974 IEEE Conf. on Decision and Control*, Phoenix, Ariz., pp. 846–853.

[11] Kudva, P., and K. S. Narendra, "An Identification Procedure for Discrete Multivariable Systems," *IEEE Trans. Automatic Control*, Vol. AC-19, No. 5, October 1974, pp. 549–552.

[12] KUDVA, P., and K. S. NARENDRA, "The Discrete Adaptive Observer," *Proc. 1974 IEEE Conf. on Decision and Control*, Phoenix, Ariz, pp. 307–312.

[13] LANDAU, I. D., "Unbiased Recursive Identification Using Model Reference Adaptive Techniques," *IEEE Trans. Automatic Control*, Vol. AC-21, No. 2, April 1976, pp. 194–203.

[14] MOORE, J. B. and G. LEDWICH, "Multivariable Adaptive Parameter and State Estimators with Convergence Analysis," submitted for publication.

[15] LJUNG, L., "Convergence of an Adaptive Filter Algorithm," submitted for publication.

[16] LEONDES, C. T., and J. O. PEARSON, "Kalman Filtering of Systems with Parameter Uncertainties—A Survey," *Int. J. Control*, Vol. 17, 1973, pp. 785–792.

[17] CAREW, B., and P. R. BELANGER, "Identification of Optimal Filter Steady-state Gain for Systems with Unknown Noise Covariance," *IEEE Trans. Automatic Control*, Vol. AC-18, No. 6, December 1973, pp. 582–587.

[18] MEHRA, R. K., "On the Identification of Variances and Adaptive Kalman Filtering," *IEEE Trans. Automatic Control*, Vol. AC-15, No. 2, April 1970, pp. 175–184.

[19] MEHRA, R. K., "On-line Identification of Linear Dynamic Systems with Applications to Kalman Filtering," *IEEE Trans. Automatic Control*, Vol. AC-16, No. 1, February 1971, pp. 12–21.

[20] BENNETT, R. J., "Nonstationary Parameter Estimation for Small Sample Situations: A Comparison of Methods," *Int. J. Systems Sci.*, Vol. 7, No. 3, 1976, pp. 257–275.

[21] SONDHI, M. M., and D. MITRA, "New Results on the Performance of a Well-known Class of Adaptive Filters," *Proc. IEEE*, Vol. 63, No. 11, November 1976, pp. 1583–1597.

[22] MANN, H. B., and A. WALD, "On the Statistical Treatment of Linear Stochastic Difference Equations," *Econometrica*, Vol. 11, 1943, pp. 173–220.

[23] LJUNG, L., "Consistency of the Least Squares Identification Method," *IEEE Trans. Automatic Control*, Vol. AC-21, No. 15, October 1976, pp. 779–780.

[24] MOORE, J. B., "On Strong Consistency of Least Squares Identification Algorithms," *Automatica*, Vol. 14, No. 5, September 1978.

[25] TSYPKIN, Y. Z., *Foundations of the Theory of Learning Systems*, Academic Press, Inc., New York, 1973.

[26] SÖDERSTROM, T., "An On-line Algorithm for Approximate Maximum Likelihood Identification of Linear Dynamic Systems," *Report 7308, Div. of Automatic Control*, Lund Inst. of Tech., Sweden, 1973.

[27] SÖDERSTROM, T., L. LJUNG and I. GUSTAVSSON, "A Comparative Study of Recursive Identification Algorithms," *Report 7427, Div. of Automatic Control*, Lund Inst. of Tech., Sweden, 1974.

[28] LJUNG, L., "The Extended Kalman Filter as a Parameter Estimator for Linear Systems," *Technical Report LiTH-ISY-I-0.154*, University of Linköping, Sweden, 1977.

CHAPTER 11

COLORED NOISE

AND SUBOPTIMAL

REDUCED ORDER FILTERS

11.1 GENERAL APPROACHES TO DEALING
WITH COLORED NOISE

In virtually all the Kalman filter theory that has gone before, we have assumed that the measurement and input noise processes have been white. We consider in this chapter what should be done when this assumption fails. As it turns out, optimal handling of the situation is normally possible, though this is generally at the expense of increased complexity of the filter. Therefore, we become interested in replacing optimal filters by less complex, suboptimal ones. Such a replacement may of course be of interest independently of whether the noises are white, and so we consider the general question of suboptimal filter design using reduced dimension filters in the last section of the chapter.

Optimal Filter Design with Colored Noise

The material on covariance factorization of Chapter 9 provides the tool for handling colored noise in an optimal fashion, as we shall now argue. Suppose that the usual signal model conditions apply, save that $\{v_k\}$ and $\{w_k\}$, the measurement and input noise, are not white. For convenience, suppose they are independent with $E[v_k v_l']$ and $E\{w_k w_l'\}$ known and given in the "separable" form

$$E[v_k v_l'] = A_k' B_l \qquad k \geq l$$

and similarly for $E[w_k w_l']$. Then we can construct finite-dimensional systems $\mathcal{S}_1$ and $\mathcal{S}_2$ with inputs white noise processes, $\{\xi_k\}$ and $\{\eta_k\}$ say, and outputs which are realizations of $\{v_k\}$ and $\{w_k\}$. To do this, we use the covariance factorization ideas of Chap. 9; in case $\{v_k\}$ and $\{w_k\}$ are stationary, we can ensure that the finite-dimensional systems just referred to are time invariant. Alternatively, of course, models $\mathcal{S}_1$ and $\mathcal{S}_2$ may be part of the a priori data.

The composite of the original signal model $\mathcal{S}$ and the finite-dimensional systems $\mathcal{S}_1$ and $\mathcal{S}_2$ together form a single linear system $\bar{\mathcal{S}}$ with white noise inputs $\{\xi_k\}$ and $\{\eta_k\}$. The usual Kalman filter can be obtained for $\bar{\mathcal{S}}$; part of the state vector of the Kalman filter will comprise an estimate of the state of $\mathcal{S}$, and a submatrix of the filter error covariance matrix will be the error covariance matrix associated with estimating the state of $\mathcal{S}$. The estimate is of course an optimal one.

In principle then, there is no difficulty about dealing with colored noise. The practical difficulty is, however, that the filter dimension will be the sum of the dimensions of $\mathcal{S}$, $\mathcal{S}_1$, and $\mathcal{S}_2$, and may therefore be uncomfortably high. Accordingly, we need to consider approaches to reduce the filter dimension.

In Secs. 11.2 and 11.3, we concentrate on using ideas stemming from properties of the output noise. Specifically, we show in Sec. 11.2 that if the output noise is Markov, then the optimal filter need be of no higher dimension than if the output noise is white. In Sec. 11.3, we show that if the measurement noise covariance is singular, the optimal filter dimension can be reduced below the usual dimension; it follows that if the measurement noise covariance is nearly singular, a suboptimal filter follows by designing as if the covariance was singular.

In Sec. 11.4, we discuss procedures for suboptimal filter design for colored input or measurement noise (or both). The noise processes are not assumed to be Markov, and the filters are of dimension equal to the dimension of the state vector of the basic signal model, i.e., the dimension which would apply were the noise processes white.

Finally, Sec. 11.5 discusses procedures for lowering the filter dimension, even when the noise processes are white.

A number of methods are heuristic, or only partly justified by the arguments to be presented. Accordingly, it is imperative *that any suboptimal filter design should have its performance compared with that of the optimal filter.* In many of the suboptimal filters presented, the performance of the suboptimal filter is evaluated as part of the design process. In virtually all of them, the performance can be calculated; it can also be determined via simulations. Incidentally, the reader should be reassured that examples as outlined in the references do verify the utility of the methods.

Filter design with Markov measurement noise is discussed in [1], whose authors refer to an original treatment in [2]. Filter design with singular measurement noise covariance is discussed in [3–7], with [5–7] emphasizing the notion that the design procedure is also a valid technique for obtaining reduced order filters when the measurement noise covariance is not singular. The ideas of Sec. 11.4 are an amalgam of various ideas scattered through [3–7], sometimes in an underdeveloped fashion; in Sec. 11.5, we carry the ideas of Sec. 11.4 forward with the aid of an approach suggested in [8].

Main Points of the Section

If the input or measurement noises are colored, one models them as the output of a linear finite-dimensional system excited by white noise. One builds a Kalman filter for the linear system comprising the original signal model and the noise models. Suboptimal filter design may be needed to reduce the dimension of this filter.

Problem 1.1 Will the ideas of this section carry through if $\{v_k\}$, $\{w_k\}$ are colored and dependent?

11.2 FILTER DESIGN WITH MARKOV OUTPUT NOISE

It turns out that when the output noise $\{v_k\}$ is Markov, there is no need to increase the dimension of the optimal filter. In this section, we shall illustrate how the filter may be found in this instance. Thus we suppose that

$$x_{k+1} = F_k x_k + G_k w_k \qquad k \geq 0 \tag{2.1a}$$

$$z_k = H'_k x_k + v_k \tag{2.1b}$$

$$v_{k+1} = A_k v_k + \eta_k \qquad k \geq 0 \tag{2.2}$$

Here, x_0, v_0, $\{\eta_k\}$, and $\{w_k\}$ are independent and gaussian. We have x_0 as $N(\bar{x}_0, P_0)$, v_0 as $N(0, R_0)$, $\{\eta_k\}$ as white with η_k as $N[0, \Xi_k]$, and $\{w_k\}$ as white with w_k as $N[0, Q_k]$.

Observe now that

$$z_{k+1} - A_k z_k = H'_{k+1} x_{k+1} - A_k H'_k x_k + v_{k+1} - A_k v_k$$
$$= (H'_{k+1} F_k - A_k H'_k) x_k + \eta_k + H'_{k+1} G_k w_k$$

Define this quantity as $\bar{z}_{k+1}$ for $k \geq 0$, and think of (2.1b) as being replaced by

$$\bar{z}_{k+1} = \bar{H}'_k x_k + \eta_k + \bar{G}_k w_k \qquad k \geq 0 \tag{2.3a}$$

where $\bar{H}'_k = H'_{k+1} F_k - A_k H'_k$ and $\bar{G}_k = H'_{k+1} G_k$. Also set $\bar{z}_0 = z_0$, so that

$$\bar{z}_0 = H'_0 x_0 + v_0 = z_0 \tag{2.3b}$$

Standard Kalman filter theory then allows us to compute $E[x_k | \bar{Z}_k]$ for each k, using a filter of dimension equal to the dimension of x_k. (Actually a minor deviation from standard procedures is required at $k = 0$.)

The definition of $\{\bar{z}_k\}$ shows that $\bar{Z}_k$ is computable from Z_k, and Z_k from $\bar{Z}_k$. Therefore $E[x_k | \bar{Z}_k] = E[x_k | Z_k]$. This means that true filtered estimates (rather than one-step-ahead prediction estimates) of x_k are produced by the Kalman filter.

From (2.1a) and (2.3a), we see that the filter will have the form

$$\hat{x}_{k+1/k+1} = F_k \hat{x}_{k/k} + K_k[\bar{z}_{k+1} - \bar{H}'_k \hat{x}_{k/k}] \qquad (2.4a)$$

for $k = 0, 1, 2, \ldots$; from (2.3b) we see that one must take

$$\hat{x}_{0/0} = P_0 H_0 (H'_0 P_0 H_0 + R_0)^{-1} z_0 \qquad (2.4b)$$

to initialize the filter. The precise sequence $\{K_k\}$ can be found by standard procedures, with a derivation being called for in the problems. What is of interest here is how one should implement (2.4a) once K_k has been found. Rewriting (2.4a) as

$$\hat{x}_{k+1/k+1} = F_k \hat{x}_{k/k} - K_k A_k z_k - K_k \bar{H}'_k \hat{x}_{k/k} + K_k z_{k+1} \qquad (2.4c)$$

allows us to see that the arrangement of Fig. 11.2-1 illustrates an implementation of (2.4c). The dimension of the linear system of Fig. 11.2-1 is the same as that of x_k; in the figure, the various quantities are those present at time k. The input to the delay is easily checked to be

$$F_k \hat{x}_{k/k} - K_k A_k z_k - K_k \bar{H}'_k \hat{x}_{k/k}$$

at time k, so its output at time k is

$$F_{k-1} \hat{x}_{k-1/k-1} - K_{k-1} A_{k-1} z_{k-1} - K_{k-1} \bar{H}'_{k-1} \hat{x}_{k-1/k-1}$$

Examination of the summing node at the delay output then allows recovery of (2.4c) with k replaced by $k - 1$.

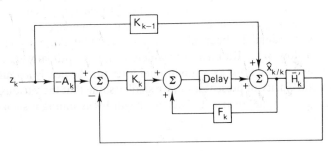

Fig. 11.2-1   Filter structure when measurement noise is Markov.

Two other minor points should be noted.

1. If v_k were the output of a linear finite-dimensional system excited by white noise, one could not carry through the above derivation unless

v_k was the same as the state of this system, and therefore possessed the Markov property.

2. The measurement noise in (2.3a) is $\eta_k + \bar{G}_k w_k$ and is not independent of the input noise in (2.1a). Therefore, the somewhat more complex formulas for the filter gain and error covariance applicable for dependent noises must be used.

Main Points of the Section

When the output noise is Markov, one can redefine the measurement equation to make the new measurement noise white. The Kalman filter dimension is, therefore, not increased.

Problem 2.1 Develop formulas for the gain matrix K_k and the associated error covariance of the filter of this section. Check what happens when $A_k = 0$.

Problem 2.2 Let $\{v_k\}$ be an autoregressive process. Show that the ideas of this section can be extended to yield a Kalman filter driven by a linear combination of z_k and past measurements.

11.3 FILTER DESIGN WITH SINGULAR OR NEAR-SINGULAR OUTPUT NOISE

In this section, we shall argue that when the output noise covariance matrix R_k has nullity m (i.e., has m zero eigenvalues) for all k, we can reduce the dimension of the Kalman filter by m. The heuristic reasoning for this is that when R_k has nullity m, there are m linear functionals of x_k known precisely once z_k is known. There is then no need to estimate them.

When the output noise covariance is nearly singular, we can derive a low order suboptimal Kalman filter by assuming that the noise is actually singular. The low order Kalman filter which would be optimal were the noise actually singular functions as a suboptimal filter for the nearly singular case.

To retain clarity of presentation, we shall assume that the various parameter matrices are time invariant. Thus we begin by assuming a signal model of the form

$$\bar{x}_{k+1} = \bar{F}\bar{x}_k + \bar{G}w_k \tag{3.1a}$$

$$\bar{z}_k = \bar{H}'\bar{x}_k + \bar{v}_k \tag{3.1b}$$

We shall assume independence and the gaussian property for $\bar{x}_0$, $\{w_k\}$, $\{\bar{v}_k\}$, the usual whiteness and zero mean assumptions for $\{w_k\}$ and $\{\bar{v}_k\}$, and $E[\bar{v}_k \bar{v}_k'] = \bar{R}$, $E[w_k w_k'] = Q$. We further assume that $\bar{R}$ has nullity m, that $\bar{z}_k$ has dimension p, and that $\bar{x}_k$ has dimension $n \geq m$.

To derive the reduced order Kalman filter, we shall adopt the following strategy:

1. We shall introduce coordinate basis changes of the output and state spaces so that part of the new output vector becomes identical with part of the new state vector.
2. We shall show how the remainder of the state vector can be estimated via a dynamical system of dimension m less than the usual Kalman filter dimension.

Coordinate basis changes. First, we set up a new output $\{z_k\}$ such that the first m entries of $\{z_k\}$ contain perfect measurements. Let T be a nonsingular matrix such that

$$T\bar{R}T' = \begin{bmatrix} 0_m & 0 \\ 0 & I_{p-m} \end{bmatrix} \tag{3.2}$$

Such T can be found by standard devices of linear algebra. Then set

$$z_k = T\bar{z}_k \qquad v_k = T\bar{v}_k \tag{3.3}$$

The first $m \times m$ entries of $\{v_k\}$ will be zero, by virtue of (3.2). Thus with

$$z_k = T\bar{H}'\bar{x}_k + v_k \tag{3.4}$$

the first m entries of $T\bar{H}'\bar{x}_k$ will be known exactly.

Next, we shall arrange for the first m entries of z_k to be the first m entries of the state vector. We must make the assumption that $\bar{H}'$ has rank equal to the number of its rows p. (This is inessential; if it were not the case, certain linear combinations of measurement vector components would be independent of $\bar{x}_k$ and could be thrown away.) Accordingly, let $\bar{H}'$ have rank equal to p. Define a nonsingular matrix S of dimension $n \times n$ by

$$S = \begin{bmatrix} T\bar{H}' \\ S_2 \end{bmatrix} \tag{3.5}$$

where S_2 is any matrix chosen to make S nonsingular. Then set

$$x_k = S\bar{x}_k \tag{3.6}$$

There results, with $F = S\bar{F}S^{-1}$, $G = S\bar{G}$,

$$x_{k+1} = Fx_k + Gw_k \tag{3.7a}$$

$$z_k = \begin{bmatrix} x_k^1 \\ [I_{p-m} \quad 0]x_k^2 \end{bmatrix} + \begin{bmatrix} 0 \\ v_k^2 \end{bmatrix} \tag{3.7b}$$

The effect of the coordinate basis change is to allow the first m entries of x_k, namely x_k^1, to be estimated without error from the first m entries, call them z_k^1, of z_k. The remaining $(p - m)$ entries of z_k are noisy measurements of the first $(p - m)$ entries of x_k^2, with x_k^2 denoting the last $(n - m)$ entries of x_k.

Estimation of x_k^2. We could write down the usual Kalman filter equation associated with (3.7). It has the form (using true filtered estimates)

$$\hat{x}_{k+1/k+1} = (I - LH')F\hat{x}_{k/k} + Lz_{k+1} \qquad (3.8)$$

Here $H' = [I_p \,\vdots\, 0]$. The first m rows of this equation must yield $\hat{x}_{k+1/k+1}^1 = z_{k+1}^1$. Therefore, for some L_2,

$$
\begin{array}{cc}
\overset{m}{\overbrace{}} & \overset{p-m}{\overbrace{}}
\end{array}
$$
$$
L = \begin{bmatrix} I_m & 0 \\ L_2 & \end{bmatrix} \begin{array}{l} \} m \\ \} n-m \end{array}
$$
$$
\underset{p}{\underbrace{}}
$$

The last $(n - m)$ rows of (3.8) yield, on identifying $\hat{x}_{k/k}^1$ with z_k^1, an equation of the form

$$\hat{x}_{k+1/k+1}^2 = A\hat{x}_{k/k}^2 + B_1 z_{k+1} + B_2 z_k \qquad (3.9)$$

This is implementable with an $(n - m)$-dimensional linear system, as shown in Fig. 11.3-1.

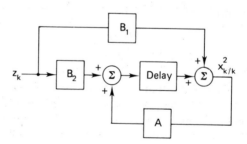

Fig. 11.3-1   Implementation of filter of equation (3.9).

Estimation of $\{\bar{x}_k\}$ from $\{\bar{z}_k\}$. It simply remains to undo the effect of the coordinate basis changes. This is best shown diagrammatically, see Fig. 11.3-2.

Before discussing the application of these ideas to suboptimal filter design problems, let us make several comments.

1. If the signal model is time varying or the noise processes non-stationary, the calculation of the basis change matrices T and S can be

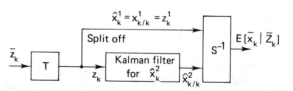

Fig. 11.3-2   Reduced order Kalman filter with singular output noise covariance.

tedious; by following more precisely the procedure of, say, [3], some of the computational burden can be cut down.

2. Some of the calculations are very reminiscent of those used in Luenberger estimator design [9], where one assumes no noise is present and accordingly can obtain an observer of dimension equal to $\dim x - \dim z = n - p$. The Luenberger observer equations are of the form (in the absence of external inputs)

$$q_{k+1} = Aq_k + Bz_k \tag{3.10a}$$

$$\hat{x}_{k/k} = Cq_k + Dz_k \tag{3.10b}$$

and q_k has dimension equal to $n - p$. If we allow q_k to have dimension $n - m$, then one can check that (3.9) and (3.10a) have the same form after manipulation.

3. The results of this section also give new perspective on the treatment of Markov measurement noise of the last section. Suppose that in (3.1), $\bar{v}_k$ is neither white nor has a singular covariance matrix but rather is described by

$$\bar{v}_{k+1} = A\bar{v}_k + \xi_k \tag{3.11}$$

for some white noise process ξ_k. Then we may regard $[\bar{x}'_k \ \ \bar{v}'_k]'$ as a new state vector [evolving according to (3.1) and (3.11)]; also, we may regard (3.1b) as stating that we have perfect measurements of a number of linear functionals of the state vector. By taking advantage of the fact that these measurements are perfect, we can derive an optimal filter of dimension equal to $\dim x_k$. This is of course what we found in the preceding section, by a different argument.

Suboptimal Filter with Nonsingular Measurement Noise Covariance

Suppose now that $E[\bar{v}_k\bar{v}'_k]$ is no longer singular. Let us suppose that the first m entries of $T\bar{z}_k$ are much more accurately known than the remaining entries. Then we can modify the previous ideas to obtain a suboptimal estimator of dimension equal to $n - m$. We follow the coordinate basis change ideas as before, but now obtain in lieu of (3.7b) the measurement equation:

$$z_k = \begin{bmatrix} x^1_k \\ [I_{p-m} \quad 0]x^2_k \end{bmatrix} + \begin{bmatrix} v^1_k \\ v^2_k \end{bmatrix} \tag{3.7c}$$

Here, $\{v^1_k\}$ has a smaller covariance matrix than $\{v^2_k\}$ and is independent of $\{v^2_k\}$.

Using (3.1) and (3.7c), it is easy to compute a joint probability density for x^1_k and z_k, and thence to derive $E[x^1_k | z_k] = L_1 z_k$ for some matrix L_1 computable from the covariance data. (Minor modification is necessary in case $E[x_k] \neq 0$.) We take $L_1 z_k$ as a suboptimal estimate of x^1_k, in lieu of $\hat{x}_{k/k} = E[x^1_k | Z_k]$. The error covariance of the suboptimal estimate is easily found.

Now in order to estimate $\hat{x}_{k/k}$ optimally, one would write down the Kalman filter equation and examine its last $(n - m)$ rows. This would yield

$$\hat{x}^2_{k+1/k+1} = A\hat{x}^2_{k/k} + B\hat{x}^1_{k/k} + Cz_{k+1} \qquad (3.12)$$

We implement instead of this optimal equation the following suboptimal equation:

$$\hat{x}^2_{k+1/k+1} = A\hat{x}^2_{k/k} + BL_1 z_k + Cz_{k+1} \qquad (3.13)$$

Conversion to the original coordinate basis proceeds as before. The arrangement of Fig. 11.3-2 still applies, save that the filter and estimates are suboptimal.

Main Points of the Section

When the measurement noise covariance matrix has nullity m, m linear functionals of the state are known exactly and the filter dimension can be reduced by an amount m. When the covariance matrix is nonsingular, the same ideas apply to yield suboptimal reduced dimension filters.

Problem 3.1 Suppose that in (3.1b), the noise process v_k is the output of the following system:

$$\xi_{k+1} = A\xi_k + B\eta_k$$
$$v_k = C\xi_k$$

Show that one can obtain an optimal filter of dimension equal to dim x + dim ξ − dim v.

Problem 3.2 Suppose that (3.1a) is replaced by $\bar{x}_{k+1} = \bar{F}\bar{x}_k + \bar{G}w_k + \Gamma u_k$ with $\{u_k\}$ a known sequence. Discuss the changes to the preceding theory.

11.4 SUBOPTIMAL DESIGN GIVEN COLORED INPUT
OR MEASUREMENT NOISE

In this section, we shall show that an idea developed originally in Chap. 3 can be used to generate a *suboptimal* filter design when either the input or measurement noise is colored. We demand a priori that the filter have the structure the optimal estimator would have were the input and measurement noise white. Thus the suboptimal filter is defined except for the filter gain matrix sequence. Then we try to find an optimal gain sequence. The idea of finding the best estimator within a class which is fixed a priori has been used earlier. In Chap. 3, we adopted such an approach in studying estimation with nongaussian noises and/or initial state.

We now fill out the above idea. Suppose the signal model is

$$x_{k+1} = F_k x_k + w_k \quad k \geq 0 \tag{4.1a}$$

$$z_k = H'_k x_k + v_k \tag{4.1b}$$

with $\{v_k\}$, $\{w_k\}$, and x_0 independent and gaussian. The process $\{v_k\}$ has zero mean and $E[v_k v'_l] = R_k \delta_{kl}$; the process $\{w_k\}$ is not white, but rather is the output of the following system:

$$\zeta_{k+1} = A_k \zeta_k + B_k \eta_k \quad k \geq 0 \tag{4.2}$$

$$w_k = C'_k \zeta_k \tag{4.3}$$

with $E[\eta_k \eta'_l] = I \delta_{kl}$ and $E[\eta_k] = 0$. (The matrix B_k can be used to accommodate changes in the covariance of η_k.) The initial state ζ_0 is $N(0, \Pi_0)$ and is independent of $\{\eta_k\}$. The initial state x_0 of (4.1) is $N(\bar{x}_0, P_0)$. The reason why no G_k appears in (4.1a) is that it can be taken up in C'_k in (4.3).

An optimal estimator would have dimension equal to the sum of the dimension of x_k and ζ_k. We shall, however, assume an estimator of dimension equal to the dimension of x_k.

Guided by the situation applicable when $\{w_k\}$ is white, we postulate that the estimator structure is to be of the form

$$\hat{x}_{0/0} = \hat{x}_{0/-1} + K_0(z_0 - H'_0 \hat{x}_{0/-1})$$

$$\hat{x}_{k+1/k+1} = F_k \hat{x}_{k/k} + K_{k+1}(z_{k+1} - H'_{k+1} F_k \hat{x}_{k/k}) \quad k \geq 0 \tag{4.4}$$

We seek K_k for each k to give the smallest possible value for the error covariance $\Sigma_{k+1/k+1}$ for a given value of $\Sigma_{k/k}$. More precisely, for given $\Sigma_{0/-1}$ we seek K_0^* such that

$$\Sigma_{0/0}(K_0^*) \leq \Sigma_{0/0}(K_0) \quad \text{for all } K_0 \tag{4.5a}$$

Then with K_0^* fixed, we seek K_1^* to minimize the resulting $\Sigma_{1/1}$:

$$\Sigma_{1/1}(K_0^*, K_1^*) \leq \Sigma_{1/1}(K_0^*, K_1) \quad \text{for all } K_1 \tag{4.5b}$$

More generally, we determine K_k^* such that

$$\Sigma_{k/k}(K_0^*, K_1^*, \ldots, K_{k-1}^*, K_k^*) \leq \Sigma_{k/k}(K_0^*, K_1^*, \ldots, K_{k-1}^*, K_k) \quad \text{for all } K_k \tag{4.5c}$$

Digression. It is important to note that when $\{w_k\}$ is colored, contrary to intuition *the sequence $K_0^*, \ldots, K_k^*$ may not minimize $\Sigma_{k/k}$ in the sense that*

$$\Sigma_{k/k}(K_0^*, K_1^*, \ldots, K_{k-1}^*, K_k^*) \leq \Sigma_{k/k}(K_0, K_1, \ldots, K_k) \tag{4.6}$$

for all $K_0, \ldots, K_k$. When the noise is white, however, then it is true that (4.5) implies (4.6). For when the noise $\{w_k\}$ is white, ordering properties on $\Sigma_{k/k}$ *propagate*, in the sense that if for two sequences $K_0^i, \ldots, K_{k-1}^i$ $(i = 1, 2)$, one has

$$\Sigma^1_{k-1/k-1}(K_0^1, \ldots, K_{k-1}^1) \geq \Sigma^2_{k-1/k-1}(K_0^2, \ldots, K_{k-1}^2)$$

then for all K_k

$$\Sigma_{k/k}(K_0^1, \ldots, K_{k-1}^1, K_k) = \Sigma_{k/k}(\Sigma_{k-1/k-1}^1, K_k)$$

$$\geq \Sigma_{k/k}(\Sigma_{k-1/k-1}^2, K_k)$$

$$= \Sigma_{k/k}(K_0^2, \ldots, K_{k-1}^2, K_k)$$

(The calculation is not hard to check.) Then if $K_0^2, \ldots, K_{k-1}^2$ is replaced by $K_0^*, \ldots, K_{k-1}^*$, it follows that

$$\Sigma_{k/k}(K_0^1, \ldots, K_{k-1}^1, K_k) \geq \Sigma_{k/k}(K_0^*, \ldots, K_{k-1}^*, K_k)$$

$$\geq \Sigma_{k/k}(K_0^*, \ldots, K_{k-1}^*, K_k^*)$$

for arbitrary $K_0^1, \ldots, K_{k-1}^1, K_k$. Thus the "propagation of ordering" property and the definition of (4.5) combine to ensure (4.6). When $\{w_k\}$ is not white, there is no guaranteed propagation of the ordering properties. This will be clearer when the actual equation for $\Sigma_{k/k}$ is presented below. Of course, though we are not achieving a global minimum for $\Sigma_{k/k}$ by our procedure for selecting $K_0^*, K_1^*, \ldots, K_{k-1}^*, K_k^*$, there is nevertheless some intuitive appeal about the choice of K_k^* sequence.

Let us now turn to the details of the calculations. We shall see that the calculations involve two steps: deriving the recursive equation for $\Sigma_{k/k}$ and finding the minimizing K_k. The details are not particularly important, though the method is.

The calculation of $\Sigma_{k/k}$ proceeds by defining a linear system with white noise input and with state covariance matrix containing $\Sigma_{k/k}$ as a submatrix. This is an important method of approaching error covariance calculation, but its importance is often not emphasized in the literature. Thus, frequently calculations of quantities like $\Sigma_{k/k}$ appear to depend on the application of much ingenuity, rather than the application of a basic technique.

Calculation of $\Sigma_{k/k}$ for Known Filter Gain Sequence

From (4.1) and (4.4), we have (with $\tilde{x}_{k+1} = x_{k+1} - \hat{x}_{k+1/k+1}$)

$$\tilde{x}_0 = (I - K_0 H_0')(x_0 - \hat{x}_{0/-1}) - K_0 v_0 \tag{4.7a}$$

$$\tilde{x}_{k+1} = (I - K_{k+1} H_{k+1}')(F_k \tilde{x}_{k/k} + w_k) - K_{k+1} v_{k+1} \qquad k \geq 0 \tag{4.7b}$$

On combining this with (4.2) and (4.3), we have, with $D_k' = (I - K_{k+1} H_{k+1}')C_k'$

$$\begin{bmatrix} \tilde{x}_{k+1} \\ \zeta_{k+1} \end{bmatrix} = \begin{bmatrix} (I - K_{k+1}H_{k+1}')F_k & D_k' \\ 0 & A_k \end{bmatrix} \begin{bmatrix} \tilde{x}_k \\ \zeta_k \end{bmatrix} + \begin{bmatrix} 0 & -K_{k+1} \\ B_k & 0 \end{bmatrix} \begin{bmatrix} \eta_k \\ v_{k+1} \end{bmatrix}$$

This is a linear system with white noise excitation, and so the state covariance can be found recursively. Thus with

$$E \begin{bmatrix} \tilde{x}_{k+1}\tilde{x}_{k+1}' & \tilde{x}_{k+1}\zeta_{k+1}' \\ \zeta_{k+1}\tilde{x}_{k+1}' & \zeta_{k+1}\zeta_{k+1}' \end{bmatrix} = \begin{bmatrix} \Sigma_{k+1/k+1} & \Xi_{k+1} \\ \Xi_{k+1}' & \Pi_{k+1} \end{bmatrix}$$

we have

$$\begin{bmatrix} \Sigma_{k+1/k+1} & \Xi_{k+1} \\ \Xi'_{k+1} & \Pi_{k+1} \end{bmatrix} = \begin{bmatrix} (I - K_{k+1}H'_{k+1})F_k & D'_k \\ 0 & A_k \end{bmatrix} \begin{bmatrix} \Sigma_{k/k} & \Xi_k \\ \Xi'_k & \Pi_k \end{bmatrix}$$

$$\times \begin{bmatrix} F'_k(I - H_{k+1}K'_{k+1}) & 0 \\ D_k & A'_k \end{bmatrix} + \begin{bmatrix} K_{k+1}R_{k+1}K'_{k+1} & 0 \\ 0 & B_kB'_k \end{bmatrix} \quad (4.8a)$$

Also, from (4.7a) we have

$$\Sigma_{0/0} = (I - K_0H'_0)P_0(I - K_0H'_0)' + K_0R_0K'_0 \quad (4.8b)$$

while also $\Xi_0 = 0$. When (4.8a) is written out term by term, there results

$$\Pi_{k+1} = A_k\Pi_kA'_k + B_kB'_k \quad (4.8c)$$

$$\Xi_{k+1} = (I - K_{k+1}H'_{k+1})(F_k\Xi_kA'_k + C'_k\Pi_kA'_k) \quad (4.8d)$$

$$\begin{aligned} \Sigma_{k+1/k+1} = {} & (I - K_{k+1}H'_{k+1})F_k\Sigma_{k/k}F'_k(I - K_{k+1}H'_{k+1})' \\ & + D'_k\Xi'_kF'_k(I - K_{k+1}H'_{k+1})' + (I - K_{k+1}H'_{k+1})F_k\Xi_kD_k \\ & + K_{k+1}R_{k+1}K'_{k+1} + D'_k\Pi_kD_k \end{aligned} \quad (4.8e)$$

Calculation of K_k^* Sequence

Suppose that $K_0^*, K_1^*, \ldots, K_k^*$ have been determined. Then we see from (4.8) that Π_k, Ξ_k, and $\Sigma_{k/k}$ are all determinable. Now with Π_k, Ξ_k, and $\Sigma_{k/k}$ all known, (4.8e) shows that $\Sigma_{k+1/k}$ can be determined after choice of K_{k+1}. One may rewrite (4.8e) as

$$\begin{aligned} \Sigma_{k+1/k+1} = {} & K_{k+1}(H'_{k+1}\Theta_kH_{k+1} + R_{k+1})K'_{k+1} \\ & - K_{k+1}H'_{k+1}\Theta_k - \Theta_kH_{k+1}K'_{k+1} + \Theta_k \end{aligned} \quad (4.9a)$$

Here,

$$\Theta_k = F_k\Sigma_{k/k}F'_k + C'_k\Xi'_kF'_k + F_k\Xi_kC_k + C'_k\Pi_kC_k \quad (4.9b)$$

from which a simple completion of the square argument shows that $\Sigma_{k+1/k+1}$ will be minimized by taking

$$K_{k+1}^* = \Theta_kH_{k+1}(H'_{k+1}\Theta_kH_{k+1} + R_k)^{-1} \quad (4.10a)$$

and one then has

$$\Sigma_{k+1/k+1} = \Theta_k - \Theta_kH_{k+1}(H'_{k+1}\Theta_kH_{k+1} + R_{k+1})H'_{k+1}\Theta_k \quad (4.11a)$$

Of course, Θ_k in (4.9b) is independent of K_{k+1}.

All the above calculations are for $k \geq 0$. We obtain K_0^* using (4.8b):

$$K_0^* = P_0H_0(H'_0P_0H_0 + R_0)^{-1} \quad (4.10b)$$

This yields

$$\Sigma_{0/0} = P_0 - P_0H_0(H'_0P_0H_0 + R_0)^{-1}H'_0P_0 \quad (4.11b)$$

It is the presence of the nonzero quantity Ξ_k in (4.9) which causes the failure of the order propagation property. In rough terms, one could conceive of a sequence $K_0, \ldots, K_k$ different to $K_0^*, \ldots, K_k^*$, with

$$\Sigma_{k/k}(K_0, \ldots, K_k) \geq \Sigma_{k/k}(K_0^*, \ldots, K_k^*)$$

but also such that the difference between

$$\Xi_k(K_0, \ldots, K_k) \quad \text{and} \quad \Xi_k(K_0^*, \ldots, K_k^*)$$

caused

$$\Sigma_{k+1/k+1}(K_0, \ldots, K_k, K_{k+1}) \leq \Sigma_{k+1/k+1}(K_0^*, \ldots, K_k^*, K_{k+1})$$

for some K_{k+1}. With Ξ_k zero, this difficulty cannot arise, and if $\{w_k\}$ is white [which is obtained by taking $A_k = 0$ in (4.2)], one sees that $\Xi_k = 0$ for all k.

Colored Output Noise

We shall now examine the situation in which in (4.1a) one has $E[w_k w_l'] = G_k Q_k G_k' \delta_{kl}$ and $\{v_k\}$ colored. That is, we shall assume that

$$\xi_{k+1} = A_k \xi_k + B_k \lambda_k \tag{4.12a}$$

$$v_k = C_k' \xi_k \tag{4.12b}$$

Here, λ_k is a zero mean, white gaussian process, independent of $\{w_k\}$, with $E[\lambda_k \lambda_l'] = I\delta_{kl}$, and ξ_0 is $N(0, \Pi_0)$, being independent of $\{\lambda_k\}$ and $\{w_k\}$.

Again, we seek an estimator of dimension equal to that of x_k, postulating that for some sequence $\{K_k\}$, the estimator is defined by (4.4). Again, we aim to choose a sequence $\{K_0^*, K_1^*, \ldots\}$ such that the error covariance minimization property (4.5c) holds.

The procedure is similar to that which applies when there is colored input noise. Thus with $\tilde{x}_k = x_k - \hat{x}_{k/k}$, one has (4.7), and tying this with (4.12) yields, with $E_{k+1} = I - K_{k+1} H_{k+1}'$,

$$\begin{bmatrix} \tilde{x}_{k+1} \\ \xi_{k+2} \end{bmatrix} = \begin{bmatrix} (I - K_{k+1}H_{k+1}')F_k & -K_{k+1}C_{k+1}' \\ 0 & A_{k+1} \end{bmatrix} \begin{bmatrix} \tilde{x}_k \\ \xi_{k+1} \end{bmatrix} + \begin{bmatrix} 0 & E_{k+1} \\ B_{k+1} & 0 \end{bmatrix} \begin{bmatrix} \lambda_{k+1} \\ w_k \end{bmatrix}$$

Set

$$E\left\{ \begin{bmatrix} \tilde{x}_{k+1} \\ \xi_{k+2} \end{bmatrix} \begin{bmatrix} \tilde{x}_{k+1} \\ \xi_{k+2} \end{bmatrix}' \right\} = \begin{bmatrix} \Sigma_{k+1/k+1} & \Xi_{k+1} \\ \Xi_{k+1}' & \Pi_{k+2} \end{bmatrix}$$

Then from this point on the calculations are the same in form as those applying to the colored input noise case. The recursive equation for $\Sigma_{k+1/k+1}$ involves K_{k+1} quadratically, and minimization for fixed $\Sigma_{k/k}$, Ξ_k, and Π_{k+1} is easily achieved. We leave the details to the reader.

It should also be clear that, in principle, one could cope with colored input and measurement noise simultaneously by the techniques of this section.

Main Points of the Section

When the input or measurement noise is colored, one can achieve a reduced order suboptimal filter of dimension equal to that of the signal model state vector. The filter structure is the same as that of the standard Kalman filter [see (4.4)] with the gain K_{k+1} being set equal to the value K_{k+1}^*, which minimizes $\Sigma_{k+1/k+1}$, the gain sequence $K_0^*, \ldots, K_k^*$ having been previously determined.

Problem 4.1 Discuss how one proceeds if in (4.1a) one has

$$x_{k+1} = F_k x_k + w_k + \Gamma_k u_k$$

where $\{u_k\}$ is a known input sequence and $\{\Gamma_k\}$ a known sequence of gain matrices.

Problem 4.2 Discuss how one could combine the ideas of this and the previous section to deal with systems with colored input noise and white measurement noise, with the measurement noise covariance being singular.

11.5 SUBOPTIMAL FILTER DESIGN BY MODEL ORDER REDUCTION

Suppose one has an n-dimensional signal model. One approach to suboptimal filter design is to replace the n-dimensional signal model by an n_1-dimensional model with $n_1 < n$; then design a Kalman filter for the n_1-dimensional model and use it on the n-dimensional model. The natural question arises as to what sort of signal model replacements should be considered. The answer turns out to be a little different, depending on whether we are interested in signal or state filtering. We shall look at these cases separately, assuming for convenience throughout this section that all processes are stationary.

Signal Filtering

Suppose that the input and measurement noises to the signal model are independent. Then it is intuitively reasonable that if the signal model is replaced by one whose output power spectrum is close to the power spectrum of the original model, then this replacement is not likely to introduce great errors when used as a basis for filter design. (The intuition can actually be checked quantitatively.) The question arises as to how one might go about the approximation procedure. To do any more than indicate the ideas behind valid approximation procedures would take us too far afield, and we therefore make only three brief comments:

1. For scalar signal processes, since the output power spectrum is the square of the model transfer function amplitude, the task is one of approximating a transfer function amplitude response by another transfer function amplitude response, the second transfer function being of lower degree. Techniques developed for network synthesis can be applied to this problem. For vector processes, this idea is harder to apply.

2. If the original physical situation giving rise to the signal model is known, one may be in a position to obtain a reduced order model by "neglect of the parasitics." Thus in an electrical network, for example, one can eliminate from consideration the stray capacitance and inductance.

3. The ideas of Chap. 10 using the Kullback information function as a measure of closeness of signal models are relevant here.

State Filtering

If one is to build a reduced order filter, one cannot necessarily expect to be able to estimate the entire state vector of the given signal model. This means that one should specify in advance of the filter design what particular linear functionals of the signal model state vector one wishes to estimate. Having done this, one then proceeds with the filter design.

The procedure we present below is an amalgam of ideas of the last section and of [8]. Suppose that the signal model is

$$\bar{x}_{k+1} = \bar{F}\bar{x}_k + \bar{G}w_k \tag{5.1a}$$

$$z_k = \bar{H}'\bar{x}_k + v_k \tag{5.1b}$$

with $\{v_k\}$, $\{w_k\}$ zero mean, independent, white gaussian processes with $E[v_k v'_k] = R$, $E[w_k w'_k] = Q$. Suppose moreover that one wishes to estimate the independent linear functionals $t'_1\bar{x}_k, t'_2\bar{x}_k, \ldots, t'_m\bar{x}_k$. The choice of these linear functionals may be dictated by the need to use them in a control law or by the fact that these are the "interesting" parts of the state vector, i.e., the parts that in some way contain useful information. Define a square matrix T by

$$T' = [t_1 \quad t_2 \quad \ldots \quad t_m \quad S]$$

where S is chosen to make T nonsingular, but is otherwise arbitrary. Then (5.1) is equivalent, under the transformation $x_k = T\bar{x}_k$, to

$$x_{k+1} = Fx_k + Gw_k \tag{5.2a}$$

$$z_k = H'x_k + v_k \tag{5.2b}$$

where $F = T\bar{F}T^{-1}$, $G = T\bar{G}$, $H' = \bar{H}T^{-1}$. Further, the first m entries of x_k are the quantities which we wish to estimate. These first m entries, denoted by x_k^1, satisfy

$$x_{k+1}^1 = F_{11}x_k^1 + F_{12}x_k^2 + G_1 w_k \tag{5.3}$$

We can regard x_k^2 in (5.3) as being an additive colored noise term and postulate an estimator of the form*

$$\hat{x}_{k+1/k+1}^1 = F_{11}\hat{x}_{k/k}^1 + K_{k+1}[z_{k+1} - H_1'F_{11}\hat{x}_{k/k}^1] \qquad (5.4)$$

In general, we should expect K_k to be time invariant. Let us retain the time variation for the moment, however. Note that (5.4) is not the only form of estimator we might consider. One could, for example, attempt to get a stochastic version of some of the ideas of [10], postulating an estimator structure

$$p_{k+1} = Ap_k + Bz_k \qquad \hat{x}_{k/k} = Cp_k + Dz_k \qquad (5.5)$$

We shall, however, consider only (5.4).

To obtain a gain-update equation, observe first from (5.3) and (5.4) that

$$x_{k+1}^1 - \hat{x}_{k+1/k+1}^1 = (I - K_{k+1}H_1')F_{11}(x_k^1 - \hat{x}_{k/k}^1)$$
$$+ \text{(terms involving } x_k^2, x_{k+1}^2, v_{k+1} \text{ and } w_k) \qquad (5.6)$$

By proceeding as in the last section, one can obtain a recursive equation for the error correlation† $\Sigma_{k+1/k+1}^{11}$ associated with using $\hat{x}_{k+1/k+1}^1$ as an estimate of x_{k+1}^1. In fact, one has

$$\Sigma_{k+1/k+1}^{11} = K_{k+1}X_kK_{k+1}' + K_{k+1}Y_k + Y_k'K_{k+1}' + Z_k \qquad (5.7)$$

for certain terms X_k, Y_k, and Z_k, which are computable provided initial state covariances are known and $K_1, \ldots, K_k$ are known. It is then trivial to select K_{k+1} to minimize $\Sigma_{k+1/k+1}^{11}$. Proceeding in this way, one obtains a sequence $K_0^*, K_1^*, \ldots, K_{k+1}^*$ with the property that for all k and K_k,

$$\Sigma_{k+1/k+1}^{11}(K_0^*, K_1^*, \ldots, K_{k+1}^*) \leq \Sigma_{k+1/k+1}^{11}(K_0^*, K_1^*, \ldots, K_k^*, K_{k+1}) \qquad (5.8)$$

Should the K_k approach a steady-state value, one obtains in this manner a time-invariant filter (5.4).

Of course, in principle one can combine this idea with those given previously for handling colored input and output noise and for reducing the estimator dimension given a small noise covariance associated with part of the measurements. In a sense, this may be equivalent to working with the structure of (5.5).

Main Points of the Section

Reduced order signal estimators can be obtained by approximating the signal model by one of reduced order and with approximately the same amplitude response. Reduced order state estimators can be obtained by treating part of the state vector as a colored noise process.

*Note that $\hat{x}_{k/k}^1$ is not generally $E[x_k^1 | Z_k]$, being a suboptimal estimate of x_k^1.

†In order that $\Sigma_{k/k}^{11} = E[(x_k^1 - \hat{x}_{k/k}^1)(x_k^1 - \hat{x}_{k/k}^1)']$ be an error covariance, we require $E[x_k^1 - \hat{x}_{k/k}^1] = 0$, which cannot always be guaranteed. For a discussion of bias, see [11].

For convenience, let us sum up the major approaches outlined in this chapter to dealing with colored noise and/or implementing reduced order filters.

1. When either noise or both noises are colored, model the input noise and output noise by finite-dimensional systems driven by white noise. Build an optimal filter. (Sec. 11.1)
2. When input noise is white and when output noise is Markov, define a new measurement process and build a filter of the normal dimension. (Sec. 11.2)
3. When both noises are white and when the measurement noise has a singular covariance, reduce the filter dimension by the nullity of the covariance matrix. (Sec. 11.3)
4. When both noises are white, extend 3 to yield a suboptimal filter when the measurement noise covariance is nonsingular. (Sec. 11.3)
5. When either noise or both noises are colored, postulate the filter structure which would apply in the white noise situation and choose the gain to optimize "one step ahead." (Sec. 11.4)
6. When both noises are white, treat part of the state vector as colored noise and proceed as in 5. (Sec. 11.5)

In all situations where a suboptimal filter is used, it is wise to compare its performance with that of the optimal filter.

REFERENCES

[1] SAGE, A. P., and J. L. MELSA, *Estimation Theory with Applications to Communications and Control*, McGraw-Hill Book Company, New York, 1971.

[2] BRYSON, A. E., and L. J. HENRIKSON, "Estimation Using Sampled-data Containing Sequentially Correlated Noise," *J. of Spacecraft and Rockets*, Vol. 5, June 1968, pp. 662–665.

[3] TSE, E., and M. ATHANS, "Optimal Minimal-order Observer-estimators for Discrete Linear Time-varying Systems," *IEEE Trans. Automatic Control*, Vol. AC-15, No. 4, August 1970, pp. 416–426.

[4] YOSHIKAWA, T., and H. KOBAYASHI, "Comments on 'Optimal Minimal-order Observer-estimators for Discrete Linear Time-varying Systems'," *IEEE Trans. Automatic Control*, Vol. AC-17, No. 2, April 1972, pp. 272–273.

[5] AOKI, M., and J. R. HUDDLE, "Estimation of the State Vector of a Linear Stochastic System with a Constrained Estimator," *IEEE Trans. Automatic Control*, Vol. AC-12, No. 4, August 1967, pp. 432–433.

[6] LEONDES, C. T., and L. M. NOVAK, "Optimal Minimal-order Observers for Discrete-time Systems—A Unified Theory," *Automatica*, Vol. 8, No. 4, July 1972, pp. 379–388.

[7] LEONDES, C. T., and L. M. NOVAK, "Reduced-order Observers for Linear Discrete-time Systems," *IEEE Trans. Automatic Control*, Vol. AC-19, No. 1, February 1974, pp. 42–46.

[8] HUTCHINSON, C. E., J. A. D'APPOLITO, and K. J. ROY, "Applications of Minimum Variance Reduced-state Estimators," *IEEE Trans. Aerospace and Electronic Systems*, Vol. AES-11, No. 5, September 1975, pp. 785–794.

[9] LUENBERGER, D. G., "An Introduction to Observers," *IEEE Trans. Automatic Control*, Vol. AC-16, No. 6, December 1971, pp. 596–602.

[10] MOORE, J. B., and G. F. LEDWICH, "Minimal-order Observers for Estimating Linear Functions of a State Vector," *IEEE Trans. Automatic Control*, Vol. AC-20, No. 5, October 1975, pp. 623–632.

[11] ASHER, R. B., K. D. HERRING, and J. C. RYLES, "Bias, Variance and Estimation Error in Reduced Order Filters," *Automatica*, Vol. 12, No. 6, November 1976, pp. 589–600.

BRIEF REVIEW OF RESULTS
OF PROBABILITY THEORY

The purpose of this appendix is to provide a concise statement of the results from probability theory which are used in this book. It is not intended as a replacement for a formal course in probability theory, and would be quite inadequate for this purpose. Nevertheless, it might serve to fill in a limited number of gaps in the reader's knowledge.

The appendix is divided into three sections, covering results from pure probability theory, results on stochastic processes, and results involving gaussian random variables and processes. For an introduction to these ideas which is suited to engineers, see, for example, [1] and [2]. For a more advanced treatment, relying on measure theory, see, for example, [3] and [4].

In our view, the material in [1] defines fairly precisely the material which is needed as background for the understanding of this text. However, in the summary that follows, we have occasionally gone beyond the level in [1] to mention ideas which we feel are particularly important.

By and large, many qualifiers, particularly existence qualifiers, are omitted in the following material.

A.1 PURE PROBABILITY THEORY

1. Sample Space, Events, Experiments, Probability

Consider an experiment with a number of possible *outcomes*. The totality of such outcomes is a *sample space* Ω. An *event A* is a subset of the sample space. A *probability measure* $P(\cdot)$ is a mapping from events into the reals satisfying the axioms

1. $P(A) \geq 0$.
2. $P(\Omega) = 1$.
3. For a countable set $\{A_i\}$ of events, if $A_i \cap A_j = \phi$ for all i, j, then $P(\cup A_i) = \sum_i P(A_i)$. (Here, ϕ denotes the empty set, and the set $\{A_i\}$ is termed *mutually disjoint*.)

Important consequences for an arbitrary countable set $\{A_i\}$ of events are

$$P(A) \leq 1, \quad P(\phi) = 0, \quad P(\bar{A}) = 1 - P(A), \quad \text{and} \quad P(\cup A_i) \leq \sum_i P(A_i)$$

with $\bar{A}$ denoting the event "not A" or "complement of A." Not all subsets of the sample space need be events, but the events must form a *sigma field*; in other words, if A is an event, $\bar{A}$ is an event, and if $\{A_i\}$ is a countable set of events, $\cup A_i$ is an event. Finally, Ω is an event. Frequently, it is also assumed that if A is an event with $P(A) = 0$ and B is any subset of A, then B is also an event with $P(B) = 0$. The probability measure is then termed *complete*.

2. Joint Probability

The joint probability of two events A and B is $P(A \cap B)$, written sometimes $P(AB)$.

3. Conditional Probability

Suppose A and B are two events and an experiment is conducted with the result that event B occurs. The probability that event A has also occurred, or the conditional probability of A given B, is

$$P(A \mid B) = \frac{P(AB)}{P(B)} \quad \text{assuming } P(B) \neq 0$$

$P(A \mid B)$ for fixed B and variable A satisfies the probability measure axioms. (Note that the definition of $P(A \mid B)$ when $P(B) = 0$ is apparently of no interest, precisely because the need to make this definition arises with zero probability.)

4. Independence

Events $A_1, A_2, \ldots, A_n$ are *mutually independent* if and only if

$$P(A_{i_1} \cap A_{i_2} \cap \cdots \cap A_{i_k}) = P(A_{i_1})P(A_{i_2})\cdots P(A_{i_k})$$

for all integers $i_1, \ldots, i_k$ selected from $1, 2, \ldots, n$ with no two the same. It is possible for three events A, B, C to have each pair mutually independent, i.e.,

$$P(AB) = P(A)P(B) \qquad P(BC) = P(B)P(C)$$

and $P(CA) = P(C)P(A)$ but not $P(ABC) = P(A)P(B)P(C)$. Two events A and B are *conditionally independent given an event C* if

$$P(AB|C) = P(A|C)P(B|C)$$

If $A_i, i = 1, 2, \ldots, n$ are mutually disjoint and $\cup A_i = \Omega$, then

$$P(B) = \sum_i P(B|A_i)P(A_i)$$

for arbitrary B.

5. Bayes' Rule

If $P(B) \neq 0$,

$$P(A|B) = \frac{P(B|A)P(A)}{P(B)}$$

If $A_i, i = 1, 2, \ldots, n$ are mutually disjoint and $\cup A_i = \Omega$,

$$P(A_j|B) = \frac{P(B|A_j)P(A_j)}{\sum_i P(B|A_i)P(A_i)}$$

6. Random Variables

It is often appropriate to measure quantities associated with the outcome of an experiment. Such a quantity is a random variable. More precisely, a random variable X is a function from the outcomes ω in a sample space Ω to the real numbers, with two properties as given below. A value of the random variable X is the number $X(\omega)$ when the outcome ω occurs. Most commonly, X can take either discrete values (X is then a discrete random variable), or continuous values in some interval $[a, b]$ (X is then a continuous random variable).

We adopt the convention that $P(X = 2)$ means $P(\{\omega \,|\, X(\omega) = 2\})$, i.e., the probability of the subset of Ω consisting of those outcomes ω for which $X(\omega) = 2$. Likewise, $P(X > 0)$ means $P(\{\omega \,|\, X(\omega) > 0\})$, etc.

For X to be a random variable, we require that:

1. $P(X = -\infty) = P(X = +\infty) = 0$.
2. For all real a, $\{\omega \,|\, X(\omega) \le a\}$ is an event, i.e.,

$$P(\{\omega \,|\, X(\omega) \le a\}) = P(X \le a)$$

is defined.

7. Distribution Function

Given a random variable X, the distribution function F_X is a mapping from the reals to the interval $[0, 1]$:

$$F_X(x) = P(X \le x)$$

The subscript X on F identifies the random variable; the argument x is simply a typical value. The distribution function is monotonic increasing, $\lim_{x \to \infty} F_X(x) = 1$, $\lim_{x \to -\infty} F_X(x) = 0$, and it is continuous from the right.

8. Density Function

It is frequently the case that $F_X(x)$ is differentiable everywhere. The probability density function $p_X(x)$ associated with the random variable X is

$$p_X(x) = \frac{dF_X(x)}{dx}$$

Then $p_X(x)dx$ to first order is $P\{x < X \le x + dx\}$. A discrete random variable only has a density function in the sense that the density function consists of a sum of delta functions.

9. Pairs of Random Variables

Let X and Y be two random variables. Then $F_{X,Y}(x, y) = P\{(X \le x) \cap (Y \le y)\}$ is the *joint distribution function*. If the derivative exists, the joint probability density function is

$$p_{X,Y}(x, y) = \frac{\partial^2}{\partial x\, \partial y} F_{X,Y}(x, y)$$

Given $F_{X,Y}(x, y)$, it follows that $F_X(x) = F_{X,Y}(x, \infty)$ and

$$p_X(x) = \int_{-\infty}^{+\infty} p_{X,Y}(x, y)\, dy$$

10. Conditional Distribution and Densities

If X and Y are discrete random variables,

$$p_{X|Y}(x_i \,|\, y_j) = P(X = x_i \,|\, X = y_j) = \frac{p_{X,Y}(x_i, y_j)}{p_Y(y_j)}$$

If X is continuous and B any event,

$$F_{X|B}(x) = P(X \leq x \,|\, B) = \frac{P\{(X \leq x) \cap B\}}{P(B)}$$

and, if the derivative exists, one has the conditional density

$$p_X(x \,|\, B) = \frac{dF_{X|B}(x)}{dx}$$

If Y is continuous, by taking B as the event $\{y < Y \leq y + \Delta y\}$ and letting $\Delta y \rightarrow 0$, we obtain [if $p_{X,Y}(x, y)$ and $p_Y(y) \neq 0$ exist]

$$p_{X|Y}(x \,|\, y) = \frac{p_{X,Y}(x, y)}{p_Y(y)}$$

Of course, $p_{X|Y}(x \,|\, y)$ is termed the conditional probability density of X given Y. Somewhat paradoxically, the conditioning here is on an event of zero probability. One also has the important formula

$$p_X(x) = \int_{-\infty}^{+\infty} p_{X|Y}(x \,|\, y) p_Y(y)\, dy$$

11. Random Vectors, Marginal and Conditional Densities*

n random variables $X_1, X_2, \ldots, X_n$ define a random n-vector X. One has

$$F_X(x) = P\{(X_1 \leq x_1) \cap \ldots \cap (X_n \leq x_n)\}$$

and

$$p_X(x) = \frac{\partial^n}{\partial x_1 \ldots \partial x_n} F_X(x)$$

Marginal densities, or densities of the form $p_{X_1,X_2,X_3}(x_1, x_2, x_3)$ can be obtained by integration:

$$p_{X_1,X_2,X_3}(x_1, x_2, x_3) = \int_{-\infty}^{+\infty} \int_{-\infty}^{+\infty} \cdots \int_{-\infty}^{+\infty} p_X(x)\, dx_4\, dx_5 \ldots dx_n$$

Conditional densities can be found as

$$p_{X_1,X_2,\ldots X_i | X_{i+1}, X_{i+2}, \ldots, X_n} = \frac{p_X}{p_{X_{i+1}, X_{i+2}, \ldots, X_n}}$$

Frequently in this book, the term *random variable* is used to cover random vectors as well as random scalars.

12. Independent Random Variables

X and Y are independent random variables if the events $\{X \leq x\}$ and

---

*Henceforth, we shall almost always omit mention of existence conditions for densities.

$\{Y \leq y\}$ are independent for all x and y; equivalently,

$$F_{X,Y}(x, y) = F_X(x)F_Y(y)$$

or

$$p_{X,Y}(x, y) = p_X(x)p_Y(y)$$

or

$$p_{X|Y}(x|y) = p_X(x)$$

There is obvious extension to random vectors and conditional independence.

13. Function of One Random Variable

Let X be a random variable and $g(\cdot)$ a reasonably behaved scalar function of a scalar variable. Then $Y = g(X)$ is a random variable, and if an experiment results in an outcome ω, Y takes the value $y = g(X(\omega))$. One has

$$F_Y(y) = P(Y \leq y) = P(x \in I_y)$$

where $I_y = \{x | g(x) \leq y\}$. In case X is continuous and g differentiable, it can be shown that

$$p_Y(y) = \sum_i \frac{p_X(x_i)}{|g'(x_i)|}$$

where x_i is a root of $y = g(x)$. If an event A has occurred, then

$$p_Y(y|A) = \sum_i \frac{p_X(x_i|A)}{|g'(x_i)|}$$

14. One Function of Two Random Variables

Let X and Y be two jointly distributed random variables and $g(\cdot, \cdot)$ a scalar function. Then $Z = g(X, Y)$ is a random variable, and

$$F_Z(z) = P(Z \leq z) = P(X, Y \in D_z) = \iint_{D_z} p_{X,Y}(x, y)\, dx\, dy$$

where $D_z = \{(x, y) | g(x, y) \leq z\}$. In case $Z = X + Y$ with X and Y independent, the convolution formula holds:

$$p_Z(z) = \int_{-\infty}^{+\infty} p_X(z - y)p_Y(y)\, dy = \int_{-\infty}^{+\infty} p_X(x)p_Y(z - x)\, dx$$

15. Two Functions of Two Random Variables

In case $U = g(X, Y)$ and $V = h(X, Y)$, one has

$$p_{U,V}(u, v) = \sum_{x_i, y_i} \frac{p_{X,Y}(x_i, y_i)}{|J(x_i, y_i)|}$$

where

$$J(x, y) = \frac{\partial g(x, y)}{\partial x}\frac{\partial h(x, y)}{\partial y} - \frac{\partial g(x, y)}{\partial y}\frac{\partial h(x, y)}{\partial x}$$

and (x_i, y_i) is a solution of $u = g(x, y)$, $v = h(x, y)$. There is an obvious extension to n functions of n random variables.

16. Functions of Independent Random Variables

If X and Y are independent random variables, so are $g(X)$ and $h(Y)$.

17. Mean, Variance, and Expectation

The *mean* or *expectation* of a random variable X, written $E[X]$, is the number $\int_{-\infty}^{+\infty} xp_X(x)dx$, where the integral is assumed absolutely convergent. If absolute convergence does not hold, $E[X]$ is not defined. *The variance σ^2 is*

$$E[(X - E[X])^2] = \int_{-\infty}^{+\infty} (x - E[X])^2 p_X(x)\,dx$$

Chebyshev's inequality states

$$P\{|X - E[X]| > K\} \leq \frac{\sigma^2}{K^2}$$

One can also show that $\sigma^2 = E[X^2] - (E[X])^2$. The definition of the mean generalizes in an obvious way to a vector. For vector X, the variance is replaced by the *covariance matrix*

$$E\{(X - E[X])(X - E[X])'\}$$

The variance is always nonnegative, and the covariance matrix nonnegative definite symmetric. If $Y = g(X)$ is a function of a random variable X, the random variable Y has expected value

$$E[g(X)] = \int_{-\infty}^{+\infty} g(x) p_X(x)\,dx$$

These notions generalize to the situation when the probability density does not exist.

18. Properties of Expectation Operator

The expectation operator is linear. Also, if X_i denote mutually independent random variables,

$$E[X_1 X_2 \ldots X_n] = E[X_1]E[X_2] \ldots E[X_n]$$

If they are also of zero mean,

$$E[(\sum_i X_i)^2] = \sum_i E[X_i^2]$$

19. Moments and Central Moments

The kth *moment* of a random variable X is $m_k = E[X^k]$. The kth *central moment* is $\mu_k = E[(X - E[X])^k]$. The *joint moments* of two random variables

X and Y are given by the set of numbers $E[X^k Y^l]$. $E[XY]$ is the *correlation* of X and Y. The *joint central moments* are defined in an obvious manner and $E[(X - E[X])(Y - E[Y])]$ is the *covariance* of X and Y.

If $E[XY] = E[X]E[Y]$, X and Y are termed *uncorrelated*, and if $E[XY] = 0$, they are termed *orthogonal*. Independent random variables are always uncorrelated.

20. Characteristic Function

With $j = \sqrt{-1}$, the *characteristic function* of a random variable X is defined by

$$\phi_X(s) = E[\exp jsX]$$

The variable s can take on complex values. Evidently $\phi_X(\cdot)$ is the Fourier transform of $p_X(\cdot)$. If $X_1, X_2, \ldots, X_n$ are n random variables, the *joint characteristic function* is

$$\phi_X(s_1, s_2, \ldots, s_n) = E[\exp \sum_i js_i X_i]$$

One has $\phi_X(0) = 1$ and $|\phi_X(s_1, s_2, \ldots, s_n)| \leq 1$ for all real s_i. From $\phi_X(s)$, $p_X(x)$ can be recovered by an inverse Fourier transform. The moments m_k of X are related to $\phi_X(\cdot)$ by

$$m_k = j^k \frac{d^k}{ds^k} \phi_X(s) \Big|_{s=0}$$

If X and Y are jointly distributed, $\phi_X(s) = \phi_{X,Y}(s_1, 0)$. If they are independent,

$$\phi_{X,Y}(s_1, s_2) = \phi_X(s_1)\phi_Y(s_2)$$

and conversely. If $\{X_i\}$ is a set of independent random variables and $Z = X_1 + X_2 + \cdots + X_n$, then

$$\phi_Z(s) = \phi_{X_1}(s)\phi_{X_2}(s) \ldots \phi_{X_n}(s)$$

21. Conditional Expectation

The conditional expected value of a random variable X, assuming occurrence of an event A, is

$$E[X|A] = \int_{-\infty}^{+\infty} x p_{X|A}(x|A)\, dx$$

Further,

$$E[g(X)|A] = \int_{-\infty}^{+\infty} g(x) p_{X|A}(x|A)\, dx$$

Suppose Y is a continuous random variable jointly distributed with X. Although $\{\omega | Y(\omega) = y\}$ is an event of zero probability, by analogy with the definition of $p_{X|Y}(x|y)$, one has

$$E[X|Y = y] = \int_{-\infty}^{+\infty} x p_{X|Y}(x|y)\, dx$$

This quantity is a number, depending on y. We can define a random variable $E[X| Y]$ as that which takes the value $E[X| Y(\omega) = y]$ when the experimental outcome ω leads to $Y(\omega) = y$. Thus $E[X| Y]$ is a *function of the random variable* Y, determined by the equation above for $E[X| Y = y]$. As a random variable, it has an expectation. It is important to note that

$$E[E[X| Y]] = E[X]$$

When X, Y, Z are jointly distributed, $E[X| Y, Z]$ is defined in the obvious manner. Then

$$E[X| Y, Z] = E[X| Y + Z, Z] = E[X| g_1(Y, Z), g_2(Y, Z)]$$

for any two functions g_1, g_2 such that to each pair of values ψ_1, ψ_2 of g_1, g_2 there is only one pair of values y, z of Y and Z for which $g_i(y, z) = \psi_i$. The intention is that the values taken by the g_i convey precisely the same information as the values taken by Y, Z. This idea obviously extends to conditioning on more than two random variables. The conditional expectation operator is linear. If X and Y are conditionally independent for a given Z, then

$$E[XY|Z] = E[X|Z]E[Y|Z]$$

If X and Y are independent, $E[X| Y] = E[X]$. In fact, $E[g(X)| Y] = E[g(X)]$ for any function $g(\cdot)$.

The number $E[g(X, Y)| X = x]$ can be evaluated as

$$\lim_{\Delta x \to 0} E[g(X, Y)|x < X \leq x + \Delta x]$$

and is

$$E[g(X, Y)| X = x] = E[g(x, Y)| X = x] = \int_{-\infty}^{+\infty} g(x, y)p_{Y|X}(y|x)\, dy$$

The random variable $E[g(X, Y)| X]$ is a function of the random variable X. If $g(X, Y) = g_1(X)g_2(Y)$, one has

$$E[g(X, Y)| X] = g_1(X)E[g_2(Y)| X]$$

Also,

$$E[E[g(X, Y)| X]] = E[g(X, Y)]$$

22. Central Limit Theorem

If the random variables X_i are independent, under general conditions the distribution of

$$Y_n = n^{-1} \sum_{i=1}^{n} X_i$$

is approximately gaussian, of mean $n^{-1} \sum_i \mu_i$ and variance $n^{-1} \sum_i \sigma_i^2$, where μ_i and σ_i^2 are the mean and variance of X_i. As $n \to \infty$, the approximation becomes more accurate.

23. Convergence

Let X_n be a sequence of random variables. We say that $X_n \rightarrow X$ everywhere as $n \rightarrow \infty$ if $X_n(\omega) \rightarrow X(\omega)$ for all $\omega \in \Omega$. This is normally too restrictive a definition, and the following three convergence concepts are the most commonly used:

C1 $X_n \rightarrow X$ almost surely, or with probability 1, if $X_n(\omega) \rightarrow X(\omega)$ for almost all ω (that is, for all $\omega \in A \subset \Omega$ where $P(A) = 1$.)

C2 $X_n \rightarrow X$ in mean square if $E[\|| X_n - X \||^2] \rightarrow 0$.

C3 $X_n \rightarrow X$ in probability if for all $\epsilon > 0$, $P[\|| X_n - X \|| > \epsilon] \rightarrow 0$.

It is known that:

(a) C1 implies C3.
(b) C2 implies C3.
(c) C3 implies that a subsequence of $\{X_n\}$ satisfies C1.
(d) C3 and $|X_n| < c$ for some c, all $n \geq$ some n_0, and almost all ω implies C2.

On occasions, the following idea is also helpful:

C4 $X_n \rightarrow X$ in νth mean if $E[\|| X_n - X \||^\nu] \rightarrow 0$.

A.2 STOCHASTIC PROCESSES

1. Discrete-time Random Process (Random Sequence)

The idea of a random process is a generalization of the idea of a random variable in the following sense. Instead of each experiment resulting in a number (the value of a random variable) or an n-tuple of numbers (the value of a random vector), the experiment results in a function mapping an underlying time set (nonnegative integers, all integers, or the reals, commonly) into the reals. A discrete-time random process has the time set comprising, usually, nonnegative integers or all integers. One has a mapping from $\omega \in \Omega$ to a set of values $x_\omega(k)$ for $k = 0, 1, 2, \ldots$ or $k = \ldots -2, -1, 0, 1, 2, \ldots$. A scalar discrete-time random process is like an infinite-dimensional random vector. Each $\{x_\omega(k)\}$ can be a sequence of vectors rather than scalars, yielding a vector random process. Normally, the notation $\{x_k\}$ will denote the process in general, or a *sample function*, that is, a particular sequence of values taken as a result of an experiment. Also, x_k will denote the random variable obtained by looking at the process at time k, as well as the value taken by that variable. Though this is a somewhat unfortunate convention, it is standard.

2. Continuous-time Random Process

The underlying time set is the interval $[0, \infty)$ or $(-\infty, \infty)$; $x(\cdot)$ denotes the process, or a particular *sample function*, and $x(t)$ denotes the random variable obtained by looking at the process at time t, or the value of that variable. By examining the values of a continuous-time random process at certain instants of time, e.g., $t = 1, 2, 3, \ldots$, one obtains a discrete-time process.

3. Probabilistic Description

Let m be arbitrary and $k_1, k_2, \ldots, k_m$ be arbitrary times in the underlying time set. Then the set of all probability densities $p_{x_{k_1} \ldots x_{k_m}}(x_{k_1}, x_{k_2}, \ldots, x_{k_m})$ (or the corresponding distribution functions) serves to define the probability structure of the random process. From these densities, one can obtain conditional densities in the usual way.

4. Mean, Autocorrelation, and Covariance of a Process

The mean m_k of a process is the time function $E[x_k]$. The autocorrelation is the set of quantities $E[x_{k_1} x'_{k_2}]$. The covariance is the set of quantities $E\{[x_{k_1} - m_{k_1}][x_{k_2} - m_{k_2}]'\}$ for all k_1 and k_2. When $k_1 = k_2$, the covariance is nonnegative definite symmetric.

5. First Order and Second Order Densities of a Process

The first order densities of a process are the set of densities $p_{x_k}(x_k)$ for all k, and the second order densities the set $p_{x_{k_1}, x_{k_2}}(x_{k_1}, x_{k_2})$ for all k_1 and k_2. The mean and covariance can be obtained entirely from the first and second order densities.

6. Uncorrelated, Orthogonal, and Independent Increment Processes

A process has uncorrelated, orthogonal, or independent increments if $x_{k_i} - x_{k_{i+1}}$ is a sequence of uncorrelated, orthogonal, or independent random variables, with $[k_i, k_{i+1}]$ a set of nonoverlapping but possibly touching intervals in the time set.

7. Uncorrelated, Orthogonal, and Independent Pairs of Processes

$\{x_k\}$ and $\{y_k\}$ are (a) uncorrelated, (b) orthogonal, or (c) independent processes according as (a) $E[x_{k_1} y'_{k_2}] = E[x_{k_1}] E[y'_{k_2}]$ for all k_1 and k_2, (b) $E[x_{k_1} y'_{k_2}] = 0$ for all k_1 and k_2, and (c) for any sets $\{k_i\}$ and $\{l_i\}$, the vector random variable $[x'_{k_1} \ldots x'_{k_n}]'$ is independent of the vector random variable $[y'_{i_1} \ldots y'_{i_m}]'$.

8. Markov Processes

Loosely, a process is Markov if, given that the present is known, the past has no influence on the future, i.e., if $k_1 > k_2 > \cdots > k_n$, then

$$p_{X_{k_1}|X_{k_2},\ldots,X_{k_n}}(x_{k_1}|x_{k_2},\ldots,x_{k_n}) = p_{X_{k_1}|X_{k_2}}(x_{k_1}|x_{k_2})$$

A Markov process is sometimes termed *first order Markov*. A *second order Markov process* is one in which, roughly, the most recent two pieces of information are all that affect the future; i.e., if $k_1 > k_2 > \cdots > k_n$, then

$$p_{X_{k_1}|X_{k_2},\ldots,X_{k_n}}(x_{k_1}|x_{k_2},\ldots,x_{k_n}) = p_{X_{k_1}|X_{k_2},X_{k_3}}(x_{k_1}|x_{k_2},x_{k_3})$$

Third and higher order Markov processes can be defined similarly.

9. Martingale Processes

A process is a martingale if, roughly, it is as likely to go up as go down at each time instant; i.e., assuming $\{X_k\}$ is defined for $k \geq 0$,

$$E[X_{k+1}|X_0, X_1, \ldots, X_k] = X_k$$

If X_k represents the stake at time k held by one of two gamblers, a game between the two gamblers is fair if and only if the martingale property holds. If

$$E[X_{k+1}|X_0, X_1, \ldots, X_k] \leq X_k \quad (\geq X_k)$$

one has a supermartingale (a submartingale). Martingales are one of the simplest kinds of stochastic processes for which a number of convergence results are available. The above definitions are sometimes generalized to replace the conditioning variables $X_0, \ldots, X_k$ by a sigma field $\mathfrak{F}_k$ related to $X_0, \ldots, X_k$, with $\mathfrak{F}_k \subseteq \mathfrak{F}_{k+1}$.

10. Stationary Processes

A process $\{x_k\}$ is *strict-sense stationary*, or simply *stationary*, if its associated probability densities are unaffected by time translation; i.e., for arbitrary integer m and times $k_1, \ldots, k_m$ and N,

$$p_{X_{k_1} \cdots X_{k_m}}(x_{k_1}, x_{k_2}, \ldots, x_{k_m}) = p_{X_{k_1+N} \cdots X_{k_m+N}}(x_{k_1+N}, \ldots, x_{k_m+N})$$

It is *asymptotically stationary* if

$$\lim_{N \to \infty} p_{X_{k_1+N} \cdots X_{k_m+N}}(x_{k_1+N}, \ldots, x_{k_m+N})$$

exists. Processes $\{x_k\}$ and $\{y_k\}$ are jointly stationary if $\{[x'_k y'_k]'\}$ is stationary. If $\{x_k\}$ is stationary, then $E[x_k] = m$, independent of k, and

$$R(k_1, k_2) = E\{[x_{k_1} - m][x_{k_2} - m]'\} = R(k_1 - k_2)$$

11. Wide-sense Stationary

A process is *wide-sense stationary* if its first and second order densities are invariant under time translation. Then its mean is constant, and its covariance $R(k_1, k_2)$ is of the form $R(k_1 - k_2)$. Stationary processes are wide-sense stationary. Both the covariance and autocorrelation $C(k_1, k_2) = C(k_1 - k_2)$ are even in the scalar case; i.e. $R(-k) = R(k)$, $C(-k) = C(k)$, and $C(0) \geq |C(k)|$ for all k. In the vector process case,

$$R(-k) = R'(k), \ C'(-k) = C(k)$$

12. Ergodic Processes

Certain stationary processes are ergodic. The basic idea behind ergodicity is that time averages can be replaced by an expectation, or averages over the set of experiment outcomes. There are two approaches. One says that a process $\{x_k\}$ is ergodic (without qualification) if for any suitable function $f(\cdot)$, the following limit exists almost surely:

$$E[f(\{x_k\})] = \lim_{K \to \infty} \frac{1}{2K + 1} \sum_{-K}^{K} f(\{x_i\})$$

If $\{x_k\}$ is gaussian with covariance sequence R_k, the following condition is sufficient for ergodicity:

$$\sum_{k=-\infty}^{+\infty} |R_k| < \infty$$

Alternatively, one seeks for a given $f(\cdot)$ conditions for the limit to exist as a mean square limit. A sufficient condition is then that

$$\sum_{k=-\infty}^{\infty} |R_k^f| < \infty$$

where R^f denotes the covariance of $f(x)$. Taking

$$f(\{x_k\}) = x_k \quad \text{and} \quad f[\{x_k\}] = x_k x_{k+l}$$

leads to the concepts of ergodicity in the mean and in the covariance function; these last two concepts have validity for processes which are wide-sense stationary.

13. Power Spectrum
(Power Spectrum Density, Spectral Density)

If $\{x_k\}$ is a discrete-time random process that is wide-sense stationary, the power spectrum is, assuming it exists for some z,

$$\Phi(z) = \sum_{k=-\infty}^{+\infty} z^{-k} R_k$$

One has $\Phi(z)$ nonnegative if a scalar, or nonnegative definite if a matrix, for all z on $|z| = 1$. Also, $\Phi(z) = \Phi'(z^{-1})$. Finally,

$$E[x_k x_k'] = \frac{1}{2\pi j} \oint \frac{\Phi(z)\, dz}{z}$$

the integration being round the unit circle.

14. White Noise

White noise processes usually have zero mean, and when stationary, are processes whose power spectrum is constant.

Constancy of the power spectrum is equivalent to

$$E[x_k x_l'] = C\delta_{kl}$$

for some constant matrix C. The discrete-time (Kronecker) delta function δ_{kl} is 0 for $k \neq l$ and 1 for $k = l$.

15. Passage through a Linear System

If a random process of power spectrum $\Phi(z)$ passes through a time-invariant, stable linear system of transfer function matrix $W(z)$, the power spectrum of the output process is $W(z)\Phi(z)W'(z^{-1})$.

A.3   GAUSSIAN RANDOM VARIABLES, VECTORS, AND PROCESSES

1. Gaussian Random Variable

X is a gaussian or normal random variable if its probability density is of the form

$$p_X(x) = \frac{1}{\sqrt{2\pi}\sigma} \exp\left[-\frac{(x - \mu)^2}{2\sigma^2}\right]$$

One can evaluate $E[X]$ and $E\{(X - E[X])^2\}$ to be μ and σ^2, respectively. In this evaluation, one can make use of the integral

$$\int_{-\infty}^{+\infty} e^{-(1/2)x^2}\, dx = \sqrt{2\pi}$$

The notation "X is $N(\mu, \sigma^2)$" is sometimes used to denote that X is gaussian with mean μ and variance σ^2. The mode of $p_X(\cdot)$, i.e., the value of x maximizing $p_X(\cdot)$, is μ. So is the median [which is that value χ for which $\Pr(x \leq \chi) = \frac{1}{2}$.

2. Sum of Independent Gaussian Variables

If X is $N(\mu_X, \sigma_X^2)$ and Y is $N(\mu_Y, \sigma_Y^2)$, with X and Y independent, then $X + Y$ is $N(\mu_X + \mu_Y, \sigma_X^2 + \sigma_Y^2)$.

3. Characteristic Function

If X is $N(\mu, \sigma^2)$, then $\phi_X(s) = \exp[j\mu s - (\sigma^2 s^2/2)]$, and conversely.

4. Gaussian Random Vector

Let X be a random n-vector. If X has a nonsingular covariance matrix, we say that X is gaussian or normal if and only if its probability density is of the form

$$p_X(x) = \frac{1}{(2\pi)^{n/2}} \frac{1}{|\Sigma|^{1/2}} \exp[-\tfrac{1}{2}(x - m)'\Sigma^{-1}(x - m)]$$

for some vector m and matrix Σ. One can evaluate

$$E[X] = m \qquad E\{[X - m][X - m]'\} = \Sigma$$

and also

$$\phi_X(s) = \exp[js'm - \tfrac{1}{2}s'\Sigma s]$$

If the covariance of X is singular, then X with probability 1 lies in a proper subspace of n-space; in fact, for any vector α in the null space of the covariance matrix Σ, one has $\alpha'(X - m) = 0$ for all X with probability 1. One cannot define the gaussian property via the probability density, but one can still define it via the characteristic function: X is gaussian if and only if for some m and Σ, $\phi_X(s) = \exp[js'm - \tfrac{1}{2}s'\Sigma s]$. As for scalar X, we write "X is $N(m, \Sigma)$." Again, m is the mode.

5. Joint Densities, Marginal Densities, and Conditional Densities

To say that X_1 and X_2 are jointly gaussian random variables is the same as saying that the random vector $X = [X_1 \quad X_2]'$ is a gaussian random vector. All marginal densities derived from a gaussian random vector are themselves gaussian; e.g., if $X = [X_1 \quad X_2 \quad \ldots \quad X_n]'$ is gaussian, then $\hat{X} = [X_1 \quad X_2 \quad \ldots \quad X_m]'$ is gaussian. All conditional densities formed by conditioning some entries of a gaussian random vector on other entries are gaussian. If $X = [X_1 \quad X_2]'$ is $N(m, \Sigma)$, then X_1 conditioned on X_2 is gaussian with mean $m_1 - \Sigma_{12}\Sigma_{22}^{-1}m_2 + \Sigma_{12}\Sigma_{22}^{-1}x_2$ and covariance $\Sigma_{11} - \Sigma_{12}\Sigma_{22}^{-1}\Sigma_{12}'$. Here the m_i and Σ_{ij} are the obvious submatrices of m and Σ. Pseudo-inverses can replace inverses if necessary.

6. Linear Transformations

Let X be $N(m, \Sigma)$ and let $Y = Ax + b$ for a constant matrix A and vector b. Then Y is $N(Am + b, A\Sigma A')$. This follows from

$$\phi_Y(s) = E\{\exp(js'Y)\} = E\{\exp(js'Ax + js'b)\}$$
$$= E\{\exp(js'b)\exp(j(As)'x)\} = \exp(js'b)\phi_x(As)$$
$$= \exp[js'Am + js'b - \tfrac{1}{2}s'A\Sigma A's]$$

In particular, if X and Y are jointly gaussian, $X + Y$ is gaussian.

7. Uncorrelated Gaussian Variables

Suppose X and Y are uncorrelated and gaussian. Then they are independent. This follows by showing that $\phi_{X,Y}(s_1, s_2) = \phi_X(s_1)\phi_Y(s_2)$.

8. Conditional Expectation

(This point is developed in the text.) Let X and Y be jointly gaussian. Then $E[X|Y]$, which is a function of Y, is of the form $AY + b$ for a constant matrix A and vector b, and is therefore gaussian. In fact, if $[X' : Y']'$ is $N(m, \Sigma)$ then $E[X|Y]$ is given by

$$m_x - \Sigma_{xy}\Sigma_{yy}^{-1}m_y + \Sigma_{xy}\Sigma_{yy}^{-1}Y$$

and is $N(m_x, \Sigma_{xy}\Sigma_{yy}^{-1}\Sigma_{xy}')$. Pseudo-inverses can replace inverses if necessary.

9. Gaussian Random Process

A random process is a gaussian or normal random process if, for any selection of points $k_1, \ldots, k_n$ in the time set, the random variables $x_{k_1}, \ldots, x_{k_n}$ are jointly gaussian, i.e.,

$$p_{x_{k_1}, \ldots, x_{k_n}}(x_{k_1}, \ldots, x_{k_n}) = \frac{1}{(2\pi)^{n/2}|\Sigma|^{1/2}} \exp[-\tfrac{1}{2}(x - m)'\Sigma^{-1}(x - m)]$$

where

$$x = [x_{k_1} \quad x_{k_2} \quad \ldots \quad x_{k_n}]' \qquad m^{(i)} = E[x_{k_i}]$$

and

$$\Sigma^{(ij)} = E\{[x_{k_i} - m^{(i)}][x_{k_j} - m^{(j)}]\}$$

A complete probabilistic description of the process is provided by $E[x_{k_i}]$ and cov $[x_{k_i}, x_{k_j}]$ for all k_i and k_j.

10. Linear Transformations

Suppose $\{x_k\}$ is a gaussian random process; define a process $\{y_k\}$ by $y_k = \sum_l a_{kl}x_l$, where $\{a_{kl}\}$ are constants. If $\{y_k\}$ is finite in a suitable sense, it is gaussian.

11. Wiener Processes

If $\{w_k\}$ is a white noise, discrete-time, gaussian process and w_k is $N[0, 1]$ for each k, the process $\{x_k\}$ defined by $x_{k+1} = x_k + w_k$, $k \geq 0$, where x_0 is $N[m_0, \Sigma_0]$, is a Wiener process.

REFERENCES

[1] BREIPOHL, A. M., *Probabilistic Systems Analysis*, John Wiley & Sons, Inc., New York, 1970.

[2] PAPOULIS, A., *Probability, Random Variables and Stochastic Processes*, McGraw-Hill Book Company, New York, 1965.

[3] TUCKER, H. G., *A Graduate Course in Probability*, Academic Press, Inc., New York, 1967.

[4] WONG, E., *Stochastic Processes in Information and Dynamical Systems*, McGraw-Hill Book Company, New York, 1971.

BRIEF REVIEW OF SOME RESULTS

OF MATRIX THEORY

The purpose of this appendix is to provide a brief statement of those particular results of matrix theory used in this book. For more extensive treatments, standard textbooks (e.g., [1-4]) should be consulted.

1. Matrices and Vectors

An $m \times n$ matrix A consists of a collection of mn quantities*

$$a_{ij} \quad (i = 1, 2, \ldots, m; j = 1, 2, \ldots, n)$$

written in an array of m rows and n columns:

$$A = \begin{bmatrix} a_{11} & a_{12} & \cdots & a_{1n} \\ a_{21} & a_{22} & \cdots & a_{2n} \\ \cdot & \cdot & & \cdot \\ \cdot & \cdot & & \cdot \\ \cdot & \cdot & & \cdot \\ a_{m1} & a_{m2} & \cdots & a_{mn} \end{bmatrix}$$

Sometimes, one simply writes

$$A = (a_{ij})$$

The quantity a_{ij} is an entry (the ijth entry, in fact) of A.

*The a_{ij} will be assumed real in most of our discussions.

An m-vector, or, more fully, a column m-vector, is a matrix with one column and m rows; thus

$$x = \begin{bmatrix} x_1 \\ x_2 \\ \cdot \\ \cdot \\ \cdot \\ x_m \end{bmatrix}$$

defines x as a column m-vector, whose ith entry is the quantity x_i. A row n-vector is a matrix with one row and n columns.

2. Addition, Subtraction, and Multiplication by a Scalar

Two matrices A and B with the *same number of rows and also the same number of columns* may be added, subtracted, or individually multiplied by a scalar. With k_1, k_2 scalar, the matrix

$$C = k_1 A + k_2 B$$

is defined by

$$c_{ij} = k_1 a_{ij} + k_2 b_{ij}$$

Thus, to add two matrices, one simply adds corresponding entries; to subtract two matrices, one simply subtracts corresponding entries; etc. Of course, addition is commutative, i.e.,

$$A + B = B + A$$

3. Multiplication of Matrices

Consider two matrices A and B, with A an $m \times p$ matrix and B a $p \times n$ matrix. Thus, the number of columns of A equals the number of rows of B. The product AB is an $m \times n$ matrix defined by

$$C = AB$$

with

$$c_{ij} = \sum_{k=1}^{p} a_{ik} b_{kj}$$

Notice that C has the same number of rows as A and the same number of columns as B.

The product of three (or more) matrices can be defined by

$$D = ABC = (AB)C = A(BC).$$

In other words, multiplication is associative. However, multiplication is not commutative; i.e., it is *not* in general true that

$$AB = BA$$

In fact, although AB can be formed, the product BA may not be capable of being formed.

For any integer p, the $p \times p$ matrix

$$I = \begin{bmatrix} 1 & 0 & \cdots & 0 \\ 0 & 1 & & 0 \\ \cdot & & \cdot & \cdot \\ \cdot & & \cdot & \cdot \\ \cdot & & \cdot & \cdot \\ 0 & 0 & \cdots & 1 \end{bmatrix}$$

possessing p rows and columns is termed the *identity matrix of order p*. It has the property that, with A any $m \times p$ matrix,

$$AI = A$$

Likewise, the identity matrix of order m has the property that

$$IA = A$$

Any matrix consisting entirely of entries that are zero is termed *the zero matrix*. Its product with any matrix produces the zero matrix, whereas if it is added to any matrix, it leaves that matrix unaltered.

Suppose A and B are both $n \times n$ matrices (A and B are then termed *square matrices*). Then AB is square. It can be proved then that

$$|AB| = |A||B|$$

where $|A|$ is the determinant of A.

[The definition of the determinant of a square matrix is standard. One way of recursively defining $|A|$ for A an $n \times n$ matrix is to expand A by its first row, thus

$$|A| = a_{11} \begin{vmatrix} a_{22} & \cdots & a_{2n} \\ \cdot & & \cdot \\ \cdot & & \cdot \\ \cdot & & \cdot \\ a_{n2} & \cdots & a_{nn} \end{vmatrix} - a_{12} \begin{vmatrix} a_{21} & a_{23} & a_{24} & \cdots & a_{2n} \\ a_{31} & a_{33} & a_{34} & \cdots & a_{3n} \\ \cdot & & & & \cdot \\ \cdot & & & & \cdot \\ \cdot & & & & \cdot \\ a_{n1} & a_{n3} & a_{n4} & \cdots & a_{nn} \end{vmatrix}$$

$$+ a_{13} \begin{vmatrix} a_{21} & a_{22} & a_{24} & \cdots & a_{2n} \\ a_{31} & a_{32} & a_{34} & \cdots & a_{3n} \\ \cdot & & & & \cdot \\ \cdot & & & & \cdot \\ a_{n1} & a_{n2} & a_{n4} & \cdots & a_{nn} \end{vmatrix} - \cdots$$

This expresses $|A|$ in terms of determinants of $(n-1) \times (n-1)$ matrices. In turn, these determinants may be expressed using determinants of $(n-2) \times (n-2)$ matrices, etc. For a scalar a, $|a| = a$.]

4. Direct Sum of Two Matrices

Let A be an $n \times n$ matrix and B an $m \times m$ matrix. The *direct sum* of A and B, written $A \dotplus B$, is the $(n + m) \times (n + m)$ matrix

$$\begin{bmatrix} A & 0 \\ 0 & B \end{bmatrix}$$

5. Transposition

Suppose A is an $m \times n$ matrix. The *transpose* of A, written A', is an $n \times m$ matrix defined by

$$B = A'$$

where

$$b_{ij} = a_{ji}$$

Thus, if

$$A = \begin{bmatrix} 1 & 3 & 2 \\ 2 & 1 & 5 \end{bmatrix}$$

then

$$A' = \begin{bmatrix} 1 & 2 \\ 3 & 1 \\ 2 & 5 \end{bmatrix}$$

It is easy to establish the important result

$$(AB)' = B'A'$$

which extends to

$$(ABC)' = C'B'A'$$

and so on. Also, trivially, one has

$$(A + B)' = A' + B'$$

6. Singularity and Nonsingularity

Suppose A is an $n \times n$ matrix. Then A is said to be *singular* if $|A|$ is zero. Otherwise, A is termed nonsingular.

7. Rank of a Matrix

Let A be an $m \times n$ matrix. The rank of A is a positive integer q such that some $q \times q$ submatrix of A, formed by deleting $(m - q)$ rows and $(n - q)$ columns, is nonsingular, whereas no $(q + 1) \times (q + 1)$ submatrix is nonsingular. For example, consider

$$A = \begin{bmatrix} 1 & 2 & 3 & 0 \\ 1 & 2 & 3 & 0 \end{bmatrix}$$

The maximum size square submatrix that can be formed is 2×2. Therefore, a priori, rank $A \leq 2$. Now the possible 2×2 submatrices are

$$\begin{bmatrix} 1 & 2 \\ 1 & 2 \end{bmatrix}, \begin{bmatrix} 1 & 3 \\ 1 & 3 \end{bmatrix}, \begin{bmatrix} 1 & 0 \\ 1 & 0 \end{bmatrix}, \begin{bmatrix} 2 & 3 \\ 2 & 3 \end{bmatrix}, \begin{bmatrix} 2 & 0 \\ 2 & 0 \end{bmatrix}, \begin{bmatrix} 3 & 0 \\ 3 & 0 \end{bmatrix}$$

These all have zero determinant. Therefore, rank $A < 2$. Of the 1×1 submatrices, two have zero determinant but six do not. Therefore, rank $A = 1$.

The rank of A is also the maximum number of linearly independent rows of A and the maximum number of linearly independent columns of A. In the example, the second row equals the first row. Furthermore, the second, third, and fourth columns are linear multiples of the first.

It can be shown that

$$\text{rank } (AB) \leq \min [\text{rank } A, \text{rank } B]$$

If rank A is equal to the number of columns *or* the number of rows of A, A is often said to have *full rank*. If A is $n \times n$, the statement "rank $A = n$" is equivalent to the statement "A is nonsingular." If, for an arbitrary matrix A, rank $A = 0$, then A is the zero matrix.

8. Range Space and Null Space of a Matrix

Let A be an $m \times n$ matrix. The range space of A, written $\mathfrak{R}[A]$, is the set of all vectors Ax, where x ranges over the set of all n-vectors. The range space has dimension equal to the rank of A; i.e., the maximal number of linearly independent vectors in $\mathfrak{R}[A]$ is rank A. The null space of A, written $\mathfrak{N}[A]$, is the set of vectors y for which $Ay = 0$.

An easily proved property is that $\mathfrak{R}[A']$ and $\mathfrak{N}[A]$ are orthogonal; i.e., if $y_1 = A'x$ for some x, and if y_2 is such that $Ay_2 = 0$, then $y_1' y_2 = 0$.

If A and B are two matrices with the same number of rows, then $\mathfrak{R}[A] \subset \mathfrak{R}[B]$ if and only if $\mathfrak{N}[A'] \supset \mathfrak{N}[B']$.

9. Inverse of a Square Nonsingular Matrix

Let A be a square matrix. If, but only if, A is nonsingular, there exists a unique matrix, call it B, termed the *inverse* of A, with the properties

$$BA = AB = I$$

The inverse of A is generally written A^{-1}. There are many computational procedures for passing from a prescribed A to its inverse A^{-1}. A formula is, in fact, available for the entries of $B = A^{-1}$, obtainable as follows.

Define the cofactor of the ij entry of A as $(-1)^{i+j}$ times the determinant of the matrix obtained by deleting from A the ith row and jth column, i.e., the row and column containing a_{ij}. Then,

$$b_{ij} = \frac{1}{|A|} \times \text{cofactor of } a_{ji}$$

It easily follows that

$$(A^{-1})' = (A')^{-1}$$

If A_1 and A_2 are two $n \times n$ nonsingular matrices, it can be shown that

$$(A_1 A_2)^{-1} = A_2^{-1} A_1^{-1}$$

10. The Pseudo-inverse of a Matrix

The pseudo-inverse $A^\#$ of a square matrix A is a useful generalization of the inverse of a matrix. There are actually a number of different pseudo-inverses [3]; here, we shall describe the Moore-Penrose pseudo-inverse.

The key is to make $A^\#A$ act as the identity matrix on as large a set of vectors as is practicable.

DEFINITION. Let A be an $n \times n$ matrix. Its pseudo-inverse $A^\#$ is uniquely defined by the following equations:

$$A^\#Ax = x \qquad \forall x \in \mathfrak{R}[A'] = \mathfrak{N}[A]^\perp$$
$$A^\#x = 0 \qquad \forall x \in \mathfrak{R}[A]^\perp = \mathfrak{N}[A']$$

Observe that $A^\#A$ is the identity on $\mathfrak{R}[A'] = \mathfrak{N}[A]^\perp$.

PROPERTIES:

1. $\mathfrak{R}[A^\#] = \mathfrak{R}[A'], \ \mathfrak{N}[A^\#] = \mathfrak{N}[A']$.
2. $(A^\#)^\# = A$.
3. $A^\#AA^\# = A^\#$.
4. $AA^\#A = A$.
5. $A^\#A$ is the orthogonal projection onto $\mathfrak{R}(A')$.
6. $AA^\#$ is the orthogonal projection onto $\mathfrak{R}(A)$.
7. $(A^\#)' = (A')^\#$.
8. Let y be an arbitrary n-vector. Then

$$\| Ax_0 - y \| \le \| Ax - y \|$$

for all x, with $x_0 = A^\#y$.

COMPUTATION:

1. For $A = \text{diag}(a_1, \ldots, a_n)$,

$$A^\# = \text{diag}(\alpha_1, \ldots, \alpha_n)$$

where $\alpha_i = a_i^{-1}$ if $a_i \ne 0$, $\alpha_i = 0$ if $a_i = 0$.

2. For A symmetric, write $A = T'\Lambda T$, where Λ is diagonal and T is nonsingular; then

$$A^{\#} = T^{-1}\Lambda^{\#}(T')^{-1}$$

3. For arbitrary A,

$$A^{\#} = (A'A)^{\#}A'$$

11. Powers of a Square Matrix

For positive m, A^m for a square matrix A is defined as $AA \ldots A$, there being m terms in the product. For negative m, let $m = -n$, where n is positive; then $A^m = (A^{-1})^n$. It follows that $A^pA^q = A^{p+q}$ for any integers p and q, positive or negative, and likewise that $(A^p)^q = A^{pq}$.

A polynomial in A is a matrix $p(A) = \sum_{i=0}^{r} a_i A^i$, where a_i are scalars. Any two polynomials in the same matrix commute, i.e., $p(A)q(A) = q(A)p(A)$, where p and q are polynomials. It follows that $p(A)q^{-1}(A) = q^{-1}(A)p(A)$, and such *rational functions* of A also commute.

12. Exponential of a Square Matrix

Let A be a square matrix. Then it can be shown that the series

$$I + A + \frac{1}{2!}A^2 + \frac{1}{3}A^3 + \cdots$$

converges, in the sense that the ij entry of the partial sums of the series converges for all i and j. The sum is defined as e^A. It follows that

$$e^{At} = I + At + \frac{1}{2!}A^2t^2 + \cdots$$

Other properties are: $p(A)e^{At} = e^{At}p(A)$ for any polynomial A, and $e^{-At} = [e^{At}]^{-1}$.

13. Differentiation and Integration

Suppose A is a function of a scalar variable t, in the sense that each entry of A is a function of t. Then

$$\frac{dA}{dt} = \left(\frac{da_{ij}}{dt}\right)$$

It follows that

$$\frac{d}{dt}(AB) = \frac{dA}{dt}B + A\frac{dB}{dt}$$

Also, from the definition of e^{At}, one has for time-invariant A

$$\frac{d}{dt}(e^{At}) = Ae^{At} = e^{At}A$$

The integral of a matrix is defined in a straightforward way as

$$\int A \, dt = \left(\int a_{ij} \, dt \right)$$

Suppose ϕ is a scalar function of a vector x. Then

$$\frac{d\phi}{dx} = \text{a vector whose } i\text{th entry is} = \frac{\partial \phi}{\partial x_i}$$

Suppose ϕ is a scalar function of a matrix A. Then

$$\frac{d\phi}{dA} = \text{a matrix whose } ij \text{ entry is } \frac{\partial \phi}{\partial a_{ij}}$$

Suppose z is a vector function of a vector x. Then

$$\frac{dz}{dx} = \text{a matrix whose } ij \text{ entry is } \frac{\partial z_i}{\partial x_j}$$

14. Eigenvalues and Eigenvectors of a Square Matrix

Let A be an $n \times n$ matrix. Construct the polynomial $|sI - A|$. This is termed the *characteristic polynomial* of A; the zeros of this polynomial are the *eigenvalues* of A. If λ_i is an eigenvalue of A, there always exists at least one vector x satisfying the equation

$$Ax = \lambda_i x$$

The vector x is termed an eigenvector of the matrix A. If λ_i is not a repeated eigenvalue—i.e., if it is a simple zero of the characteristic polynomial—to within a scalar multiple x is unique. If not, there *may* be more than one eigenvector associated with λ_i. If λ_i is real, the entries of x are real, whereas if λ_i is complex, the entries of x are complex.

If A has zero entries everywhere off the main diagonal, i.e., if $a_{ij} = 0$ for all i, j, with $i \neq j$, then A is termed *diagonal*. (*Note:* Zero entries are still permitted on the main diagonal.) It follows trivially from the definition of an eigenvalue that the diagonal entries of the diagonal A are precisely the eigenvalues of A.

It is also true that for a general A,

$$|A| = \prod_{i=1}^{n} \lambda_i$$

If A is singular, A possesses at least one zero eigenvalue.

The eigenvalues of a rational function $r(A)$ of A are the numbers $r(\lambda_i)$, where λ_i are the eigenvalues of A. The eigenvalues of e^{At} are $e^{\lambda_i t}$.

If A is $n \times m$ and B is $m \times n$, with $n \geq m$, then the eigenvalues of AB are the same as those of BA together with $(n - m)$ zero eigenvalues.

15. Trace of a Square Matrix A

Let A be $n \times n$. Then the trace of A, written tr $[A]$, is defined as

$$\text{tr}\,[A] = \sum_{i=1}^{n} a_{ii}$$

An important property is that

$$\text{tr}\,[A] = \sum_{i=1}^{n} \lambda_i$$

where the λ_i are eigenvalues of A. Other properties are

$$\text{tr}\,[A + B] = \text{tr}\,[B + A] = \text{tr}\,[A] + \text{tr}[B]$$

and, assuming the multiplications can be performed to yield square product matrices,

$$\text{tr}\,[AB] = \text{tr}\,[B'A'] = \text{tr}\,[BA] = \text{tr}\,[A'B']$$

$$\text{tr}\,[A'A] = \sum_{i=1}^{n} \sum_{j=1}^{n} a_{ij}^2$$

16. Companion Matrix

A matrix A of the form

$$\begin{bmatrix} 0 & 1 & 0 & \cdots & 0 \\ 0 & 0 & 1 & \cdots & 0 \\ \cdot & \cdot & \cdot & \cdot & \cdot \\ \cdot & \cdot & \cdot & \cdot & \cdot \\ \cdot & \cdot & \cdot & \cdot & \cdot \\ 0 & 0 & 0 & \cdots & 1 \\ -a_n & -a_{n-1} & -a_{n-2} & \cdots & -a_1 \end{bmatrix}$$

is termed a companion matrix. One has

$$|sI - A| = s^n + a_1 s^{n-1} + \cdots + a_n$$

17. Orthogonal, Symmetric, and Skew-Symmetric Matrices and Their Eigenvalue Properties

If a square matrix A is such that $AA' = I$, and thus $A'A = I$, A is termed *orthogonal*. The eigenvalues of A then have a magnitude of unity. If $A = A'$, A is termed *symmetric* and the eigenvalues of A are all real. Moreover, if x_1 is an eigenvector associated with λ_1, x_2 with λ_2, and if $\lambda_1 \neq \lambda_2$, then $x_1' x_2 = 0$. The vectors x_1 and x_2 are termed *orthogonal*. (*Note:* Distinguish between an orthogonal matrix and an orthogonal pair of vectors.) If $A = -A'$, A is termed *skew*, or *skew symmetric*, and the eigenvalues of A are pure imaginary.

18. The Cayley-Hamilton Theorem

Let A be a square matrix, and let

$$|sI - A| = s^n + \alpha_1 s^{n-1} + \cdots + \alpha_n$$

Then

$$A^n + \alpha_1 A^{n-1} + \cdots + \alpha_n I = 0$$

The Cayley-Hamilton theorem is often stated, rather ambiguously, as "the matrix A satisfies its characteristic equation."

From the Cayley-Hamilton theorem, it follows that A^m for any $m > n$ and e^A are expressible as a linear combination of $I, A, \ldots, A^{n-1}$.

19. Minimum Polynomial

For square A, the minimum polynomial is the unique monic polynomial $m(\cdot)$ of least degree such that $m(A) = 0$. If $p(\cdot)$ is any polynomial for which $p(A) = 0$, then $m(\cdot)$ divides $p(\cdot)$; in particular, $m(\cdot)$ divides the characteristic polynomial.

20. Similar Matrices and Diagonalizability

Let A and B be $n \times n$ matrices. If there exists a nonsingular $n \times n$ matrix T such that $B = T^{-1}AT$, the matrices A and B are termed *similar*. Similarity is an equivalence relation. Thus:

1. A is similar to A.
2. If A is similar to B, then B is similar to A.
3. If A is similar to B and B is similar to C, then A is similar to C.

Similar matrices have the same eigenvalues. This may be verified by observing that

$$sI - B = T^{-1}sIT - T^{-1}AT = T^{-1}(sI - A)T$$

Therefore,

$$|sI - B| = |T^{-1}||sI - A||T| = |sI - A||T^{-1}||T|$$

But $T^{-1}T = I$ so that $|T^{-1}||T| = 1$. The result is then immediate.

If, for a given A, a matrix T can be formed such that

$$\Lambda = T^{-1}AT$$

is diagonal, then A is termed *diagonalizable*, the diagonal entries of Λ are eigenvalues of A, and the columns of T turn out to be eigenvector of A. Both Λ and T may be complex.

Not all square matrices are diagonalizable. If a matrix has no repeated eigenvalues, it is diagonalizable; it may or may not be diagonalizable if it has repeated eigenvalues. If it is orthogonal, symmetric, or skew, it is diagonalizable. It is diagonalizable if and only if its minimal polynomial has no repeated roots.

21. Jordan Form

Though not all square matrices are diagonalizable, it is always possible to get very close to a diagonal matrix via a similarity transformation. In fact, there always exists a matrix T such that

$$T^{-1}AT = \begin{bmatrix} \lambda_1 & 1 & & & & & & & \\ & \lambda_1 & & & & & & & \\ & & \lambda_1 & & & & & & \\ & & & \lambda_2 & 1 & & & & \\ & & & & \lambda_2 & & & & \\ & & & & & \lambda_3 & 1 & & \\ & & & & & & \lambda_3 & 1 & \\ & & & & & & & \lambda_3 & \\ & & & & & & & & \cdot \\ & & & & & & & & & \cdot \\ & & & & & & & & & & \cdot \end{bmatrix}$$

or something similar. Here, all blank entries are zero, the eigenvalues of A occur on the main diagonal, and there *may* or *may not* be entries of 1 above and to the right of repeated eigenvalues, i.e., on the superdiagonal. For any A, the distribution of 1's and 0's on the superdiagonal is fixed, but different A matrices yield different distributions. The preceding almost-diagonal matrix is called the *Jordan canonical form* of A. The *Jordan blocks* of A are the matrices

$$\begin{bmatrix} \lambda_1 & 1 \\ 0 & \lambda_1 \end{bmatrix}, \quad [\lambda_1], \quad \begin{bmatrix} \lambda_2 & 1 \\ 0 & \lambda_2 \end{bmatrix}, \quad \begin{bmatrix} \lambda_3 & 1 & 0 \\ 0 & \lambda_3 & 1 \\ 0 & 0 & \lambda_3 \end{bmatrix}, \text{ etc.}$$

In general, T and the λ_i are complex. By allowing diagonal blocks

$$\begin{bmatrix} \alpha & -\beta \\ \beta & \alpha \end{bmatrix}$$

to replace diagonal elements $\alpha \pm j\beta$, one obtains a "real Jordan form" for a real A with T real. If A is skew, symmetric, or orthogonal, T may be chosen orthogonal.

22. Positive and Nonnegative Definite Matrices

Suppose A is $n \times n$ and symmetric. Then A is termed *positive definite*, if for all nonzero vectors x the scalar quantity $x'Ax$ is positive. Also, A is termed *nonnegative definite* if $x'Ax$ is simply nonnegative for all nonzero x. Negative definite and nonpositive definite are defined similarly. The quantity $x'Ax$ is termed a *quadratic form*, because when written as

$$x'Ax = \sum_{t,j=1}^{n} a_{ij}x_i x_j$$

it is quadratic in the entries x_i of x.

There are simple tests for positive and nonnegative definiteness. For A to be positive definite, all *leading principal minors* must be positive, i.e.,

$$a_{11} > 0 \quad \begin{vmatrix} a_{11} & a_{12} \\ a_{12} & a_{22} \end{vmatrix} > 0 \quad \begin{vmatrix} a_{11} & a_{12} & a_{13} \\ a_{12} & a_{22} & a_{23} \\ a_{13} & a_{23} & a_{33} \end{vmatrix} > 0, \text{ etc.}$$

For A to be nonnegative definite, all minors whose diagonal entries are diagonal entries of A must be nonnegative. That is, for a 3×3 matrix A,

$$a_{11}, a_{22}, a_{33} \geq 0 \quad \begin{vmatrix} a_{11} & a_{12} \\ a_{12} & a_{22} \end{vmatrix}, \begin{vmatrix} a_{11} & a_{13} \\ a_{13} & a_{33} \end{vmatrix}, \begin{vmatrix} a_{22} & a_{23} \\ a_{23} & a_{33} \end{vmatrix} \geq 0$$

$$\begin{vmatrix} a_{11} & a_{12} & a_{13} \\ a_{12} & a_{22} & a_{23} \\ a_{13} & a_{23} & a_{33} \end{vmatrix} \geq 0$$

A symmetric A is positive definite if and only if its eigenvalues are positive, and it is nonnegative definite if and only if its eigenvalues are nonnegative.

If D is an $n \times m$ matrix, then $A = DD'$ is nonnegative definite; it is positive definite if and only if D has rank n. An easy way to see this is to define a vector y by $y = D'x$. Then $x'Ax = xDD'x = y'y = \Sigma y_i^2 \geq 0$. The inequality becomes an equality if and only if $y = 0$ or $D'x = 0$, which is impossible for nonzero x if D has rank n. One terms D a square root of A.

If D is a square root with number of columns equal to rank A, all other square roots are defined by DT, where T is any matrix for which $TT' = I$.

If A is nonnegative definite, there exists a matrix B that is a *symmetric square root* of A; it is also nonnegative definite. It has the property that

$$B^2 = A$$

If A is nonsingular, so is B.

If A and B are nonnegative definite, so is $A + B$; and if one is positive definite, so is $A + B$. If A is nonnegative definite and $n \times n$, and B is $m \times n$, then BAB' is nonnegative definite.

If A is a symmetric matrix and λ_{max} is the maximum eigenvalue of A, then $\lambda_{max}I - A$ is nonnegative definite.

23. Norms of Vectors and Matrices

The norm of a vector x, written $\| x \|$, is a measure of the size or length of x. There is no unique definition, but the following postulates must be satisfied.

1. $\|x\| \geq 0$ for all x, with equality if and only if $x = 0$.
2. $\|ax\| = |a| \|x\|$ for any scalar and for all x.
3. $\|x + y\| \leq \|x\| + \|y\|$ for all x and y.

If $x = (x_1, x_2, \ldots, x_n)$, three common norms are

$$\|x\| = \left[\sum_{i=1}^{n} x_i^2\right]^{1/2}, \quad \|x\| = \max_i |x_i| \quad \text{and} \quad \|x\| = \sum_{i=1}^{n} |x_i|$$

The Schwartz inequality states that $|x'y| \leq \|x\| \|y\|$ for arbitrary x and y, with equality if and only if $x = \lambda y$ for some scalar λ.

The norm of an $m \times n$ matrix A is defined in terms of an associated vector norm by

$$\|A\| = \max_{\|x\|=1} \|Ax\|$$

The particular vector norm used must be settled to fix the matrix norm. Corresponding to the three vector norms listed, the matrix norms become, respectively

$$[\lambda_{\max}(A'A)]^{1/2}, \quad \max_i \left(\sum_{j=1}^{n} |a_{ij}|\right), \quad \text{and} \quad \max_j \left(\sum_{i=1}^{n} |a_{ij}|\right)$$

Important properties of matrix norms are

$$\|Ax\| \leq \|A\| \|x\| \qquad \|A + B\| \leq \|A\| + \|B\|$$

and

$$\|AB\| \leq \|A\| \|B\|$$

24. Linear Matrix Equations

Let A, B, C be prescribed $m \times m$, $n \times n$, and $m \times n$ matrices. One can form the following equation for an unknown $m \times n$ matrix X:

$$X - AXB = C$$

The equation has a unique solution if and only if $\lambda_i(A)\lambda_j(B) \neq 1$ for any eigenvalues of A, B. In this case, the solution may be found by rearranging the equation as $Dx = e$, where D is a square matrix formed from A and B, and x and e are vectors with entries consisting of the entries of X and C, respectively.

If $B = A'$ and $|\lambda_i(A)| < 1$ for all i, the equation always has a solution which is symmetric if and only if C is symmetric. The linear equation

$$AX + XB + C = 0$$

is also sometimes encountered and has a unique solution if and only if

$$\lambda_i(A) + \lambda_j(B) \neq 0$$

for any i and j. In case $B = A'$ and $\operatorname{Re} \lambda_i(A) < 0$, X always exists; also if $C = C'$, then $X = X'$, and if $C = C' \geq 0$, then $X = X' \geq 0$.

25. Unitary and Hermitian Matrices

To this point, almost all results have been for real matrices, though many ideas apply to matrices with complex entries with little or no change. For example, a complex matrix A is unitary if $AA'^* = A'^*A = I$, hermitian if $A = A'^*$, and skew hermitian if $A = -A'^*$. Matrices with any of these properties may be diagonalized with a unitary matrix. Hermitian matrices have all eigenvalues real.

26. Common Differential Equations Involving Matrices

The equation

$$\frac{d}{dt} x(t) = A(t)x(t) \qquad x(t_0) = x_0$$

commonly occurs in system theory. Here, A is $n \times n$ and x is an n-vector. If A is constant, the solution is

$$x(t) = \exp[A(t - t_0)]x_0$$

If A is not constant, the solution is expressible in terms of the solution of

$$\frac{dX(t)}{dt} = A(t)X(t) \quad X(t_0) = I$$

where now X is an $n \times n$ matrix. The solution of this equation cannot normally be computed analytically, but is denoted by the *transition matrix* $\Phi(t, t_0)$, which has the properties

$$\Phi(t_0, t_0) = I \qquad \Phi(t_2, t_1)\Phi(t_1, t_0) = \Phi(t_2, t_0)$$

and

$$\Phi(t, t_0)\Phi(t_0, t) = I$$

The vector differential equation has solution

$$x(t) = \Phi(t, t_0)x_0$$

The solution of

$$\frac{dx(t)}{dt} = A(t)x(t) + B(t)u(t) \qquad x(t_0) = x_0$$

where $u(t)$ is a forcing term, is

$$x(t) = \Phi(t, t_0)x_0 + \int_{t_0}^{t} \Phi(t, \tau)B(\tau)u(\tau) \, d\tau$$

The matrix differential equation

$$\frac{dX}{dt} = AX + XB + C(t) \qquad X(t_0) = X_0$$

also occurs commonly. With A and B constant, the solution of this equation may be written as

$$X(t) = \exp\left[A(t - t_0)\right]X_0 \exp\left[B(t - t_0)\right]$$
$$+ \int_{t_0}^{t} \exp\left[A(t - \tau)\right]C(\tau) \exp\left[B(t - \tau)\right] d\tau$$

A similar result holds when A and B are not constant.

27. Several Manipulative Devices

Let $f(A)$ be a function of A such that

$$f(A) = \sum_{i=0}^{\infty} a_i A^i \qquad (a_i \text{ is constant})$$

[In other words, $f(z)$, where z is a scalar, is analytic.] Then,

$$T^{-1}f(A)T = f(T^{-1}AT)$$

This identity suggests one technique for computing $f(A)$, if A is diagonalizable. Choose T so that $T^{-1}AT$ is diagonal. Then $f(T^{-1}AT)$ is readily computed, and $f(A)$ is given by $Tf(T^{-1}AT)T^{-1}$. It also follows from this identity that the eigenvalues of $f(A)$ are $f(\lambda_i)$, where λ_i are eigenvalues of A; the eigenvectors of A and $f(A)$ are the same.

For n-vectors x and y and A any $n \times n$ matrix, the following trivial identity is often useful:

$$x'Ay = y'A'x$$

If A is $n \times m$, B is $m \times n$, I_m denotes the $m \times m$ unit matrix, and I_n the $n \times n$ unit matrix, then

$$|I_n + AB| = |I_m + BA|$$

If A is a column vector a and B a row vector b', then this implies

$$|I + ab'| = 1 + b'a$$

Next, if A is nonsingular and a matrix function of time, then

$$\frac{d}{dt}\left[A^{-1}(t)\right] = -A^{-1}\frac{dA}{dt}A^{-1}$$

(This follows by differentiating $AA^{-1} = I$.)

If F is $n \times n$, G and K are $n \times r$, the following identity holds:

$$[I + K'(zI - F)^{-1}G]^{-1} = I - K'(zI - F + GK')^{-1}G$$

Finally, if P is an $n \times n$ symmetric matrix, we note the value of grad $(x'Px)$, often written just $(\partial/\partial x)(x'Px)$, where the use of the partial derivative occurs since P may depend on another variable, such as time. As may be easily checked by writing each side in full,

$$\frac{\partial}{\partial x}(x'Px) = 2Px$$

REFERENCES

[1] GANTMACHER, F. R., *The Theory of Matrices*, Vols. 1 and 2, Chelsea Publishing Co., New York, 1959.

[2] FADEEVA, V. N., *Computational Methods in Linear Algebra*, Dover Publications, Inc., New York, 1959.

[3] BARNETT, S., *Matrices in Control Theory*, Van Nostrand Reinhold Company, London, 1971.

[4] BELLMAN, R. E., *Introduction to Matrix Analysis*, 2nd ed., McGraw-Hill Book Company, New York, 1970.

BRIEF REVIEW OF SEVERAL

MAJOR RESULTS OF LINEAR

SYSTEM THEORY

This appendix provides a summary of several facts of linear system theory. A basic familiarity is, however, assumed. Source material may be found in, e.g., [1–3].

1. z-transforms

Given a sequence $\{g_n\}$, define the z-transform $G(z)$ as

$$G(z) = \sum_{-\infty}^{\infty} g_n z^{-n}$$

Frequently, the lower limit of the summation is taken to be $n = 0$.

2. Convolution and z-transforms

Let $\{g_n\}$ be an impulse response, $\{u_n\}$ an input sequence, and $\{y_n\}$ an output sequence. Then

$$y_n = \sum_{k=0}^{\infty} g_k u_{n-k}$$
$$Y(z) = G(z)U(z)$$

3. Passage from State-space Equations to Impulse Response and Transfer Function Matrix

Suppose that

$$x_{k+1} = Fx_k + Gu_k$$
$$y_k = H'x_k + Ju_k \qquad \text{(C.1)}$$

The associated impulse response is the sequence $\{g_k\}$ defined by $g_k = 0$ for $k < 0$, $g_0 = J$, $g_k = H'F^{k-1}G$ for $k \geq 1$. The transfer function matrix is $J + H'(zI - F)^{-1}G$.

4. Stability

With $u_k \equiv 0$ in (C.1), the system is asymptotically stable if $|\lambda_i(F)| < 1$. Then bounded input sequences produce bounded output sequences. If $|\lambda_i(F)| < 1$, then $\| F^k \| \to 0$ as $k \to \infty$.

5. Complete Reachability

The pair $[F, G]$ is completely reachable with F an $n \times n$ matrix and G an $n \times r$ matrix if any of the following equivalent conditions hold.

1. Rank $[G\ FG \ldots F^{n-1}G] = n$.
2. $w'F^iG = 0$ for $i = 0, 1, \ldots, n - 1$ implies $w = 0$.
3. $w'G = 0$ and $w'F = \lambda w'$ for some constant λ implies $w = 0$.
4. There exists an $n \times r$ matrix K such that the eigenvalues of $F + GK'$ can take on arbitrary values.
5. Given $x_0 = 0$ in (C.1), there exists $\{u_k\}$ for $k \in [0, n - 1]$ such that x_n takes an arbitrary value.

If the pair $[F, G]$ is not completely reachable, there exists a nonsingular T such that

$$TFT^{-1} = \begin{bmatrix} F_{11} & F_{12} \\ 0 & F_{22} \end{bmatrix} \qquad TG = \begin{bmatrix} G_1 \\ 0 \end{bmatrix}$$

with $[F_{11}, G_1]$ completely reachable.

6. Complete Controllability

The pair $[F, G]$ is completely controllable if any of the following equivalent conditions hold:

1. Range $[G\ FG \ldots F^{n-1}G] \supset$ range F^n
2. $w'F^iG = 0$ for $i = 0, 1, \ldots n - 1$ implies $w'F^n = 0$.

3. $w'G = 0$ and $w'F = \lambda w'$ implies $\lambda = 0$ or $w = 0$.
4. Given arbitrary x_0 in (C.1), there exists $\{u_k\}$ for $k \in [0, n-1]$ such that $x_n = 0$.

7. Complete Stabilizability

The pair $[F, G]$ is completely stabilizable if any of the following equivalent conditions holds:

1. $w'G = 0$ and $w'F = \lambda w'$ for some constant λ implies $|\lambda| < 1$ or $w = 0$.
2. There exists an $n \times r$ matrix K such that $|\lambda_i(F + GK')| < 1$ for all i.
3. If

$$TFT^{-1} = \begin{bmatrix} F_{11} & F_{12} \\ 0 & F_{22} \end{bmatrix}, \qquad G = \begin{bmatrix} G_1 \\ 0 \end{bmatrix}$$

with $[F_{11}, G_1]$ completely controllable, then $|\lambda_i(F_{22})| < 1$.

8. Complete Observability, Constructibility, and Detectability

One says that the pair $[F, H]$ is completely observable, constructible, or detectable according as $[F', H]$ is completely reachable, controllable, or stabilizable, respectively. If (C.1) is completely observable [or completely constructible], knowledge of $\{u_k\}$ and $\{y_k\}$ for $k \in [0, n-1]$ suffices to determine x_0 [or x_n].

9. Minimality

If a transfer function matrix $W(z)$ is related to a matrix triple F, G, H by

$$W(z) = H'(zI - F)^{-1}G$$

then F has minimal dimension if and only if $[F, G]$ is completely reachable and $[F, H]$ is completely observable. The triple F, G, H is termed a minimal realization of $W(z)$. Given two minimal realizations of $W(s)$—call them F_1, G_1, H_1 and F_2, G_2, H_2—there always exists a nonsingular T such that

$$TF_1T^{-1} = F_2 \qquad TG_1 = G_2 \qquad (T^{-1})'H_1 = H_2$$

10. Passage from Transfer Function to State-Space Equations

The determination of state-space equations corresponding to a scalar transfer function is straightforward. Given

$$W(z) = \frac{b_1 z^{n-1} + b_2 z^{n-2} + \cdots + b_n}{z^n + a_1 z^{n-1} + \cdots + a_n}$$

one may take

$$F = \begin{bmatrix} 0 & 1 & 0 & \cdots & 0 \\ 0 & 0 & 1 & \cdots & 0 \\ \vdots & & & \ddots & \\ & & & & 1 \\ -a_n & -a_{n-1} & -a_{n-2} & \cdots & -a_1 \end{bmatrix},$$

$$g = \begin{bmatrix} 0 \\ 0 \\ \vdots \\ \cdot \\ 1 \end{bmatrix}, \qquad h = \begin{bmatrix} b_n \\ b_{n-1} \\ \cdot \\ \cdot \\ b_1 \end{bmatrix}, \qquad j = 0 \qquad \text{(C.2)}$$

or

$$F = \begin{bmatrix} 0 & 0 & \cdots & -a_n \\ 1 & 0 & \cdots & -a_{n-1} \\ 0 & 1 & \cdots & -a_{n-2} \\ \cdot & & \cdot & \cdot \\ \cdot & & \cdot & \cdot \\ 0 & 0 & \cdots & 1 & -a_1 \end{bmatrix}, \qquad g = \begin{bmatrix} b_n \\ b_{n-1} \\ \cdot \\ b_1 \end{bmatrix}, \qquad h = \begin{bmatrix} 0 \\ 0 \\ \cdot \\ 1 \end{bmatrix}, \qquad j = 0$$

The first realization is always completely reachable, the second always completely observable.

In case $\lim_{z \to \infty} W(z) \neq 0$, one takes $W(\infty) = j$, and F, g, h as a realization of $W(z) - j$.

11. Passage from Rational Transfer Function Matrix to State-space Equations

A number of techniques are available (see the references). We describe a technique in [4]. Suppose $W(\infty) = 0$ [otherwise consider $W(z) - W(\infty)$]. Write

$$W(z) = \begin{bmatrix} \dfrac{w_1(z)}{d_1(z)}, \dfrac{w_2(z)}{d_2(z)}, \ldots, \dfrac{w_q(z)}{d_q(z)} \end{bmatrix}$$

Here, with $W(z)$ a $p \times q$ matrix, $w_i(z)$ is a p-vector and $d_i(z)$ is the greatest common denominator of entries of the ith column of $W(z)$. Take as a preliminary realization

$$\hat{F} = \begin{bmatrix} \hat{F}_1 & 0 & \cdots & 0 \\ 0 & \hat{F}_2 & \cdots & 0 \\ \cdot & \cdot & \cdot & \cdot \\ \cdot & \cdot & & \cdot \\ \cdot & \cdot & & \cdot \\ 0 & 0 & \cdots & \hat{F}_q \end{bmatrix},$$

$$\hat{G} = \begin{bmatrix} \hat{g} & 0 & \cdots & 0 \\ 0 & \hat{g} & \cdots & 0 \\ \cdot & \cdot & & \cdot \\ \cdot & \cdot & & \cdot \\ \cdot & \cdot & & \cdot \\ 0 & 0 & \cdots & \hat{g} \end{bmatrix}, \qquad \hat{H}' = [\hat{H}'_1 \quad \hat{H}'_2 \quad \cdots \quad \hat{H}'_q]$$

with $\hat{F}_i$ the companion matrix associated with $d_i(z)$ [as F in (C.2) is associated with $z^n + a_1 z^{n-1} + \cdots + a_n$], with $\hat{g} = g$ of (C.2), and with the rth row of $\hat{H}_i$ comprising the numerator coefficients of the rth entry of $w_i(z)/d_i(z)$. The realization $\{\hat{F}, \hat{G}, \hat{H}\}$ is completely reachable. Then one eliminates the unobservable states. This involves finding a nonsingular T such that

$$T\hat{F}T^{-1} = \begin{bmatrix} F_{11} & 0 \\ F_{21} & F_{22} \end{bmatrix}, \qquad T\hat{G} = \begin{bmatrix} G_1 \\ G_2 \end{bmatrix}$$

$$\hat{H}'T^{-1} = [H'_1 \quad 0]$$

with $[F_{11}, H_1]$ completely observable. The triple $\{F_{11}, G_1, H_1\}$ is minimal.

12. Passage from Markov Parameters to State-space Equations

Suppose the rational $p \times m$ matrix $W(z)$ has $W(\infty) = 0$ and is expanded as

$$W(z) = \frac{A_1}{z} + \frac{A_2}{z^2} + \frac{A_3}{z^3} + \cdots$$

when the A_i (termed Markov parameters) are known; then state-space equations for $W(z)$ can be determined from the A_i using an algorithm due to Ho and Kalman [5]. The A_i are arranged to form *Hankel* matrices H_N as follows:

$$H_N = \begin{bmatrix} A_1 & A_2 & \cdots & A_N \\ A_2 & A_3 & \cdots & A_{N+1} \\ \cdot & \cdot & & \cdot \\ \cdot & \cdot & & \cdot \\ \cdot & \cdot & & \cdot \\ A_N & A_{N+1} & \cdots & A_{2N-1} \end{bmatrix}$$

The next step requires the checking of the ranks of H_N for different N to determine the first integer r such that rank

$$H_r = \text{rank } H_{r+1} = \text{rank } H_{r+2} = \cdots$$

If $W(z)$ is rational, there always exists such an r. Then nonsingular matrices P and Q are found so that

$$PH_rQ = \begin{bmatrix} I_n & 0 \\ 0 & 0 \end{bmatrix}$$

where $n = \text{rank } H_r$. The following matrices "realize" $W(z)$, in the sense that $W(z) = H'(zI - F)^{-1}G$:

$$G = n \times m \text{ top left corner of } PH_r$$

$$H' = p \times n \text{ top left corner of } H_rQ$$

$$F = n \times n \text{ top left corner of } P(\sigma H_r)Q$$

where

$$\sigma H_r = \begin{bmatrix} A_2 & A_3 & \cdots & A_{r+1} \\ A_3 & A_4 & \cdots & A_{r+2} \\ \vdots & \vdots & & \vdots \\ A_{r+1} & A_{r+2} & \cdots & A_{2r} \end{bmatrix}$$

Moreover, $[F, G]$ is completely reachable and $[F, H]$ is completely observable.

13. Sampling of a Continuous-time Linear State Equation

Suppose that $\dot{x} = Fx + Gu$ is a continuous-time, state-variable equation. Then

$$x(\overline{k+1}T) = \Phi(\overline{k+1}T, kT)x(kT) + \int_{kT}^{(k+1)T} \Phi(\overline{k+1}T, \tau)G(\tau)u(\tau) \, d\tau$$

or

$$x_{k+1} = F_k x_k + w_k$$

with obvious definitions. If $u(\cdot)$ is vector white noise process in continuous time, and $E[u(t)u'(s)] = Q(t)\delta(t - s)$ with $\delta(\cdot)$ the Dirac delta function, then

$$E[w_k w_l] = \delta_{kl} \int_{kT}^{(k+1)T} \Phi(\overline{k+1}T, \tau)G(\tau)Q(\tau)G'(\tau)\Phi'(\overline{k+1}T, \tau) \, d\tau$$

The relation between $E[w_k]$ and $E[u(t)]$ is easy to obtain.

14. Sampling of a Continuous-time Output Process

Suppose that $z(t) = H'x(t) + v(t)$. A sampling process is usually defined by

$$z_k = \frac{1}{\Delta} \int_{kT}^{kT+\Delta} z(t) \, dt$$

The result is

$$z_k = H'_k x_k + v_k$$

where

$$H'_k = \frac{1}{\Delta} \int_{kT}^{kT+\Delta} H'(t) \, dt$$

$$v_k = \frac{1}{\Delta} \int_{kT}^{kT+\Delta} \left[H'(t) \int_{T}^{t} \Phi(t, \tau) G(\tau) u(\tau) \, d\tau + v(t) \right] dt$$

If $u(\cdot)$ and $v(\cdot)$ are white noise processes, so is $\{v_k\}$. The mean of v_k is easily obtained. If $u(\cdot)$ and $v(\cdot)$ are independent, then one finds as $\Delta \to 0$ that $\{w_k\}$ and $\{v_k\}$ also approach independence, while

$$E[v_k v'_k] \longrightarrow \frac{1}{\Delta^2} \int_{kT}^{kT+\Delta} R(\tau) \, d\tau$$

The fact that this quantity is infinite when $\Delta \to 0$ is the reason that instantaneous sampling of $z(t)$ is not postulated.

One takes $E[x(t) | z_0, z_1, \ldots, z_k]$ to be $\hat{x}_{k/k}$ for $t = kT$ and to be $\Phi(t, kT) \hat{x}_{k/k}$ for, $kT \leq t < \overline{k+1}T$, with the latter formula involving neglect of the correlation between w_k and v_k.

REFERENCES

[1] BROCKETT, R. W., *Finite-dimensional Linear Systems*, John Wiley & Sons, Inc., New York, 1970.

[2] CHEN, C. T., *Introduction to Linear System Theory*, Holt, Rinehart and Winston, Inc., New York, 1970.

[3] CADZOW, J. A., and H. R. MARTENS, *Discrete-time and Computer Control Systems*, Prentice-Hall, Inc., Englewood Cliffs, N.J., 1970.

[4] MUNRO, N., and R. S. McLEOD, "Minimal Realisation of Transfer Function Matrices using the System Matrix," *Proc. IEE*, Vol. 118, No. 9, September 1971, pp. 1298–1301.

[5] HO, B. L., and R. E. KALMAN, "Effective Construction of Linear State-variable Models from Input/Output Functions," *Regelungstechnik*, Vol. 14, No. 12, 1966, pp. 545–548.

LYAPUNOV STABILITY

1. Stability Definitions

Lyapunov theory is a technique for studying the stability of free or unforced equations. As references, we quote [1–3]. Consider

$$x_{k+1} = f(x_k, k) \qquad (D.1)$$

in which it is assumed that $f(0) = 0$, so that $x_e = 0$ is an equilibrium state.

DEFINITION: The equilibrium state x_e is called *stable* if for arbitrary k_0 and $\epsilon > 0$, there exists a $\delta(\epsilon, k_0)$ such that $\| x_{k_0} - x_e \| < \delta$ implies $\| x_k - x_e \| < \epsilon$ for all $k \geq k_0$.

(Provided the initial state deviation is kept small enough, the trajectory deviation can be kept arbitrarily small.)

DEFINITION: The equilibrium state x_e is called *asymptotically stable* if it is stable and if the following *convergence* condition holds: For arbitrary k_0, there exists a $\delta_1(k_0)$ such that $\| x_{k_0} - x_e \| < \delta_1$ implies $\lim_{k \to \infty} \| x_k - x_e \| = 0$.

DEFINITION: The equilibrium state is called *bounded* or *Lagrange stable* if there exists a $\delta_2(x_{k_0}, k_0)$ such that $\| x_k \| \leq \delta_2(x_{k_0}, k_0)$ for all $k \geq k_0$.

An important specialization of the above definitions is when they hold uniformly. This occurs when δ, δ_1, and δ_2 can be chosen independently of k_0. If (D.1) is specialized to the autonomous (undriven, time-invariant) system

$$x_{k+1} = f(x_k) \tag{D.2}$$

the uniformity property naturally holds.

Global asymptotic stability arises when δ_1 can be taken arbitrarily large. Uniform global asymptotic stability arises when there is uniform stability and uniform boundedness, and when δ_1 can be taken independent of k_0 and arbitrarily large.

Exponential asymptotic stability is a special case of uniform global asymptotic stability, arising when

$$\|x_k - x_e\| < \alpha(x_{k_0})\rho^{k-k_0}$$

for some $\alpha > 0$ and $\rho \in (0, 1)$.

2. Lyapunov Theorems for Autonomous Systems

Let $V(x)$ be a real scalar function of the n-vector x and $\mathbb{S}$ be a closed bounded region in R^n containing the origin.

DEFINITION: $V(x)$ is positive definite (semidefinite) in $\mathbb{S}$, written $V > 0$ ($V \geq 0$) if $V(0) = 0$, $V(x) > 0$ ($V(x) \geq 0$) for all $x \neq 0$ in $\mathbb{S}$.

Along motions or trajectories of (D.2), one can compute the change of V. Thus

$$\begin{aligned} \Delta V[x_k] &= V(x_{k+1}) - V(x_k) \\ &= V[f(x_k)] - V(x_k) \end{aligned} \tag{D.3}$$

THEOREM 2.1 (Stability). If there exists in some $\mathbb{S}$ containing the origin a $V > 0$ with $\Delta V \leq 0$ along motions of (D.2), then $x_e = 0$ is stable.

THEOREM 2.2 (Asymptotic Stability). If there exists in some $\mathbb{S}$ containing the origin a $V > 0$ with $\Delta V < 0$, then $x_e = 0$ is asymptotically stable.

THEOREM 2.3 (Asymptotic Stability). If there exists in some $\mathbb{S}$ containing the origin a $V > 0$ with $\Delta V \leq 0$ and if ΔV is not identically zero along any trajectory in $\mathbb{S}$ except the zero trajectory, then $x_e = 0$ is asymptotically stable.

THEOREM 2.4 (Global Asymptotic Stability). If in Theorems 2.2 or 2.3 one has $\mathbb{S} = R^n$ and if $V(x) \to \infty$ as $\|x\| \to \infty$, then $x_e = 0$ is globally asymptotically stable.

Because autonomous systems are being considered, all these results hold uniformly.

A function $V(x)$ which allows proof of a stability result via one of these theorems is termed a *Lyapunov function*.

3. Lyapunov Theory for Time-varying Systems

The definitions and theorems need to be modified slightly. We consider real scalar functions $W(x, k)$ of the n-vector x and time argument k defined for x in a closed bounded region $\mathcal{S}$ containing the origin.

DEFINITION: $W(x, k)$ is positive definite in $\mathcal{S}$, written $W > 0$, if $W(0) = 0$, and $W(x, k) \geq V(x)$ for some positive definite $V(\cdot)$ and all x in $\mathcal{S}$; it is nonnegative definite in $\mathcal{S}$, written $W \geq 0$, if $W(0) = 0$ and $W(x, k) \geq 0$ for all x in $\mathcal{S}$.

One varies (D-3) to

$$\Delta W[x_k, k] = W[f(x_k, k), k + 1] - W[x_k, k]$$

and one defines W as *decrescent* if $W(x, k) \leq V_1(x)$ for some positive definite V_1.

Theorem 2.1 holds with V replaced by W and (D.2) by (D.1). If W is decrescent, stability is uniform. Theorem 2.2 with the same replacements; and a decrescent assumption implies uniform asymptotic stability. Theorem 2.3 cannot be generalized easily. Theorem 2.4 is generalized in the following way. If for all x, $W > 0$, $\Delta W < 0$, W is decrescent and $W(x, k) \to \infty$ uniformly in k as $\| x \| \to \infty$, (D.1) is globally, uniformly, asymptotically stable.

REFERENCES

[1] HAHN, W., *Theory and Application of Lyapunov's Direct Method*, Prentice-Hall, Inc., Englewood Cliffs, N.J., 1963.

[2] WILLEMS, J. L., *Stability Theory of Dynamical Systems*, Thomas Nelson and Sons, Ltd., London, 1970.

[3] KALMAN, R. E., and BERTRAM, J. E., "Control Systems Analysis and Design via the 'Second Method' of Lyapunov: Part II, Discrete Systems," *J. Basic Eng., Trans. ASME*, Vol. 82, No. 2, June 1960, pp. 394–400.

AUTHOR INDEX

SUBJECT INDEX